T0091659

WHERE TO WATCH BIRDS IN WALES

5TH EDITION

DAVID SAUNDERS AND JON GREEN

H E L M
LONDON • OXFORD • NEW YORK • NEW DELHI • SYDNEY

HELM
Bloomsbury Publishing Plc
50 Bedford Square, London, WC1B 3DP, UK
29 Earlsfort Terrace, Dublin 2, Ireland

BLOOMSBURY, HELM and the Helm logo are
trademarks of Bloomsbury Publishing Plc

First published in the UK 2000
This fifth edition published 2022

© Bloomsbury Publishing, 2022

David Saunders and Jon Green have asserted their rights under the Copyright,
Designs and Patents Act, 1988, to be identified as Authors of this work

All rights reserved. No part of this publication may be reproduced or
transmitted in any form or by any means, electronic or mechanical,
including photocopying, recording, or any information storage or
retrieval system, without prior permission in writing from the publishers

Bloomsbury Publishing Plc does not have any control over, or responsibility
for, any third-party websites referred to or in this book. All internet addresses
given in this book were correct at the time of going to press. The author
and publisher regret any inconvenience caused if addresses have changed or
sites have ceased to exist, but can accept no responsibility for any such changes

A catalogue record for this book is available from the British Library
Library of Congress Cataloguing-in-Publication data has been applied for

ISBN: PB: 978-1-4729-7951-3; ePub: 978-1-4729-8066-3;
ePDF: 978-1-4729-8067-0

2 4 6 8 10 9 7 5 3 1

Typeset in the UK by Mark Heslington
Maps by Brian Southern

Printed and bound in Great Britain by CPI Group (UK) Ltd, Croydon CR0 4YY

To find out more about our authors and books visit
www.bloomsbury.com and sign up for our newsletters

Cover photographs. Front: (t) Red Kite, Steve Littlewood/Getty;
(b) Pied Flycatcher; Danny Green/naturepl.com
Spine: Puffin; Jon Green
Back: (l–r) Redstart, Manx Shearwater, Gannet; Lyndon Lomax

CONTENTS

ACKNOWLEDGEMENTS

We must straight away extend our thanks to the hundreds of birdwatchers, named and unnamed, who for the past century or more have watched and enjoyed birds in Wales. Their observations, their records and their writings were the basis of the *Birds of Wales 2021*, the county avifaunas and the numerous bird reports – all essential reading for those who wish to follow, and to learn about, the birds of the Principality. To all these pioneers, both present and future birdwatchers owe a considerable debt. Let us repay this by our continued efforts not only to watch birds in Wales, but to ensure that the habitats which birds use are properly protected, managed and enhanced. Many sites are protected as nature reserves; add to this what seems a plethora of designations – Sites of Special Scientific Interest, National Nature Reserves, Ramsar Sites to mention but a few – and one might be forgiven for believing all is well with the birds of Wales. We must not, however, be complacent as we move forward in this century, whose changes may be more dramatic than anything we have witnessed before, as climate change begins to have an effect on habitats, weather patterns, ocean currents and creatures. Our efforts must not diminish. So, enjoy your birdwatching, but at the same time encourage others and keep those records coming in; they are the warp and weft of this most glorious hobby and the means by which we can together provide a better future for the birds of Wales – our birds.

Our thanks are due to Brian Southern, who so carefully prepared the maps for this book; his expertise is greatly appreciated.

This fifth edition was written during the COVID-19 pandemic, when repeated lockdowns restricted site visits and prevented annual censuses and many breeding counts. As in previous editions, local knowledge has been sought from many of the county recorders and birdwatchers who know the sites far better than we do. We owe them a great debt. They include Rob Sandham (Anglesey), Marc Hughes (Caernarfonshire), Steve Stansfield (Bardsey), Gary Harper (Carmarthenshire), Arfon Williams (Ceredigion), Chris Jones (Denbighshire & Flintshire), Phil Bristow (East Glamorgan), Jim Dustow (Meirionnydd), Mike Haigh and Simon Boyes (Montgomery), David Astins, Richard Brown and Giselle Eagle (Skokholm), Richard Ellis and Lisa Morgan (Pembrokeshire), Rob Davies and Robert Taylor (West Glamorgan & Gower) and Julian Hughes (RSPB).

Since the previous edition of our book was published in 2008, too many of our leading ornithologists, each a good friend across more years than we care to remember, have passed on. Each made a singular contribution to knowledge and recording of birds in Wales, and everyone is sorely missed. We dedicate our book to their memory: Peter Davis MBE, Stuart Devonald, Jack Donovan MBE, Peter Hope-Jones, Graham Rees, Mike Shrubb, Hywel Roderick, Derek Moore OBE.

Finally, we would like to thank our families for all the support they have given us over the years, without which we would not have been able to enjoy the birds of Wales.

INTRODUCTION

Wales is a small, mainly upland country of 8,018 square miles (20,766 square km), of which about 60% lies above the 500-foot (150-m) contour. At the same time, it is a maritime land, bounded on three sides by the sea: to the south by the Bristol Channel, to the west by St George's Channel and to the north by the Irish Sea. For the geologist, Wales is a place of almost unrivalled opportunities. Here are the oldest rocks, here also some of the youngest. The action of ice is everywhere recorded: glacial lakes, the screes and moraines, overflow channels and other features are a constant reminder of how much of this spectacular landscape was moulded. The rivers are mostly short and fast-flowing, terminating in estuaries of major importance for their birdlife. Three much longer rivers, the Dee, Severn and Wye, murmur into life high on the hills of Wales, and form in parts its eastern marches with England. This is a country, then, of contrasts, for you do not have to travel far in any direction to encounter a great variety and range of habitats. It is, for instance, but a short journey through the industrial valleys of Glamorgan to the uplands of the Black Mountains, the Brecon Beacons and the Carmarthen Vans. Leave the coast, flatlands and lakes of Anglesey and within half an hour you can be high in Snowdonia. The coastline is equally variable, with sand dunes, saltmarshes, rocky shores, impressive sea cliffs and then the islands – and what islands they are: Bardsey in the north, and Grassholm, Ramsey, St Margaret's, Skokholm and Skomer in the south.

Despite the amount of high ground and spectacular landscape features of great interest, there is not the variety nor the numbers of birds that one encounters in Scotland. The threats to the uplands of Wales are very real, especially those of large-scale afforestation and of agricultural improvements to high-altitude areas. Nevertheless, there are important populations of key species such as Hen Harrier, Merlin, Peregrine, Golden Plover and Dunlin. Pride of place among all upland birds, indeed among all the birds of Wales, is without question taken by the Red Kite, for well over a century found only here and nowhere else in Great Britain. Its survival is assured, as recounted elsewhere in this book, only by much dedicated effort – surely the longest-running protection scheme for a single species anywhere in the world. Good news that egg and chick theft has almost ceased, although this is tempered by an RSPB report covering 1990–2019 which demonstrates that raptor persecution by shooting, trapping and poisoning has increased in the past thirty years with no fewer than 52 cases of poisoning confirmed in the last decade. Such persecution is three times higher in areas where driven shooting of gamebirds is practised.

The lakes, marshes and especially the woodlands of Wales are of interest and importance, supporting a rich variety of birds, including significant populations of such species as Wood Warbler and Pied Flycatcher. The greatest bird spectacles of Wales, however, belong to the coast: wildfowl and waders by the thousand – flashing wings and magical calls on estuaries right around the coast. Then the clamorous seabird colonies, several being of international importance, where the odour of guano ledges drifts upwards, while Fulmars sweep and glide as only they can along the cliff faces in the summer breeze. Of all the sights, there is probably none more magnificent in the whole of Wales than the massed ranks of Gannets, now 36,011 AON (apparently occupied nests) in 2015, on lonely Grassholm. And who can fail to be captivated by the Puffins at places such as Skokholm and Skomer, where they are literally at your feet. These same two islands contain perhaps half the world population (342,000–393,000 pairs) of Manx Shearwaters, an international treasure – so make sure you spend at least one night with them, an experience without compare.

It is often to our remote islands that rare birds come, vagrants from the forest and tundra of Siberia, from the steppes of Central Asia, from North America, seabirds from all the world's oceans. As this book shows, rare birds are by no means restricted to remote islands; they are likely to be seen anywhere, even in the most unexpected locations, as demonstrated by the records of a Dartford Warbler in Singleton Park, Swansea, an Arctic Redpoll in Goodwick, a Baltimore Oriole in Roch and a Black-throated Thrush in a Swansea garden. There is always the chance of the unexpected. Imagine the amazement of the bird recorder when someone asks what they know about Hoopoes breeding in the county – the answer: nothing. Further enquiries reveal that this is exactly what had occurred in Montgomeryshire in 1996, when a pair of Hoopoes reared three young in a derelict farm building without a single birdwatcher knowing until after the event. Birdwatching is continually full of surprises, so it pays to be vigilant whenever you go forth.

The changing patterns of bird distribution make for fascinating study. Turn the pages of the avifaunas and you will see many examples. For instance, the first Fulmars did not nest in Pembrokeshire until the 1940s, while it is easy to forget that the first Collared Doves arrived in Britain only as recently as 1955, and reached Wales several years later, where the population now numbers perhaps 87,000 pairs. Meanwhile, on the debit side, after a long decline, the Turtle Dove no longer breeds in Wales. Such shifts in bird distribution are continually taking place – the decline and loss of the Woodlark, the spread of Goosander and Cetti's Warbler, and the colonisation of our wetlands by Little Egret. Who would have forecast the dramatic changes in our populations of Lesser Black-backed Gulls (increased from 11,529 in 1969–70 to 20,722 in 1998–2002 before declining to 13,434 in 2015–19) and Herring Gulls (declined from 48,576 AON in 1969–70 to 7,988 in 2015–19), or realised that Dippers would disappear from some rivers as a result of acid waters reducing the number of invertebrates on which they feed? What of Green Woodpecker? Described in *Birds in Wales* as recently as 1994 as 'A common breeding resident throughout lowland Wales', there have of late been serious declines in some areas, with the county bird report for Ceredigion, for instance, noting: 'The future for this species in the county looks bleak.' Remaining on the debit side, the reduction and loss of birds such as Ringed Plovers and terns from many Welsh beaches is disheartening, while even in the 34 years since the first edition of this book was published the calamitous decline of species such as Lapwing, Snipe, Curlew and Redshank now reaches crisis proportions. But there are human-induced gains to report as well. Canada Geese are well established in many areas, some would say too well established with approaching 10,000 pairs, while Greylag Geese now number some 1,650 pairs in addition to which Barnacle Geese now breed at two locations. As our knowledge increases so new, or possibly new, species are recorded, the result of detailed scientific investigation supplemented by observation in the field. Yellow-browed Warbler, is now divided into the Yellow-browed Warbler and Hume's Leaf Warbler; the humble and usually overlooked Herring Gull is now Herring Gull, Yellow-legged Gull and Caspian Gull. There is going to be much more of this, we imagine. With forecasts of rising temperatures as this century unfolds, what other surprises still await as southern species move north and northern ones are lost? The answer is to keep watching, keep recording; who knows what is in store.

For the birdwatcher, the fact that many classic sites are now nature reserves, managed by one of the at-times confusing variety of conservation organisations, is something to be applauded. Some areas have been designated Sites of Special Scientific Interest under the provisions of the Wildlife and Countryside Act 1981, and thus are given a greater degree of protection from damaging activities than in the past. Access to many parts of Wales is now much easier than it was even a few years ago, not just as the result of road improvements but also because of the creation of or improvements to footpaths and bridleways. Long-distance footpaths such as that which follows Offa's Dyke along the border with England, or the Pembrokeshire Coast Path, enable birdwatchers to reach previously difficult or inaccessible points. Wherever you are, please ensure that you do not disturb the very birds you are watching, otherwise they will suffer, as will other birdwatchers. Always put the birds' welfare first; you will then have the satisfaction and knowledge of having your own enjoyment without causing detriment or harm.

The aim of this book is to help and encourage birdwatchers, especially beginners, to set forth, to learn more about this most delightful of hobbies, and at the same time to contribute to our knowledge. If we derive enjoyment, if we glean knowledge, if we get satisfaction from our watching, then we should try to put something back. Please send records of your observations to the county bird recorder, join the Welsh Ornithological Society and your local bird club or county wildlife trust – please support, please participate. The rewards of doing so can only add to your further enjoyment, and to the further enjoyment of others.

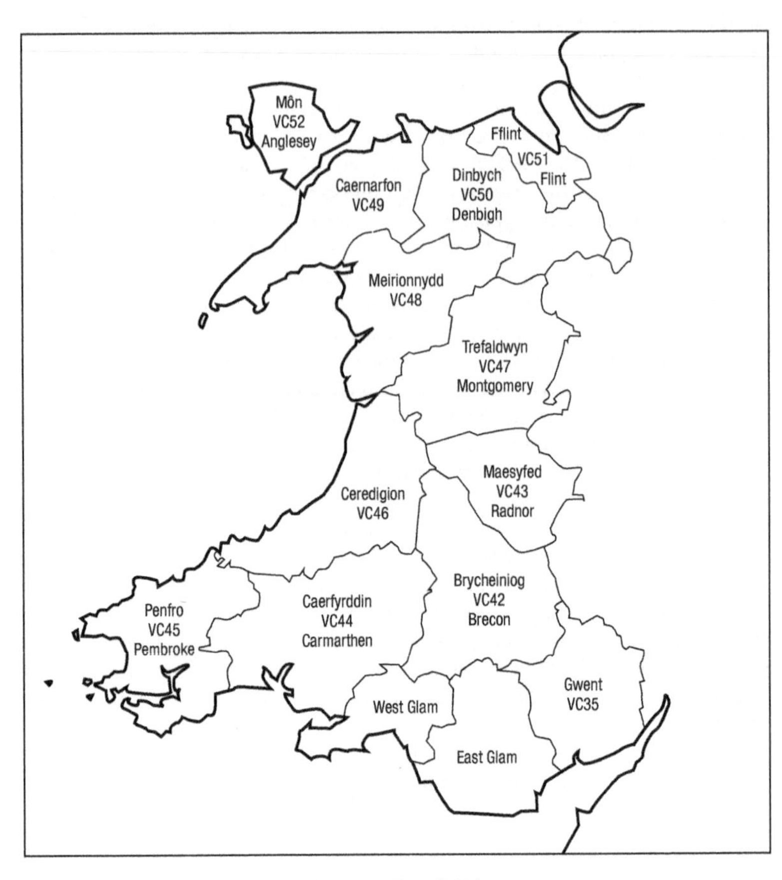

Vice counties of Wales

HOW TO USE THIS BOOK

Few birdwatchers have watched birds throughout the length and breadth of Wales, so from this book hopefully even the most expert will be able to learn of new sites of interest to explore and of the birds that they support. Local knowledge is often all important, and there is a vast amount contained among the 142 site descriptions. Follow this, and join your local bird club or wildlife trust, and your birdwatching opportunities will be much improved. However, there is a further pleasure, and that is to discover new opportunities, new places to watch, and by so doing add further to our knowledge about birds in Wales.

In 1973, one of the effects of changes in our system of local government was that some county boundaries were amended, while in Wales there was a complete revision of the county names. In most cases this resulted in a linking of counties, save for Glamorgan where the opposite took place. Then in 1996 there were further changes; some of the former counties such as Pembrokeshire fortunately reappeared and Dyfed disappeared. Despite this, birdwatchers, and indeed all naturalists, still follow the old and much-loved Watsonian vice-county system which retains the necessary historical continuity; this is how the bird clubs and societies operate, and long may it remain so!

Measurements

We have given measurements in traditional units: distances are given in miles, areas in acres and altitude in feet. Since most modern maps use metric units, however, the metric equivalent is given in parentheses after each measurement.

Habitat

Descriptions of each site are given under this heading in an attempt, hopefully successful, to provide some background for the birdwatcher. Where appropriate, we have also added items of historical interest. Brief references to other features of natural history interest which may be encountered or which warrant further inquiry are also included.

Species

The main bird interest is described, together with snippets of background information which may help add to the knowledge of the reader, enabling everyone to get the most out of the visit. No-one can fail to be moved to some excitement by the sight of an unusual bird, a rare visitor, perhaps one with the accolade 'first for Wales' or, for those extremely fortunate, a 'first for Britain' or even a 'first for the Western Palearctic'. One of the authors is fortunate to have discovered one of the latter, a Blackburnian Warbler, on Skomer in October 1961. Indeed, some birdwatchers will travel great distances to witness such an event, so we have drawn attention to many of the scarce visitors seen in Wales. Hopefully, this will help inspire further careful watching, and many more such sightings. At the same time, do not neglect more common species, information on these is always required – and, after all, a bird widespread today may not be so in several decades' time. Changes are continually taking place and these need to be known.

Timing

If you can be flexible in your arrangements, then follow the broad suggestions given in the respective sections. Generally speaking, on estuaries one is governed by the state of the tide, while for seawatching the hours immediately after dawn are often by far the most productive, though do not neglect later in the day if the opportunity presents itself. The early morning is in fact the best whether you are in the depths of woodland or tramping the high sheep-walks.

Access

Maps are provided for each site and these will enable the birdwatcher to locate the area with reference to that essential item of equipment, the Ordnance Survey Landranger or Explorer map. The relevant Ordnance Survey map numbers are given beside the heading for each locality.

Calendar

This provides a quick reference section. The first part covers birds resident throughout the year, though during severe weather many, even all, of some resident species will desert an area, while in the dog days of late July and August, with the breeding season for many species over, small passerines can at times become surprisingly elusive; it also covers birds such as Turnstone, which do not breed in Britain but may be seen at any time throughout the year. The rest of the year is generally divided into four sections. The period December–February is when all winter visitors are present, numbers often reaching a peak early in the year, especially if the rest of Europe experiences severe weather. March–May sees the departure of winter visitors, the arrival of summer migrants and the passages of waders and terns; by the end of the period, breeding species will be fully active. June–July covers the end of the breeding season for most species, and the commencement of return passage, waders usually the first to be seen. August–November is a period spanning the whole of the autumn migration and the arrival of winter visitors on our shores.

To facilitate quick reference, the sequence of bird names follows the British List published by the British Ornithologists' Union in 1998. However, we have not adopted the new names as advocated in this and some other checklists. To most of us a Robin is, and always will be, a Robin, not a European Robin!

GLOSSARY OF WELSH PLACE NAMES

It is appropriate and, we hope, of assistance to include a short glossary of the Welsh names that readers will encounter in the text, or when consulting other works or maps.

aber	mouth (of a river)
aderyn	bird
adwy	gap, pass
afon	river
agen	cleft
allt	hillside
bach	small
bala	stream outlet
bannau	peaks
berllan	orchard
big	peak
blaen	head of valley
bont	bridge
bren	wood
brig	peak
bron	rounded hill
bwlch	pass, gap
bychan	small
cader	seat
cae	field, enclosure
caer	fort
camlas	canal
capel	chapel
carnedd	mountain
carreg	rock
castell	castle
cefn	ridge
cerrig	rocks
ceunant	ravine, gorge
clawdd	embankment
clegyr	rock
clun	meadow
cnwc	hillock
coed	woodland
comin	common
copa	summit
cors	bog
craig	rock
cwm	valley
cymer	confluence
darren	bill (narrow promontory)
ddol	meadow
ddu/du	black
dinas	fort
draeth	shore
drws	pass
dyffryn	valley
eglwys	church
esgair	ridge
fach	small
fawr	large
felin	mill
fford	road
ffridd	upland fringe
ffrwd	torrent
ffynnon	spring, well
figyn	bog
foel	bare hill
fynydd	mountain
gaer	fort
gallt	hillside
garn	mountain
garth	hill, ridge
glyn	glen
gogof	cave
gopa	summit
gors	bog
graig	rock
gwastad	level place
gwaun	moor
gwern	marsh
gweunydd	moors
heol	road
isaf	lower
llan	church
llannerch	glade
llechwedd	hillside

llethr	slope
llwyn	grove
llyn	lake
maen	stone
maes	field
mawnog	peaty
mawr	large
melin	mill
migyn	bog
morfa	coastal marsh
mynydd	mountain
nant	stream, valley
odyn	kiln
ogof	cave
pant	valley
pen	top
penmaen	rocky promontory
pennant	head of valley
penrhyn	promontory
pistyll	waterfall
plas	mansion
pont	bridge
porth	landing place, harbour
pwll	pit, pool
rhaeadr	waterfall
rhiw	hill
rhos	moor, bog
rhyd	ford
sarn	causeway, route
sgwd	waterfall
tarren	knoll
ton (pl. tonnau)	wave
traeth	shore
trwyn	promontory
twll	chasm
twyn	hillock
tywyn	seashore, dune
uchaf	highest
waun	moor
ynys	island
ystrad	valley floor
ystwyth	winding

ANGLESEY (YNYS MÔN)

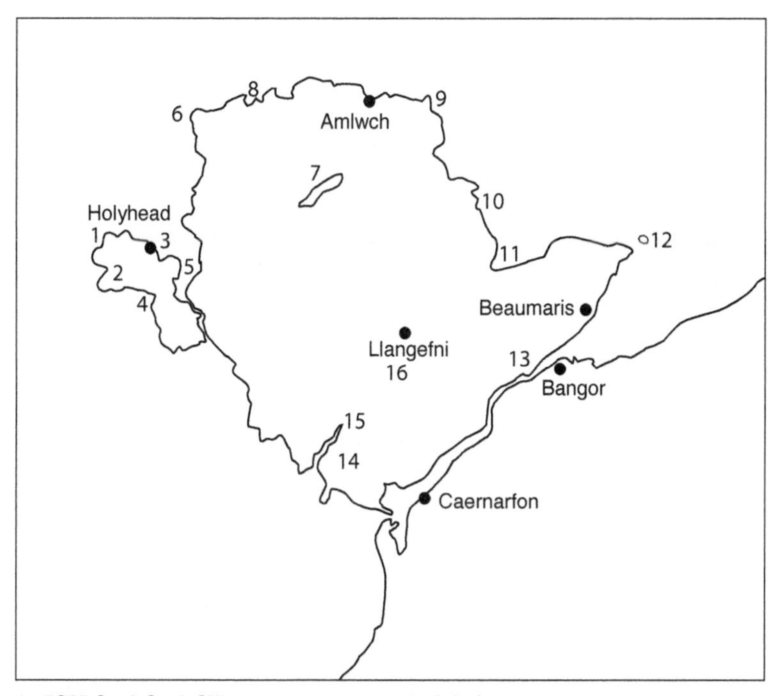

1 RSPB South Stack Cliffs
2 The Range
3 Holyhead Harbour, Soldier's Point and Breakwater Country Park
4 Trearddur Bay
5 Inland Sea and Beddmanarch Bay
6 Carmel Head
7 Anglesey Wetlands
8 Cemlyn Lagoon
9 Point Lynas
10 Traeth Dulas
11 Red Wharf Bay
12 Puffin Island and the Penmon Peninsula
13 Church Island and Coed Cymol
14 Newborough Warren and forest
15 Malltraeth Cob and Llyn Parc Mawr
16 RSPB Cors Ddyga

Anglesey is a large (276 square miles) island off the north-west coast, separated from Caernarfonshire by the Menai Straits, but connected by two roads and a rail link to the port of Holyhead – the main vehicular ferrylink to Ireland. In the centre of the island there are numerous lakes and wetlands under intensive management by the RSPB, North Wales Wildlife Trust and Natural Resources Wales as they work to create and improve wetland habitats. In this area bird-watchers should be able to find good numbers of wintering wildfowl, with smaller numbers remaining to breed.

Around the coast there are numerous bays and valleys, all of which are attractive to wintering and passage birds. Recent concentrated effort at certain sites on Anglesey, particularly South Stack and Cemlyn, has made it one of the most productive counties in Wales for finding rarities. For stunning coastal cliffs and views of breeding seabirds, RSPB South Stack reserve is a must. Cemlyn Lagoon offers excellent views of breeding terns, even Roseate in some years, while seawatchers have tended to congregate at Point Lynas to view the autumn passage, although in recent years South Stack and Carmel Head have been equally productive.

1 RSPB SOUTH STACK CLIFFS

OS Landranger map 114
OS Explorer 262
Grid Ref: SH203823
Website:
https://www.rspb.org.uk/reserves-and-events/reserves-a-z/south-stack-cliffs/

Habitat

South Stack Cliffs contain some of the most spectacular coastal scenery in North Wales and extend for nearly 2 miles (3.2 km) along the northwest coast of Holy Island, Ynys Gybi, the westernmost extremity of Anglesey. The main section of cliffs, together with a hinterland of maritime and moorland heath extending onto Holyhead Mountain, is an RSPB reserve. At Gogarth Bay the cliffs rise to about 360 feet (110 m), gradually descending as one moves south. In places they are much eroded, with numerous gullies, sea caves and offshore stacks, while the major folding near South Stack of these ancient Pre-Cambrian rocks is of enormous geological interest. Cliff-climbers not surprisingly find the whole area a great challenge, though fortunately most observe the voluntary ban on climbing during the seabird breeding season; there are descriptive signs indicating the areas to be avoided from the beginning of February to the end of July. The main stack, South Stack, has a lighthouse (built in 1809) which is reached by a long flight of steps and a narrow footbridge.

In spring and early summer the cliff edges are a blaze of colour, with plants such as Kidney Vetch, Scurvygrass, Spring Squill and Thrift in full flower. The heaths come into their own later in the summer when the golden carpets of Western Gorse contrast with three species of heather. Other notable plants include Bog Pimpernel, Devil's-bit Scabious, Heath Spotted-orchid, Pale Dog-violet and Heath Dog-violet. Two rarities occur here: the British Isles endemic Spotted Rock-rose, found only on Anglesey, at one site in Caernarvonshire and at a few localities in the west of Ireland and on Coll, Scotland, and a maritime variety of the Field Fleawort, found only at South Stack Cliffs.

Species

Seabirds are one of the main attractions at South Stack, with over 6,000 Guillemots and almost 1,200 Razorbills but only a handful of pairs of Puffins. Fulmars, Shags, Kittiwakes (less than 20 pairs) and the three large gulls also nest on the cliffs. Choughs are another speciality, the close-cropped swards and heaths providing good feeding areas for up to 10 pairs and a few non-breeding immatures. The RSPB manages the heathland for Choughs and Silver-studded Blue butterflies, keeping areas of heathland short or bare. During the autumn, Choughs often gather in loose flocks which may comprise up to 30 individuals. One, sometimes two pairs of Peregrines nest; no doubt the seabird colonies and Feral Pigeons that share the cliffs provide at least part of this raptor's food supply. Other cliff-nesting birds include several pairs of Kestrels, Stock Doves, Rock Pipits, Jackdaws, Carrion Crows and Ravens. The maritime heath, especially where pockets of scrub have become established, supports birds such as Meadow Pipit, Stonechat, Whitethroat and Linnet.

Seabird passage can be noteworthy, the west coast of Holy Island being a key vantage point for observing the southward passage of birds which may have been driven into Liverpool Bay by stormy weather. Manx Shearwaters and Gannets are regular offshore, but each autumn such birds as Sooty Shearwater, Arctic and Great Skuas together with large passages of auks are observed.

Regular watching by a small band of observers has paid off in recent years. A Black-browed Albatross was seen passing in February 2005, while the cliffs and heaths have played host to Ring-necked Duck on one of the small ponds nearby, as well as Gyr Falcon, Red-rumped Swallow, Bluethroat, Yellow-browed and Barred Warblers and Red-backed Shrike. Two species stand head and shoulder above the rest. A Grey Catbird, the first for Britain, was found here in October 2001. Unfortunately, it was extremely difficult to see; however, one lucky observer found a Red-eyed Vireo while attempting to sight the Catbird. The second highlight was the discovery of Britain's second Black Lark, a species that has a restricted global range, in the steppes of Kazakhstan, in June 2003 and which was easily seen by a multitude of birdwatchers from all over Britain and Europe.

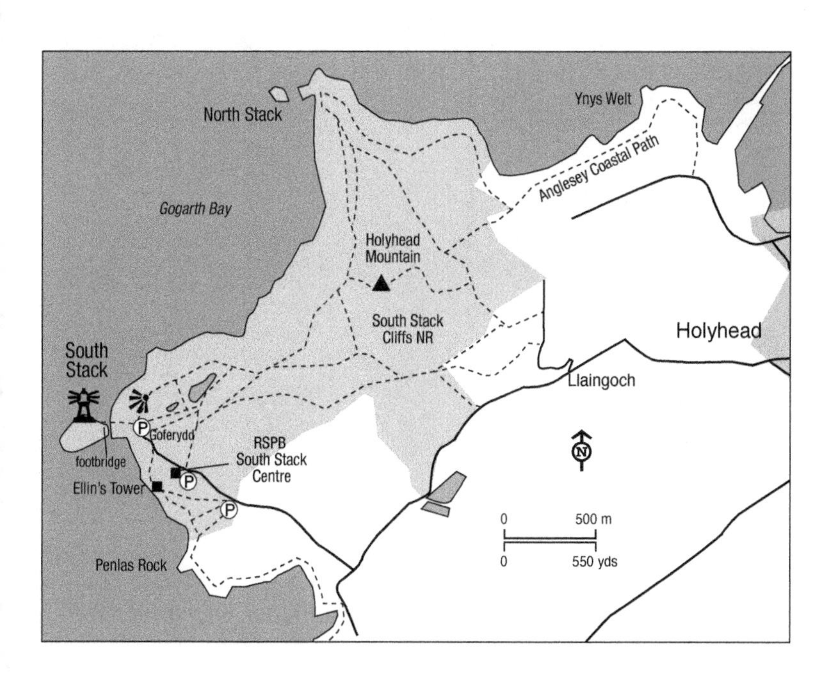

Timing

For a general visit, no particular time is better than another, but for seabird passage in late summer, or scarce visitors at any time, you must be there in the early morning. Do not forget that here, and indeed at other seabird colonies, Guillemots and Razorbills have all departed the cliffs by the end of July and Puffins shortly afterwards.

Access

Two unclassified roads lead from Holyhead, the southern one going via Penrhosfeilw, and both continue towards South Stack. There are three car parks near the reserve and a network of footpaths throughout the whole area. The RSPB has a visitor centre, with café, shop and toilets next to the middle car park and is open all year round. Ellin's Tower is open throughout the summer months and contains a first-floor viewing room overlooking the seabird colonies. Remote-controlled video cameras complete with windscreen-wipers have been installed on the cliffs close to a Guillemot ledge; this enables visitors to watch the birds at really close quarters on a screen in the tower. Choughs and Stonechats can often be seen from the car park. The lighthouse on South Stack itself is open during the summer months; for admission pay at the ticket booth near the middle car park.

CALENDAR

Resident: Cormorant, Shag, Buzzard, Kestrel, Peregrine, Oystercatcher, Turnstone, large gulls, Stonechat, Chough, Jackdaw and Raven.

December–February: Great Northern and Red-throated Divers, Fulmars return to cliffs and Gannets appear offshore. Hen Harrier, Merlin usually present, while both Guillemot and Razorbill make sporadic visits to the cliffs and Kittiwakes are virtually resident by end of this period.

March–May: A handful of pairs of Puffins arrive about the beginning of April, though are not present continuously until around the end of the month. Most Guillemots and Razorbills commence egg laying in early May. Passage migrants throughout whole period, Wheatear usually the first.

June–July: Manx Shearwater passage daily offshore, terns also seen, all seabirds now feeding young. By the end of the period the auks have left the cliffs, save for a few Puffins which remain into early August.

August–November: Seabird passage throughout, with the greatest range of species in September. Wader passage under way, including Dotterel, Whimbrel and Greenshank. Passerines move through, and late in period large influxes of Skylarks, Meadow Pipits, Fieldfares, Redwings, Starlings and Chaffinches may occur.

2 THE RANGE

OS Landranger map 114
OS Explorer 262
Grid Ref: SH215800

Habitat

When visiting South Stack do not overlook The Range, the low coastal headland with cliffs and offshore rocks immediately to the south of the quaintly named Abraham's Bosom bay. Although seabirds, save for Fulmars, are absent

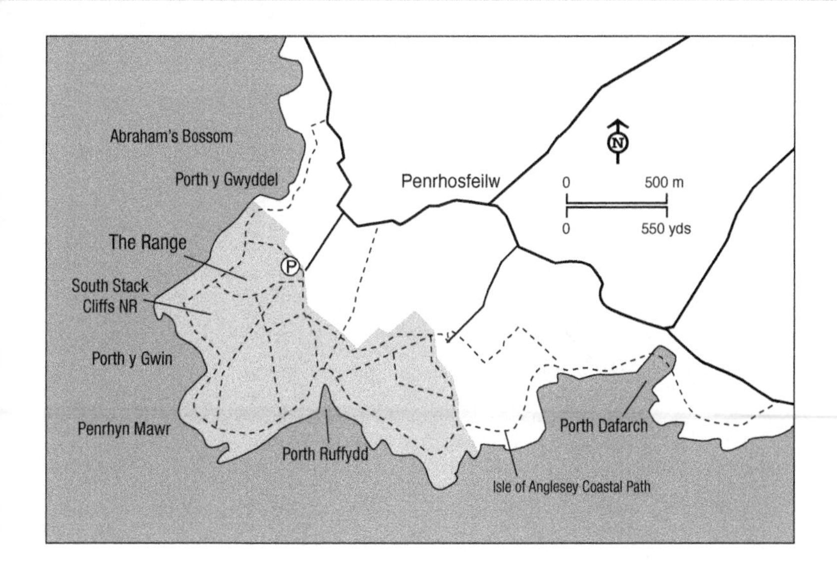

from the cliffs, they are home to Choughs and Ravens while Peregrines never seem far away. The offshore waters are frequented by Grey Seals, Harbour Porpoise, Bottlenose and Common Dolphins.

Species

Both Guillemot and Razorbill breed on the cliffs, with over 300 of the former and over 80 of the latter. Good passages of Manx Shearwaters on midsummer feeding forays, the nearest colony being that on Bardsey, but you could well be watching some from the great colonies on the Pembrokeshire islands 100 miles to the south. Fast forward to September and October and more Manx Shearwaters; now hurrying south from the Inner Hebrides colonies, their destination the waters off Brazil, Uruguay and Argentina. Terns and Kittiwakes also and as the autumn days pass, increasing numbers of Guillemots and Razorbills head south.

Choughs, Ravens, Skylarks, Meadow and Rock Pipits, Stonechats and Linnets are resident, while passage migrants commencing in spring and concluding in autumn with Wheatears and the occasional Black Redstart include Fieldfares, Redwings, Starlings and Chaffinches. Scarce visitors have included Black Kite, Dotterel, Stone-curlew, Tawny Pipit, Shorelark and Bluethroat.

Timing

Throughout the year, but spring and autumn most productive.

Access

Open throughout the year.

CALENDAR

Resident: Cormorant, Shag, Buzzard, Peregrine, Oystercatcher, Herring Gull, Great Black-backed Gull, Skylark, Meadow Pipit, Rock Pipit, Stonechat, Chough, Jackdaw, Raven, Linnet.

December–February: Fulmars back on the cliffs, Great Northern and Red-throated Divers offshore, Hen Harrier and Merlin over the heaths.

March–May: Breeding Guillemots and Razorbills can be seen on the cliffs. Wheatears in early March usually the first summer visitor to arrive, to be followed by warblers on passage and from mid-April into May some spectacular movements north of Swallows and martins, and of course, always the chance of the unexpected.

June–July: Manx Shearwater passage daily offshore, terns on fishing expeditions and Guillemots and Razorbills from the colonies here and nearby at South Stack. Listen out early in the morning – before they move too far offshore – for the plaintive piping of Guillemot and Razorbill chicks, each accompanied by their attentive father.

August–November: The last Fulmars fledge in early September, after which the cliffs are vacated. Swallows and martins heading south in September, while seabird passage throughout the period and always the chance of land bird migrants to enchant the watcher.

3 HOLYHEAD HARBOUR, SOLDIER'S POINT AND BREAKWATER COUNTRY PARK

OS Landranger Map 114
OS Explorer 262
Grid Ref: SH231836
Website:
https://www.anglesey.gov.uk/documents/Dogfennau-Cy/Cefn-Gwlad/Llwybr-Natur-Parc-Gwledig-Morglawdd-Caergybi.pdf

Habitat

This large harbour is sheltered by the longest breakwater in Great Britain, constructed between 1847 and 1853, close to its base Soldier's Point and the Breakwater Country Park leading up to Holyhead Mountain. The Country Park has good facilities and includes in one roofless building splendid reproductions of Charles Tunnicliffe paintings. Fine opportunities for birdwatching with easy access on land, within the harbour and offshore.

Species

During spring and autumn, if conditions are right, the area can be awash with migrants. Good numbers of 'phylloscopus' Warblers, Blackcaps, Spotted Flycatchers, winter thrushes and pipits occur, while rare and scarce birds have included the first record of Isabelline Shrike for Wales in 1985, Killdeer in 1993, Hoopoe, Wryneck, Bee-eater, Richard's Pipit, Barred, Pallas's, Yellow browed and Dusky Warblers, Serin and Common Rosefinch. The nearby harbour is certainly worth checking, especially in winter and autumn. All four species of diver have occurred here, including a White-billed Diver in 1991, with Great Northern being the most regular. Black Guillemots that nest in the area are often seen near the breakwater and offer excellent views.

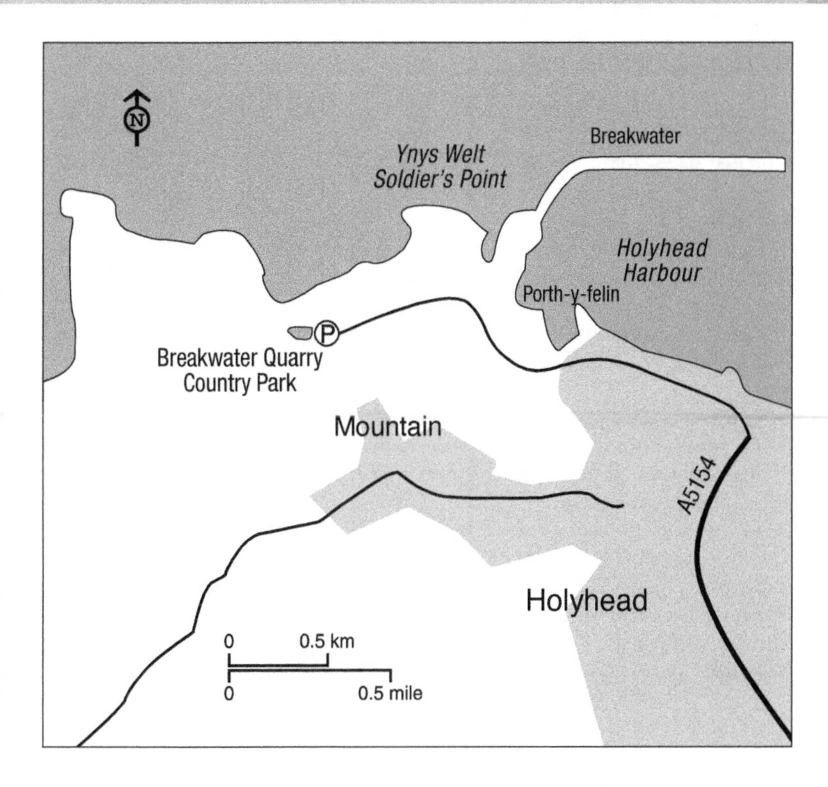

Timing

Winter, spring and autumn. Best viewing is up to two hours either side of high tide, and a telescope is certainly beneficial.

Access

Park in a car park of the Country Park, then go on foot to either of the two migration sites, while the Harbour can be viewed from many places.

CALENDAR

Resident: Cormorant, Shag, Peregrine, large gulls, Black Guillemot, Stonechat, Meadow and Rock Pipit and Raven.

December–February: Great Northern and Red-throated Diver.

March–May: Passage migrants throughout whole period, Wheatear usually the first, with the Breakwater Country Park being the best option.

June–July: Breeding Stonechat, Whitethroat, Meadow and Rock Pipits, Wheatear.

August–November: Passerines move through, and late in period Fieldfares, Redwings. This is the time for scarce and rare visitors with a little patience needed by the observer, who with good fortune will be blessed by scarce visitors and rarities like Red-breasted Flycatcher, Yellow-browed and Barred Warblers; even the chance of adding a new bird to the county list or daydream of a new species for Wales.

4 TREARDDUR BAY

OS Landranger Map 114
OS Explorer 262
Grid Ref: SH254788

Habitat

A secluded bay on a rocky and rugged coastline, which gives shelter from the almost constant southwesterly wind. The bay is surrounded by fields and coastal scrub. The area offers observers the chance of seeing the usual Welsh coastline specialties all year as well as Purple Sandpiper and sea-ducks during the winter and the chance of finding unusual migrants on passage.

Species

From spring to autumn, Manx Shearwaters and Gannets pass the coastline, sometimes close inshore, while during winter Purple Sandpipers roost on the rocks in the bay, sometimes up to 20. Goldeneye and Red-breasted Merganser can be seen in the calmer waters of the bay while terns feed here during the summer months. During passage periods, the surrounding area holds Cuckoo, Whinchat, Black Redstart, Sedge Warbler and Whitethroat, while almost anything can turn up. Scarcities have included Red-footed Falcon and Stone-curlew while the bay has held Surf Scoter. Breeding specialties include Stonechat and Chough which feed in the surrounding fields.

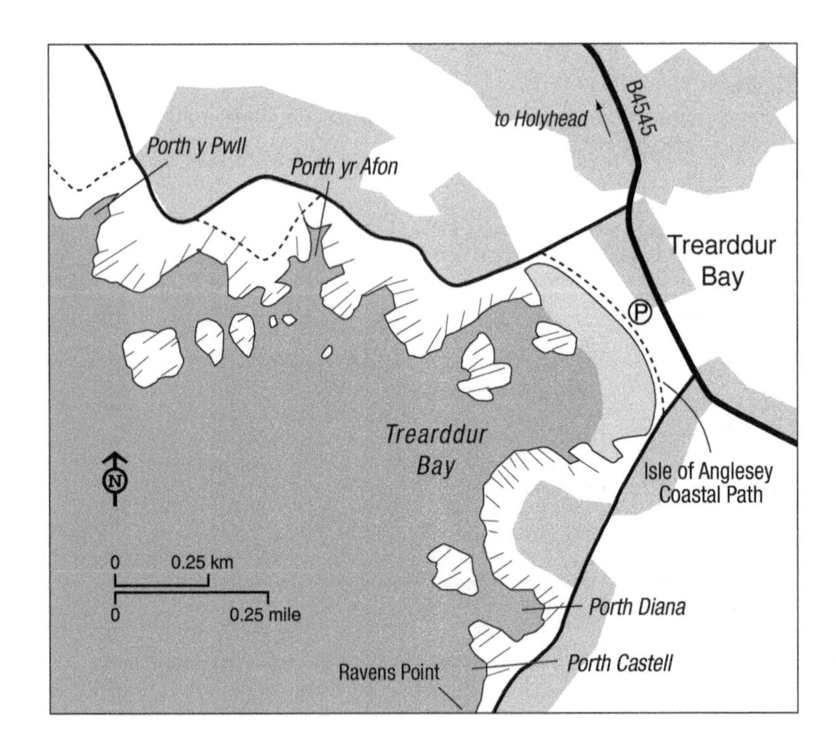

Timing
All year, though as this is a very popular holiday beach it comes into its own for birdwatchers in winter.

Access
Parking is available near the caravan site and along the B4545, OS Ref: SH255776. The coast path offers access in both directions. It is also worth checking the bays along the coast to the north-west as far as Porth Dafarch.

CALENDAR

Resident: Cormorant, Shag, Buzzard, Kestrel, Peregrine, Oystercatcher, large gulls, Rock Pipit, Stonechat, Chough and Raven.
December–February: Great Northern and Red-throated Divers, Fulmars return to cliffs and Gannets appear offshore. Goldeneye and Red-breasted Merganser in the bay, Turnstone and Purple Sandpiper on the rocks. Possibility of Black Redstart.
March–May: Passage migrants throughout period, Wheatear usually the first, with Sedge Warbler and Whitethroat later in May.
June–July: Manx Shearwater passage daily offshore, terns also seen, Sedge Warblers and Whitethroats in scrub.
August–November: Seabirds offshore usually include Manx Shearwaters, with numbers decreasing in October, and Common Scoters.

5 INLAND SEA AND BEDDMANARCH BAY

OS Landranger Map 114
OS Explorer 262
Grid Ref: SH275795

Habitat
The main road and rail links across mainland Anglesey cross onto Holy Island, splitting a large bay in half – into the Inland Sea and Beddmanarch Bay. Together they attract large numbers of ducks, waders and seabirds, especially during the winter and migration periods.

Species
During winter Great Northern Diver and Slavonian Grebe are present until spring, while a large flock of Pale-bellied Brent Geese can be found on each side of the embankment that separates the Bay and Inland Sea. Ducks include Wigeon, Scaup and Red-breasted Merganser, while waders can be found in good numbers. Black and Bar-tailed Godwits, Grey and Golden Plovers, Greenshank and Sanderling are all regular, as are Mediterranean Gulls.

Timing
Winter, spring and autumn. Best viewing is up to two hours either side of high tide, and a telescope is certainly beneficial.

Access
The area is best viewed from two locations; the first at Penrhos Nature reserve off the A5 just as you reach Holy Island, while the second from Four Mile

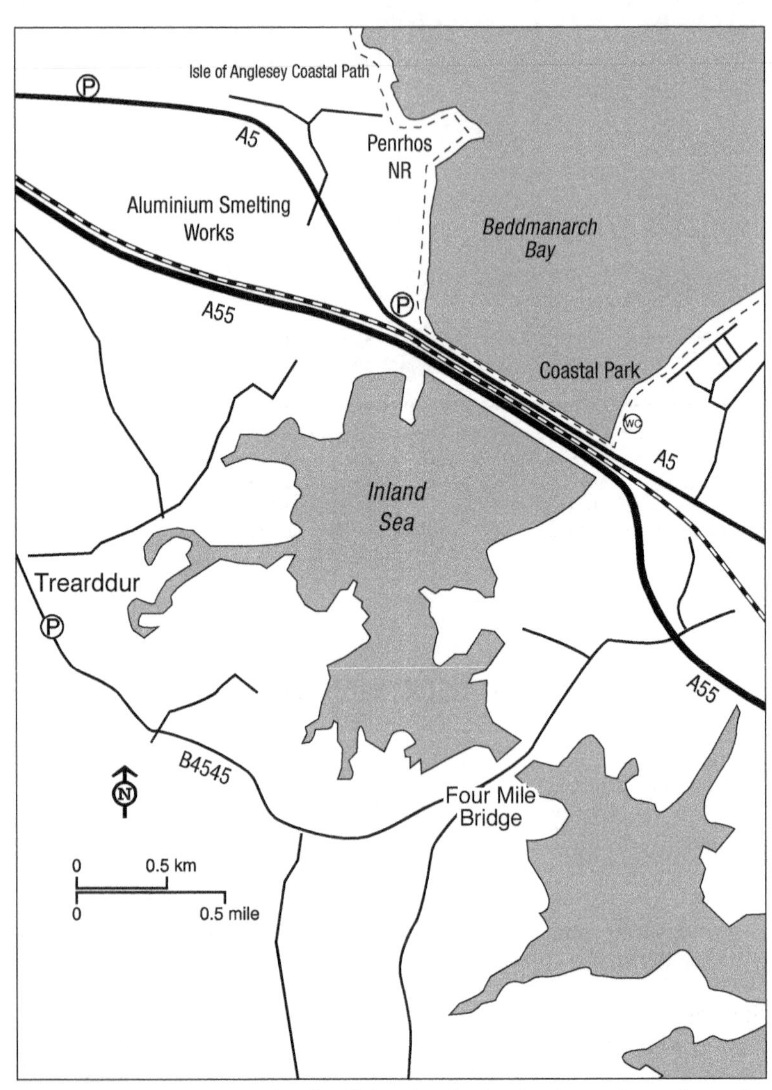

Bridge on the B4545 south of Holyhead and take the footpath north on the mainland side.

CALENDAR

Resident: Cormorant, Shag, Peregrine, Oystercatcher, Curlew, Turnstone, Herring, Lesser Black-backed and Great Black-backed Gulls, Stonechat, Linnet and Reed Bunting.
December–February: Great Northern and Red-throated Divers, Whooper Swan, Brent Geese, Wigeon, Teal, Gadwall, Pintail, Shoveler, Great Crested and Slavonian Grebes in either the Inland Sea or Beddmanarch Bay.
March–May: Passage waders throughout period, Ringed Plover, Dunlin, Sanderling, Redshank, Greenshank, Bar-tailed Godwit and Whimbrel.

June–July: Shelduck depart on their moult migration and will not return until late October. Terns from colonies elsewhere on Anglesey feed on both sides of the embankment.

August–November: Brent Geese return from September onwards. Wader passage under way, including Whimbrel, Bar-tailed and Black-tailed Godwits and Greenshank.

6 CARMEL HEAD

OS Landranger Map 114
OS Explorer 262
Grid Ref: SH291926

Habitat

This must be the remotest part of Anglesey and requires that extra bit of effort to reach. But your exertions are well rewarded by spectacular cliffs and hidden coves, moulded from some of the oldest rocks in Wales. Why Carmel? The name in Welsh – Pen Bryn-yr-Eglwys – translates as Hill or Mount of the Church. Now there is no trace of a church but it does not seem unreasonable to suggest there was one in the past, perhaps used by the miners seeking copper; the abandoned shafts and a chimney confirm an industry that probably commenced here in Roman times. This is a perilous coast for mariners: a marker painted on the West Mouse rock is matched by two on the Head erected in the 1860s by the Mersey Harbour Board – known as the White Ladies, they warn of the Coral Rock reef, only visible at low water on spring tides.

Species

Two miles offshore are The Skerries with the largest Arctic Tern colony in Great Britain, at least until 2019 when there were 2,814 pairs. In 2020 none nested, one suggestion being that the presence of a pair of Peregrines had caused the birds to move to nearby Cemlyn Lagoon (Site 8).

The Carmel Head cliffs provide nesting sites for Kestrels, Peregrines, Ravens and Choughs while Hooded Crows have occasionally been reported. Rock Pipits along the cliff line replaced by Meadow Pipits inland, while Wheatears, Stonechats and Whitethroats can be prominent. Easily overlooked but well worth the wait, a Black Guillemot or two is a distinct possibility, after all they breed not so far away on Anglesey. The wooded valley to the south of the Head offers nesting for Buzzards, while Grey Herons might breed, as they are regularly seen, the marshy lagoon a favourite feeding area. Woodland birds include Chiffchaff, Willow Warbler, Greenfinch and Goldfinch.

Scarce visitors have included Black Stork, Black Kite, Buff-breasted Sandpiper, Dotterel, Wryneck, Wood Lark, Subalpine Warbler, Richard's and Tawny Pipits, Isabelline Wheatear and Lapland Bunting. Enough here to whet the appetite while encouraging more observations.

Timing

Throughout the year, but spring and autumn most productive.

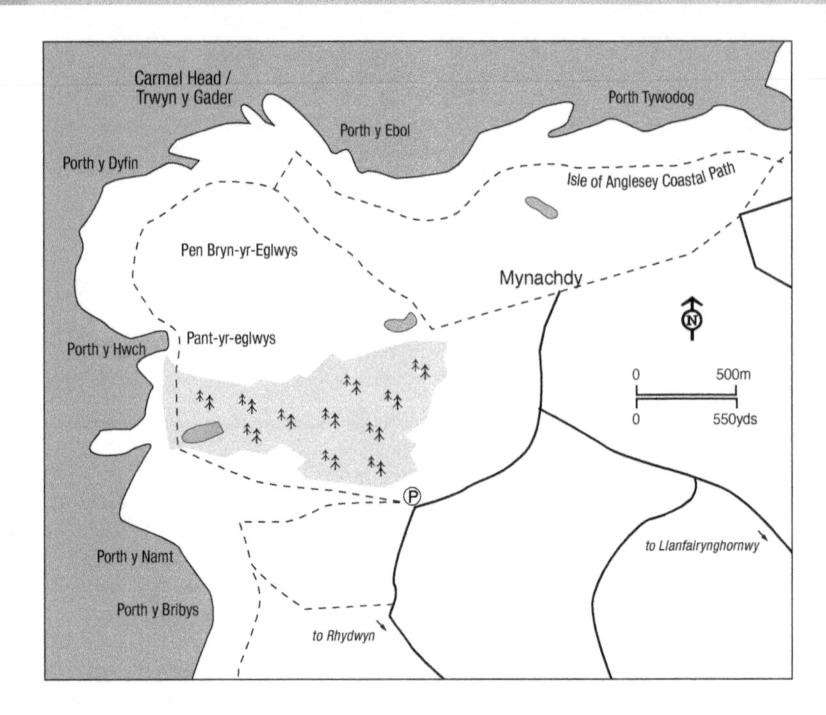

Access

Open throughout the year, except a section of the Isle of Anglesey Coastal Path south of the Head may be closed for Pheasant shooting between 14 September and 1 February.

CALENDAR

Resident: Grey Heron, Cormorant, Shag, Buzzard, Peregrine, Oystercatcher, Herring Gull, Great Black-backed Gull, Little Owl, Skylark, Meadow Pipit, Rock Pipit, Stonechat, Chough, Jackdaw, Raven, Linnet.
December–February: Great Northern and Red-throated Divers, Hen Harrier, Merlin, Purple Sandpiper, Turnstone, Short-eared Owl, Woodcock, Fieldfare, Redwing, Goldcrest, Brambling, Siskin, Lesser Redpoll, Snow Bunting.
March–May: Terns and Manx Shearwaters on passage at sea, Wheatears are usually the first summer visitor to arrive, while the Greenland race can be encountered in late April and May. Good passages of Swallows and martins and from early May there are Swifts.
June–July: Terns and Manx Shearwaters on feeding forays, the nearest shearwater colony that on Bardsey Island nearly 50 miles distant – though that is no distance for these great travellers. Gannets, perhaps from the nearest colony, that on Ireland's Eye off Howth, Co. Dublin.
August–November: Seabird passage throughout this period and well worth that extra effort required to reach the headland; all four skuas are likely with Arctic and Great Skuas most prominent. Passages of Manx Shearwaters and Kittiwakes and later in the period Guillemots and Razorbills.

7 ANGLESEY WETLANDS

OS Landranger map 114
OS Explorer 262
Website:
https://www.rspb.org.uk/reserves-and-events/reserves-a-z/valley-wetlands/

Habitat

Anglesey abounds in freshwater sites, from reservoirs such as Llyn Alaw and Llyn Cefni to the natural lakes and marshes in the shallow valleys that intersect the island on parallel lines to the Menai Strait. Indeed, the island contains some of the best areas of lowland open water and marsh in the whole of western Britain, and these lime-rich fens are unique in Wales and comparable in importance with the fens of East Anglia. Some 1,545 acres (625 ha) have been designated a Ramsar site, a wetland of international importance, while the reedbeds, mesotrophic lakes, fens, Purple Moor-grass and rush pastures are key habitats. In the first edition of this book, disappointment was expressed that so few were nature reserves: it is a pleasure to inform the reader that much has changed, with a number now managed because of their conservation importance, while footpaths and viewpoints enable good birdwatching on the remainder. Furthermore, the Anglesey Wetlands Strategy launched in 1998 has brought together the key organisations, both statutory and voluntary, to conserve the wetland biodiversity of the island through communication and cooperation and the promotion of good wetland management. Most have a rich aquatic flora, including rare plants such as Frogbit, Flowering-rush, Eight-stamened Waterwort, Lesser Bulrush, Great Fen-sedge, Blunt-flowered Rush, Marsh Fern and Narrow-leaved Marsh-orchid, to mention but a few. Good plants, and good invertebrates too, the lakes being excellent for dragonflies, Hairy Dragonfly and Variable Dragonfly among them. Less obvious are scarce water beetles and strong populations of Medicinal Leech at one of its few remaining locations in Wales.

Species

Breeding waterfowl include Great Crested and Little Grebes, Mute Swan, Greylag Goose (introduced), Canada Goose, Wigeon, Gadwall, Teal, Mallard, Shoveler, Pochard, for which Anglesey is the Welsh stronghold, Tufted Duck, Moorhen, Coot and Black-headed Gull. Garganey have nested in the past and small but variable numbers occur each spring, some lingering longer, giving rise to hopes of successful breeding.

Wildfowl numbers increase in winter when Pochard and Tufted Ducks may each rise to 400 at Llyn Alaw, while on Llyn Coron the number of Wigeon reaches 1,600, and Shoveler can exceed 200 – mostly on Llyn Penrhyn. Other visitors include Whooper Swans, on Llyn Alaw and in meadows at Llanddeusant a little to the west, Greenland White-fronted Goose, Gadwall, Goldeneye, Water Rail, Jack Snipe, Snipe and Woodcock. Wader passage can be good with Little Stint, Curlew Sandpiper, Dunlin, Ruff, Black-tailed Godwit, Greenshank, Green and Common Sandpipers. Casual visitors have included Night Heron, Spoonbill, Mandarin Duck, Ring-necked Duck, Smew, Osprey, Hobby, Quail, Pectoral Sandpiper, Red-necked Phalarope, Little Gull, Common, Whiskered,

Black and White-winged Black Terns. The only Welsh record of Falcated Duck was of a male on Llyn Traffwll in May 2008.

Often occurring along the shore or within close vicinity of the lakes, and depending upon the season, are birds such as Hen Harrier, Kestrel, Merlin, Peregrine, Oystercatcher, Lapwing, Snipe, Redshank, Barn and Short-eared Owls, Kingfisher, Whinchat, Stonechat, Grasshopper, Cetti's, Sedge and Reed Warblers, Lesser Whitethroat, Long-tailed Tit, Linnet, Lesser Redpoll, Siskin, Yellowhammer and Reed Bunting.

Timing

Early morning visits are best, as there is much less chance of birds having been disturbed. At the same time, however, ensure that you yourself do not cause disturbance.

Access

Cors Goch (SH500815) a National Nature Reserve, owned by the North Wales Wildlife Trust, is a fine area of acid heath, limestone grassland, fen and open water noted for its rich plant communities including orchids. Good for birds in summer and winter. Footpaths cross the reserve from limited parking on the minor road at SH503815 and provide some views of Llyn Cadarn.

Llyn Alaw (SH380865). At 777 acres (314.45 ha) this, the largest area of open fresh water on Anglesey, is the flooded valley of the Afon Alaw and principal

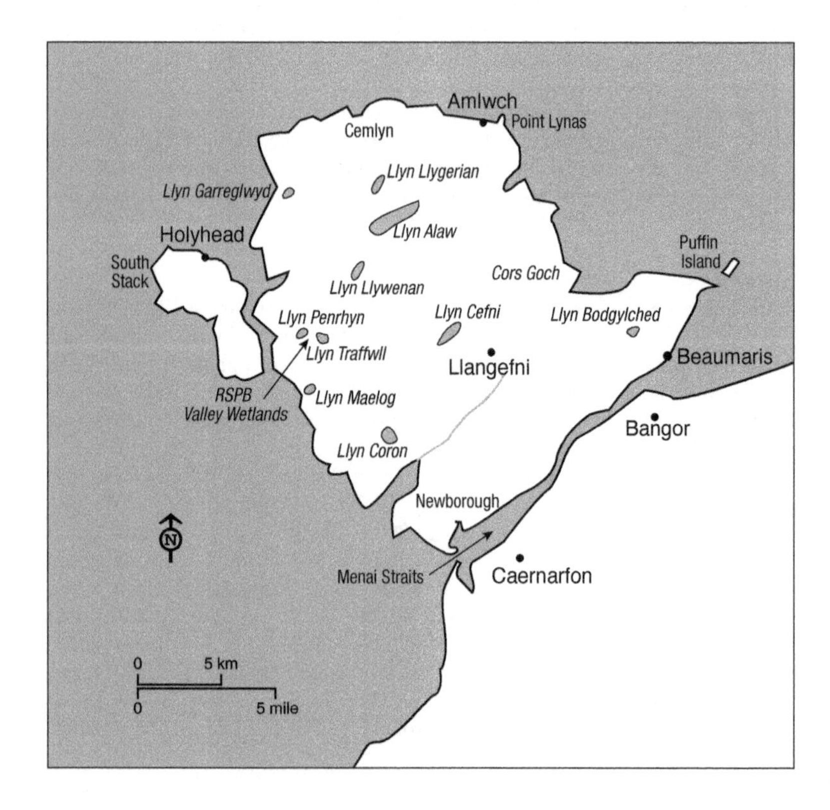

Anglesey reservoir for Dŵr Cymru – Welsh Water. Two hides overlook the north-east section of the reservoir and a small visitor centre with toilets is situated close to the dam at the south-west. With the highest waterfowl population of any inland water in Wales and a fine herd of Mute Swans. Low water levels in late summer prove attractive to passing waders, while Black Tern and Little Gulls are seen most years. There is only limited access, with viewing of the reservoir restricted to public footpaths.

Llyn Bodgylched (SH585770). May be observed from the B5109 west of Beaumaris or from the footpath which runs from the road towards the Bulkeley Memorial.

Llyn Cefni (SH445775). Surrounded by mature forestry plantations, the second-largest area of fresh water on Anglesey has easy access for viewing. Up to 80 Whooper Swans may use the lake to roost on in winter. Park at SH452783, just off the B5111 Llangefni to Amlwch road, from where a footpath leads to a hide. Alternatively, use the car park at SH444771 close to the waterworks on the southern shore from where the dam may be reached. The best birdwatching is on the northern half, but note that the reservoir can sometimes be closed in winter when shooting is taking place.

Llyn Coron (SH378700). A footpath from near Bodorgan Station runs close to the northern shore, while two others run to the lake from the A4080 Aberffraw road west of Llangadwaladr. Be especially careful not to cause disturbance to the wildfowl which often congregate in fields close to the lake. Visit here in spring or early summer.

Llyn Garreglwyd (SH309878). A fragment of water, surrounded by woodland to the southwest of Church Bay, from where a footpath leads to the lake.

Llyn Llygerian (SH346898). Park at the viewpoint SH346902, while a footpath further east leads south to a causeway at the eastern side of the lake.

Llyn Llywenan (SH346815). An unclassified road runs northward close from the B5109, half a mile (0.8 km) west of Bodedern to the west shore of this the largest natural freshwater lake on Anglesey. Access is limited and views should be obtained from the road, or from the car parks close to the north and south ends of the lake.

Llyn Maelog (SH325730). A footpath skirts much of the lakeshore, leaving the A4080 on the eastern outskirts of Rhosneigr and further east at SH327726. Alternatively use the lay-by on the road for a good general view.

RSPB Valley Wetlands, a wetland complex which includes Llyn Penrhyn (SH314770) and Llyn Traffwll (SH325770). Small car park and footpaths around the lakes.

CALENDAR

Resident: Little and Great Crested Grebes, Cormorant, Grey Heron, Mute Swan, Greylag Goose, Shelduck, Wigeon, Gadwall, Teal, Mallard, Shoveler, Pochard, Tufted Duck, Moorhen, Coot, Oystercatcher, Lapwing, Black-headed Gull and Cetti's Warbler.

December–February: Numbers of winter visitors peak, especially if there is hard weather in England and in Western Europe.

Additional species are likely to include Whooper Swan, Long-tailed Duck and Little Gull. On the heaths around some of the lakes there are wintering Hen Harriers, Merlins, Peregrine and Short-eared Owls, while resident Barn Owls are likely to be seen just before dusk. Woodcock frequent the plantations.
March–May: Winter visitors depart, summer migrants arrive throughout period, breeding species such as Black-headed Gull, Grasshopper Warbler, Sedge Warbler, Reed Warbler and Lesser Whitethroat.
June–July: Summer visitors and residents present. Breeding season draws to a close and wader passage commences, often the first species seen being Green Sandpiper.
August–November: Wader passage during August and some southward-moving terns call briefly. Duck numbers begin to increase during September, and by the end of period many winter visitors have arrived, including Bittern, Whooper Swan, White-fronted Goose, Pintail, Goldeneye and Water Rail.

8 CEMLYN LAGOON

OS Landranger map 114
OS Explorer 262
Grid Ref: SH330931
Website:
http://www.northwaleswildlifetrust.org.uk/reserves/cemlyn

Habitat

Cemlyn Lagoon on the north coast of Anglesey has been formed by a shingle-ridge storm beach which has effectively sealed off, save for an overflow channel, the inner reaches of Cemlyn Bay. Behind the beach, an area of brackish water is retained, being flooded by the sea at high spring tides. It has been a nature reserve since the early 1930s, first as part of the Hewitt Estate; then, on the death of Captain V. Hewitt – an extraordinary man, a pioneer aviator, racing car driver and a collector whose vast purchases included four mounted specimens and no fewer than thirteen eggs of the Great Auk, like Citizen Kane he used his money simply to 'buy things' – the lagoon was purchased by the National Trust and subsequently has been managed by the North Wales Wildlife Trust. The shingle ridge has an interesting maritime flora, with Sea Kale, Sea Beet, Thrift, Sea Aster, Sea Milkwort, Spring Squill and Sea Purslane.

Species

The highlight of Cemlyn is without question the tern colonies. Sandwich, Common and Arctic Terns all nest, though numbers of each can vary considerably from year to year. Over the last few years between 1,200 and 1,500 pairs of Sandwich Terns nested, the highest number ever recorded breeding in Wales, an estimated 11% of the total British and Irish breeding population. Some 10 pairs of Common and 25 pairs of Arctic Terns are also present. Roseate Terns have nested in the past, but not recently, while Little Terns only occur on passage. Among these rarities have been discovered a Bridled Tern in 1988, Sooty Terns in 2005 and 2020, and an Elegant Tern in 2021.

Other breeding species include Shelduck, Mallard, Red-breasted Merganser, Oystercatcher, Ringed Plover, Redshank and Black-headed Gull. A wide range of waders uses the muddy fringe of the lagoon and the nearby beach, including

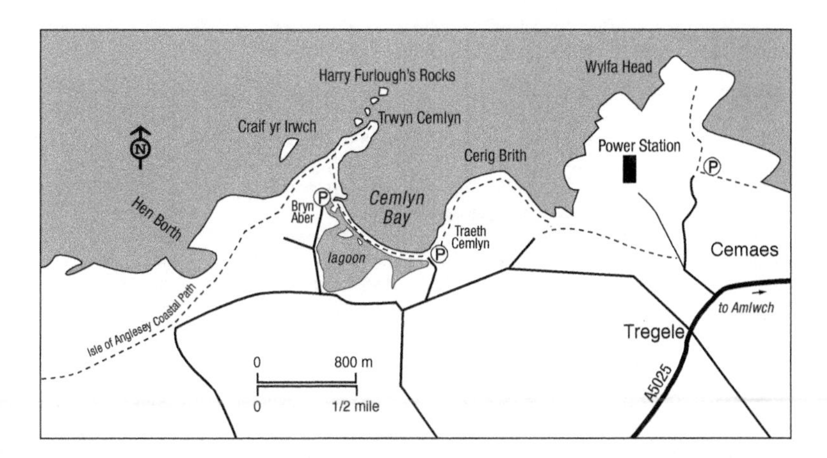

Grey Plover, Lapwing, Sanderling, Curlew Sandpiper, Purple Sandpiper, Dunlin, Curlew, Common Sandpiper and Turnstone. In winter, flocks of Wigeon and Mallard may rise to about 400 and there are always smaller numbers of Teals, Shovelers, Pochards, Tufted Ducks and Goldeneyes present. The sheltered waters of the bay throughout the winter months often contain Red-throated Diver, Great Crested Grebe, Common Scoter, Guillemot and Razorbill. Seabird passage offshore can be good in autumn and can include Manx Shearwaters in large numbers, petrels, skuas and terns. Occasionally Twite, Tree Sparrow, Lapland and Snow Buntings. More unusual species have been Squacco Heron, Spoonbill, Garganey, American Golden Plover, Terek Sandpiper, Little Gull, Caspian Tern, White-winged Black Tern, Shorelark, Richard's Pipit, Great Grey Shrike, Isabelline Shrike, Black-headed Wagtail and Wales's first Eastern Yellow Wagtail in September 2019.

Timing

Any daylight hours should yield sightings, but during the summer months the early morning may mean that you have the shingle ridge to yourself – at least apart from other readers of this book – plus hopefully a good selection of birds. Winter visits are probably best when the wind is in the south-west or west and Cemlyn Bay provides a sheltered area. Autumn seawatches off the Trwyn are worthwhile and should produce European and Leach's Storm Petrels and Sabine's Gull in suitable conditions.

Access

Leave the A5025 at Tregele (SH355927) on the unclassified but signed road. There are two car parks near the lagoon, at Traeth Cemlyn on the east side and near Bryn Aber on the west, and the best views may be had from the shingle ridge, but please keep below the crest to avoid disturbance. Please note that there are access restrictions in the interests of the tern colonies. A warden is resident in summer.

CALENDAR

Resident: Cormorant, Shag, Grey Heron, Shelduck, Mallard, Red-breasted Merganser, Kestrel, Coot, Oystercatcher, Ringed Plover, Lapwing, Redshank, Turnstone, Black-headed, Herring and Great Black-backed Gulls, Rock Pipit, Linnet, Reed Bunting.

December–February: Red-throated and Great Northern Divers, Great Crested Grebe, Wigeon, Teal, Pochard, Tufted Duck, Eider, Common Scoter, Hen Harrier, Peregrine, Purple Sandpiper, Grey Plover, Twite and Snow Bunting.

March–May: Passage waders include Sanderling, Whimbrel and Common Sandpiper. Terns arrive during April, Sandwich Tern usually being the first noted. Swallows, Sand and House Martins often feed over the lagoon in large numbers during April and May. Other summer migrants include Cuckoo, Wheatear, Whinchat and Yellow Wagtail.

June–July: All breeding species present, first return wader passage commences during July.

August–November: Wader passage well underway by late August, when a wide range of species can be expected. Terns usually present in variable numbers until late September. Manx Shearwaters pass offshore together with occasional other petrels, Common Scoter and skuas. Winter ducks begin to arrive by early October, and by the end of period midwinter species such as divers and swans usually much in evidence.

9 POINT LYNAS

OS Landranger Map 114
OS Explorer 263
Grid Ref: SH479934

Habitat

Point Lynas, a narrow promontory rising to almost 120 feet, extending northwards for half a mile (0.8 km) from the main coastline at the northeast tip of Anglesey seems made for seawatching, a fact recognised over two centuries ago, though for other purposes, when the Liverpool Pilotage Authority established a pilot station there (this is still in operation today). There is a good access road but no parking on the track beyond the mini roundabout; however, there is a new car park on the right of the road just before the roundabout. The lighthouse complex is a private house with ancillary self-catering accommodation. Further interest for the watchers at Point Lynas derives from regular sightings of Harbour Porpoise, while several species of dolphins are not infrequent.

Species

Seawatching first carried out in 1967 and 1968 quickly showed the possibilities here, especially following stormy conditions. Manx Shearwaters are frequently seen, being especially numerous on summer evenings with birds returning from fishing expeditions further north to their colonies on Bardsey and even the Pembrokeshire islands. The larger movements of September could well involve birds from the enormous colonies on Rum and other islands in the Inner Hebrides and from the smaller ones on Rathlin Island and the Copeland Islands off Northern Ireland. Undoubtedly, those from the tiny colony on the Calf of Man will come this way. It was from the Calf that the first Manx Shearwaters were described for science in the mid-17th century, hence the name Manx.

Small numbers of Fulmars pass, with peak movements in August and September. Like the Manx Shearwaters, Gannets passing here have come from the north, probably from the great colony on Ailsa Craig, sentinel of the Clyde, and from the small colony on Scar Rocks, Wigtownshire; others may well be wanderers from the south, from Grassholm, the only gannetry in Wales. Kittiwakes are the most frequent species, perhaps not surprisingly as this attractive gull is one of the most numerous seabirds in Britain. Many thousands pass Point Lynas each autumn, which is quite remarkable considering the decline in UK breeding numbers since around the turn of the millennium.

Most exciting of all to observe is the passage of auks, mainly Guillemots and

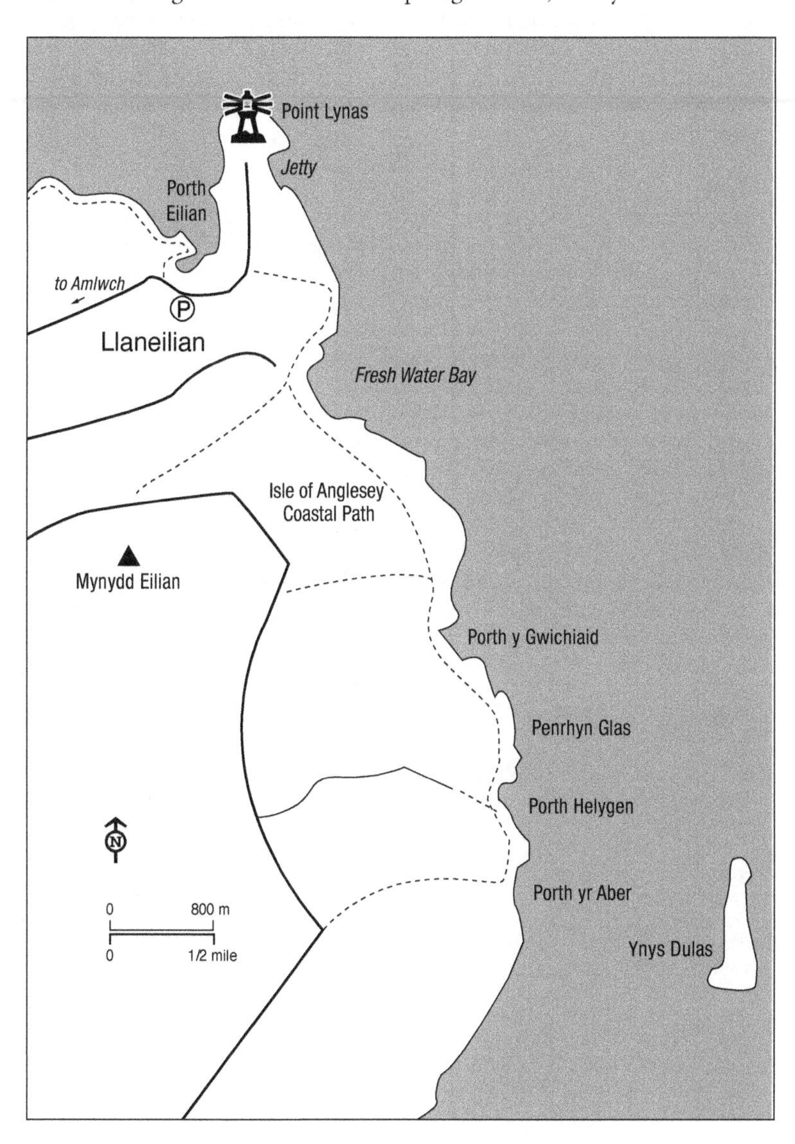

Razorbills, though very occasionally a Puffin or a Black Guillemot may be seen, or, if you are very lucky, even a Little Auk. Peak numbers are observed from October to December, by which time divers should be present.

Terns can be seen throughout the summer, birds from breeding colonies elsewhere on Anglesey and indeed throughout Liverpool Bay. The main passage south takes place during late August and September. Although most skuas, mainly Arctic and Great, pass at the same time, there are records of birds as early as mid-July while others may be seen as late as November. Other species seen at Port Lynas, though only in small numbers, include grebes, Sooty Shearwater, Leach's Petrel, Scaup, Eider, Long-tailed Duck, Common Scoter, Goldeneye, Red-breasted Merganser, Black Tern. More unusual records have been of Cory's Shearwater, Velvet Scoter, Little and Sabine's Gulls. Movements of passerines, especially in autumn when larks, pipits, thrushes and finches move through in numbers, with Lapland Buntings being among the scarce visitors recorded.

Timing

The first essential of seawatching is to ensure that you arrive and are ready to commence at daybreak. Most movement takes place in the hours immediately after dawn, and although, depending on the weather, there will be some movement throughout the day, it will generally not compare with that observed in the early hours. The weather is all important, not just on the morning of the watch but also during the preceding days, and at Point Lynas the essential ingredients are a period of stormy and unsettled conditions, with the wind in the northwest or the northern quarter. One of the reports laments what can happen when these conditions fail to materialise, as it recalls how 'observer enthusiasm fell in consequence'. As with any seawatching, very warm clothing is a must, together with suitable quantities of food and drink.

Access

Take any of the several unclassified roads off the A5025 south of Amlwch, or from the town itself, pass through the hamlet of Llaneilian, and continue to Porth Eilian and so to the point. A new car park is available on the right of the hill just before the mini roundabout.

CALENDAR

Resident: Cormorant, Shag, Oystercatcher, large gulls, Peregrine, Rock Pipit, Stonechat, Raven and Linnet.

December–February: Divers and Great Crested Grebe usually offshore, occasional sea-ducks, Fulmars return to nearby cliffs. Some passage of auks through whole period.

March–May: Terns noted from early April onwards. Passerine passage develops from April onwards, most noticeable being the movements north and east of Swallows, smaller numbers of Sand Martins and a few House Martins. Pay attention to the bay on the western side of the headland, Black Guillemots are a possibility.

June–July: Terns still present, passage waders seen late in period, regular feeding movements of Manx Shearwaters.

August–November: Peak passage time for Common Scoter, all seabirds including skuas and best time of year for scarce species. Large Swallow and martin passage in mid-September, Skylarks, Meadow Pipit, Fieldfare, Redwing, Starling and Chaffinch movement from mid-October. Divers are now likely and so are Purple Sandpipers, though these will need an extra effort to locate.

10 TRAETH DULAS

OS Landranger Map 114
OS Explorer 263
Grid Ref: SH482886

Habitat

An area of extensive saltmarsh and mudflats which can be viewed from your vehicle.

Species

The area holds good numbers of typical estuarine species such as Shelduck, Wigeon and Red-breasted Merganser. Careful scrutiny should produce Bar and Black-tailed Godwit, Grey Plover and Turnstone, while Curlew Sandpiper and Little Stint are annual during autumn. The saltmarsh attracts good numbers of Snipe and Jack Snipe in winter. A walk around the surrounding scrub and fields should yield Stonechat, Sedge and Grasshopper Warblers, Whitethroat and Lesser Whitethroat, Yellowhammer and Reed Bunting. This is also one of the most reliable sites in North Wales for Red-legged Partridge.

Timing

Close views of waders and duck can be had on the rising tide, during winter and passage periods.

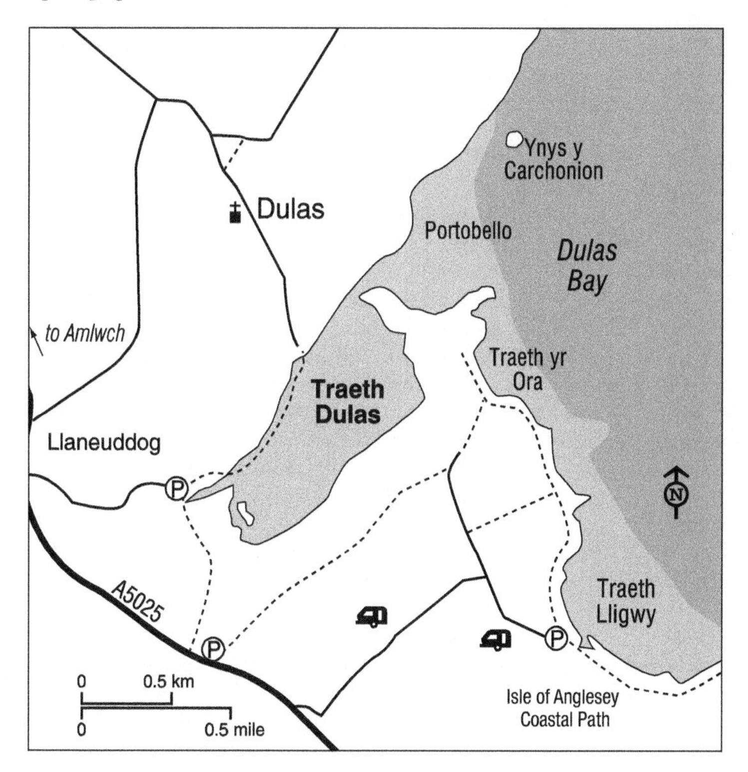

Access
Ample parking is available at the end of the single-track road (SH476881), reached from the A5025.

CALENDAR

Resident: Grey Heron, Little Egret, Red-legged Partridge, Stonechat, Yellowhammer, Reed Bunting.
November–February: Shelduck, Wigeon, Red-breasted Merganser, Grey Plover, Ringed Plover, Redshank, Dunlin, Turnstone, Snipe and Jack Snipe.
March–May: Wader passage, Grey Plover, Redshank, Greenshank, Dunlin, Curlew, Whimbrel, Bar-tailed and Black-tailed Godwit and Turnstone.
June–July: Stonechat, Sedge and Grasshopper Warblers, Whitethroat and Lesser Whitethroat.
August–October: Wader passage including Bar and Black-tailed Godwit, Whimbrel, Greenshank, Dunlin, Ringed Plover.

11 RED WHARF BAY

OS Landranger Map 114
OS Explorer 263
Grid Ref: SH540800

Habitat
Red Wharf Bay is an extensive mudflat backed by scrub and fields. When the tide is in, it creates a shallow bay that attracts divers and sea-ducks, while when the tide is out, the area is a haven for waders.

Species
During the winter the bay can be alive with waders and duck, especially at low tide. Large flocks of Wigeon, Teal and Pintail feed in the channels, while waders include Greenshank and Black-tailed Godwit. The site was made famous by the first ever Green Heron recorded in Wales, in November 2005, which spent its time feeding alongside the numerous Little Egrets that are resident here. Peregrine, Sparrowhawk and Merlin hunt the saltings for waders and small birds including Linnet, Reed Bunting and occasional Yellowhammer. At high tide, Red-throated Diver and Common and occasionally Velvet Scoter feed in the bay along with occasional Great Northern Divers and Slavonian Grebes among the numerous Red-breasted Mergansers and Great Crested Grebes. Snow Buntings are almost annual around the car park in winter.

Timing
Medium tides, for at low water most birds will be the best part of a mile distant. High tide in winter for sea-ducks and divers.

Access
By narrow lanes from Llanddona and Pentraeth with car parks beside the Bay or in the village of Red Wharf Bay on the western shore, here a public house adds to the attraction while the nearby Trwyn Dwlban offers a good vantage point on all tides. The Wales Coast Path provides access along the whole shore.

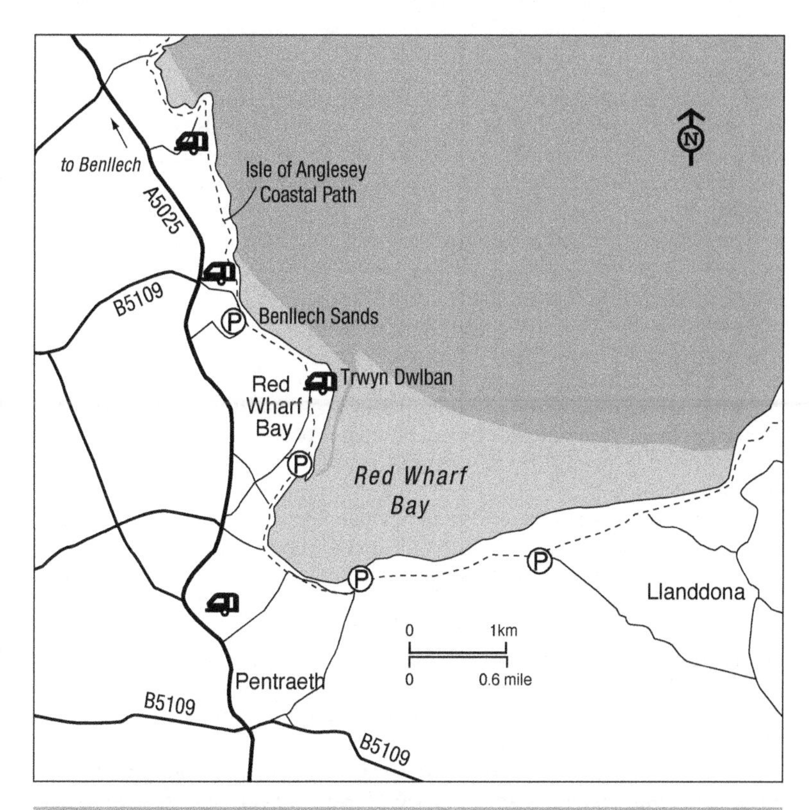

CALENDAR

Resident: Cormorant, Little Egret, Heron, Oystercatcher, Curlew, Black-headed and Herring Gulls, Linnet and Reed Bunting.
November–February: Great Northern and Red-throated Divers, Great Crested and Slavonian Grebes, Wigeon, Teal, Pintail, Red-breasted Merganser and Common Scoter. Waders include Jack Snipe, though these are remain well hidden in the saltmarsh. Raptors include Peregrine, Merlin and Sparrowhawk.
March–May: Passage migrants throughout whole period, waders such as Greenshank and Black-tailed Godwit.
August–October: Wader passage, including Redshank, Greenshank, Black-tailed Godwit, Curlew, Whimbrel, Ringed Plover and Dunlin.

12 PUFFIN ISLAND AND THE PENMON PENINSULA

OS Landranger Map 114
OS Explorer 263
Grid Ref: SH651821

Habitat

Puffin Island, also known as Priestholm and Ynys Seiriol, the island of Saint Seiriol, who lived there in the 6th century, is some 70 acres (28 ha) in extent and rises to 163 feet (50 m). It is separated from the Penmon Peninsula, the easternmost part of mainland Anglesey, by a sound nearly half-a-mile (0.8 km) wide. John Ray, greatest of all field naturalists visited the island in 1662 and noted the Puffin colony. Unlike on many of our other islands, there is a veritable forest of Elder covering up to a third of the plateau; other prolific plants include Alexanders up to 10-feet (over 3 m) high, Burdock, Hemlock and Henbane.

The Penmon Peninsula, especially its north coast, is of interest for its seabird colonies. There are extensive quarry workings, the Ordovician limestone having been shipped away in large quantities by sea. When visiting the peninsula, look for the remarkable dovecote close to the church of St Seiriol at Penmon. Although the church dates from the 12th century, the dovecote was built by Sir Richard Bulkeley in around 1600; it contains about 1,000 nest holes, which were reached by a ladder from a centre pillar with corbelled steps. What a sight and sound the dovecote must have been when fully occupied!

Species

John Ray, on his visit in 1662, found huge numbers of Puffins on Puffin Island; a century or so later, Thomas Pennant likened the birds flighting over the island to bees; while at the beginning of the 19th century it was thought 50,000 pairs nested. Large numbers were taken for food, not just locally but for sale far afield, even into England. Lewis Morris (1701–1765) recorded how 'The young ones before they are quite feathered, are fledged and opened and strewed with pepper and salt and broiled and eat pleasant enough. But the nice way of managing them is to pickle them, and these are sent as rarities for ye Tables of the Great. Puffins which pickled bring profit.'

Brown Rats scrambled ashore from a shipwreck in 1817 and quickly decimated the Puffin colony, so that by 1836 all had been exterminated. By around

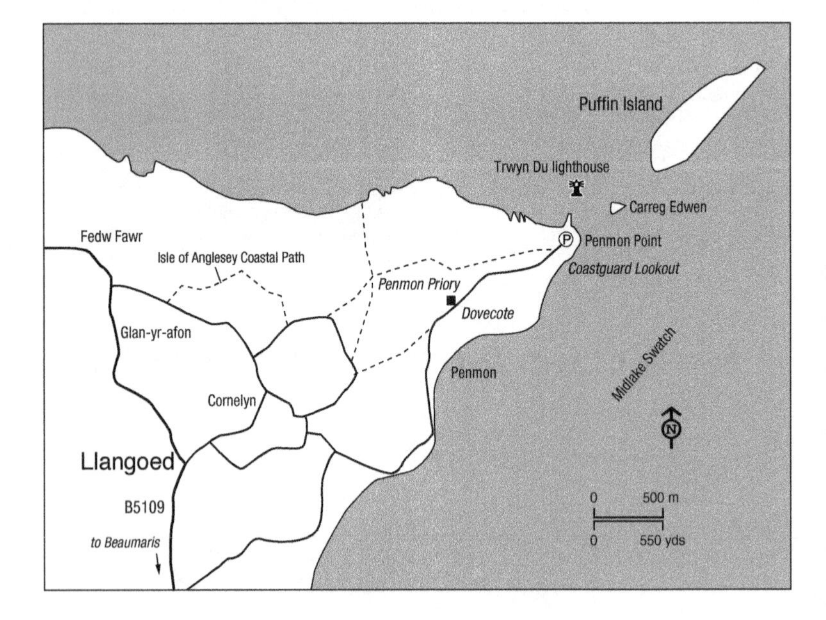

1880 the rats were eradicated and the Puffins rapidly returned, with 2,000 pairs in 1907. Then, although how is unknown – one speculates a deliberate action – the rats returned about 1960 and, once again, Puffins disappeared; another extermination programme in 1998 proved successful and slowly, this iconic seabird is returning, with 29 apparently occupied burrows in 2019.

Few Black Guillemots nest in Wales and those that do are mainly confined to Anglesey, with occasional sightings in Caernarvonshire and at least one pair breeding in Pembrokeshire. They returned to Anglesey to nest in the late 1940s, having been recorded there previously by Thomas Pennant in the 18th century. The population remains quite small, and their stronghold (though can one call a handful of pairs a stronghold?) is the north side of the Penmon Peninsula and Holyhead Harbour. Why so few in Wales? On the other side of the Irish Sea this species occurs in many localities, extending to the far southwest in Co. Cork and Co. Kerry. It seems strange that it has not become more widely established on the coast of Wales: a question of avian distribution still to be answered but possibly due to sea depth and nesting opportunities.

The Cormorant colony on Puffin Island is the largest in Wales, with around 700 pairs. Other seabirds of Puffin Island and the peninsula include Fulmars, Shags, Kittiwakes (some 300), Guillemot (3,600–4,200 birds) and Razorbills (fewer than 500). The discovery of an incubating Eider Duck on Puffin Island in May 1997 was the first breeding record for Wales. Pairs bred for a couple of years after this, but unfortunately, a breeding population has not established itself.

A Forster's Tern, a North American species, took up residence in North Wales in late summer 1984, during which time it was seen at a number of locations for the next five years, including Penmon. Wales's only record of Chimney Swift occurred here, following Hurricane Wilma, in November 2005.

Timing

No particular time of day is the best for a visit, though the north side of the peninsula is best visited when the wind is in the southerly quarter.

Access

Take the B5109 eastwards from Beaumaris to Llangoed, beyond which an unclassified road meanders close to the south shore of the peninsula to Penmon and beyond to the easternmost part of Anglesey. The western section of the peninsula is best reached by going north from Llangoed to Glan-yr-afon and thence to Fedw Fawr. Landing is not permitted but several companies offer boat trips to the island. Enquiries should be made in Beaumaris or online.

CALENDAR

Resident: Cormorant, Shag, Buzzard, Kestrel, Peregrine, Oystercatcher, large gulls, Black Guillemot, Rock Pipit, Stonechat, Chough, Jackdaw, Raven, Linnet, Yellowhammer.

December–February: Divers usually offshore, especially on south side of the peninsula, also various sea-ducks. The first Fulmars return. Waders on some sections of southern shore, Purple Sandpiper and Turnstone on the rocky areas. First Guillemots and Razorbills return to the cliffs, Kittiwakes gather at their colonies by end of period.

March–May: Returning auks more frequent, but not resident until April. Terns are seen

during early April, Sandwich Tern usually the first to arrive. Puffins from early April. Wheatears and Ring Ouzels are among the early migrants, the former nesting in the coastal areas. Other summer visitors include Cuckoo, Grasshopper and Sedge Warblers, Lesser Whitethroat.
June–July: All breeding species present. Terns regular offshore, although the nearest colonies are some 24 miles (40 km) away. Other offshore visitors, though these will have travelled greater distances, are Manx Shearwater and Gannet.
August–November: Auks have departed by early August, though late in period many moving offshore. Wader and seabird passage under way, followed by the arrival of winter ducks and the occasional diver and grebe. Although seawatching at Penmon may not be so productive as elsewhere, scarce species have included Cory's Shearwater, Velvet Scoter and Sabine's Gull.

13 CHURCH ISLAND AND COED CYMOL

OS Landranger Map 114
OS Explorer 263
Grid Ref: SH551717
Websites:
https://angleseyisle.co.uk/church-island
https://www.anglesey-hidden-gem.com/coed-cyrnol-menai-bridge.html

Habitat

At the very heart of the Menai Strait and squeezed between the two bridges linking Anglesey with the rest of Wales: the eastern one opened in 1826, while the Britannia Bridge to the west built 1850 now carries both road and rail traffic. Church Island, in Welsh Ynys Tysilio after the 6th-century Saint Tysilio is less than three acres in extent and entirely taken up by a 15th-century church and graveyard. On the summit, a fine memorial to mostly local men who did not return to these parts from two World Wars. Between the two bridges are the ominously named Swellies, known since medieval times for their shoals and rocks with whirlpools on this the most treacherous section of the Menai Straits.

Species

It was here in 2002 that a pair of Little Egrets first nested in North Wales, successfully rearing four young before returning the following year with others to become part of a heronry near Bangor. Grey Herons stalk the shallows while Cormorants and Shags from colonies on the outer coasts of Anglesey are regularly seen, being joined in the summer months by terns from the coastal colonies. Shelduck are always present save when on their late summer moult migration, while winter visiting duck include Wigeon, Teal and Red-breasted Mergansers. Oystercatchers, Curlews and Redshank forage along the winter shoreline; Common Sandpipers pause during spring and autumn migration. Do not ignore the bridges, Ravens which have nested and Peregrines find these convenient vantage points. The woodlands, though small, make a pleasing contrast, with Great Spotted Woodpecker, Nuthatch and Treecreeper joined in summer by Blackcap, Garden Warbler, Chiffchaff and Willow Warbler.

Timing

All year.

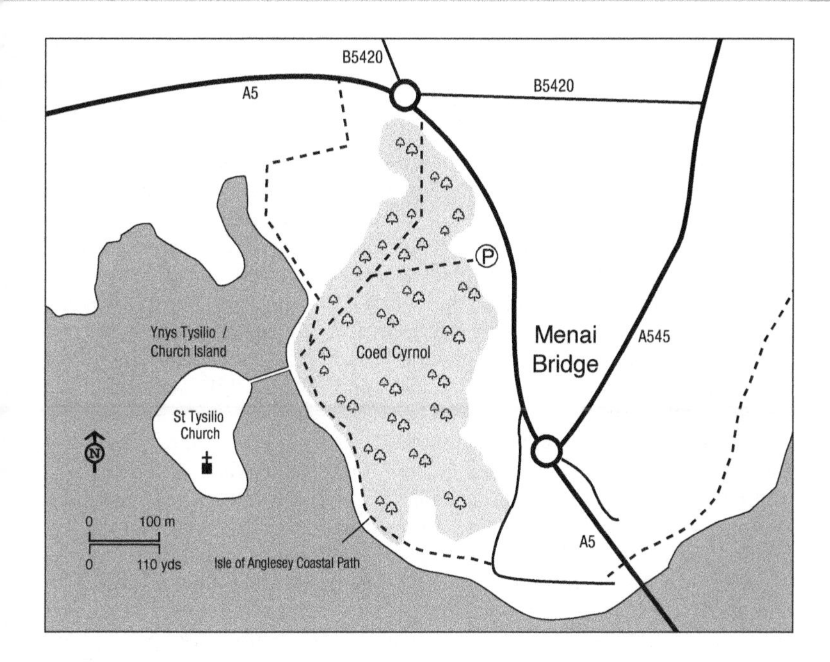

Access

Use the Coed Cymol pay-and-display car park. The Wales Coast Path provides access to both shores, while there is disabled access to Church Island.

CALENDAR

Resident: Greylag Goose, Shelduck except in late summer, Cormorant, Little Egret, Grey Heron, Buzzard, Sparrowhawk, Peregrine, Oystercatcher, Herring Gull, Great Spotted Woodpecker, Grey Wagtail, Wren, Robin, Blackbird, Song Thrush, Mistle Thrush, Goldcrest, Coal Tit, Blue Tit, Great Tit, Nuthatch, Treecreeper, Chaffinch, Goldfinch.
December–February: Wigeon, Teal, Red-breasted Merganser, Shag, Curlew, Redshank, Fieldfare, Redwing.

March–May: Common Sandpipers on passage, summer visitors arrive.
June–July: Terns from colonies elsewhere in Anglesey forage here.
August–November: Common Sandpipers commence return passage south while winter visitors arrive from early October onwards.

14 NEWBOROUGH WARREN AND FOREST

OS Landranger Map 114
OS Explorer 263
Grid Ref: SH398644
Website:
https://naturalresources.wales/days-out/places-to-visit/north-west-wales/newborough/?lang=en

Habitat

The southwest extremity of Anglesey is dominated by a huge area of sandhills, saltmarshes and dune grasslands, between the mouth of the Menai Straits and the estuary of the Afon Cefni. Prior to the 14th century this was a prosperous agricultural area, even with its own harbour at Abermenai. A series of great storms changed all of this as vast quantities of sand were blown inland. The fields and trackways quickly became choked, crops could no longer be grown, there was little grazing, houses and farm buildings began to disappear, and the population, unequal to the struggle, was forced to move inland. Subsequent centuries saw the wind mould the dunes into four main ridges, separated by flat hollows known as slacks, a wild remote area which for long periods was the haunt only of those who grazed their stock and of the rabbiters – for, as the name suggests, Newborough held a vast population of this animal until the arrival of myxomatosis.

In 1948, the Forestry Commission began planting the northern section of the dunes, the area now known as Newborough Forest, which with its serried ranks of Corsican pines extends for some 2,000 acres (810 ha), the largest forest on Anglesey. Fortunately, the whole of the dunes was not taken over for forestry, and in 1955 a National Nature Reserve was established, so that now 3,590 acres (1,453 ha) of one of the finest dune systems in Britain is fully protected. In addition to the huge plantations, there are three distinct areas: the eastern dunes between Llyn Rhos-ddu, the Braint Estuary and the sea; the rocky promontory of Ynys Llanddwyn; and the saltmarshes fringing the eastern shore of the Cefni Estuary, including Malltraeth Pool (Site 15).

Species

Birds of prey were, in years gone by, a speciality among the sandhills, where both Montagu's Harrier and Merlin nested. Now, alas, the former is but a rare visitor, indeed in some years none is recorded in Wales. They last nested at Newborough in 1964 after being harried by egg-collectors, while the growth of trees was now such as to make their breeding sites untenable. The shores of Ynys Llanddwyn provide a range of rocky habitats not found elsewhere at Newborough. Cormorants and Shags nest on Ynys Adar – Bird Island off the south-west shore – while Turnstones can be seen throughout the year and may at times number up to 70; the smaller numbers of Purple Sandpipers are restricted to the winter months. The sheltered waters of Llanddwyn Bay attract a number of birds, mainly as winter visitors, including Red-throated and Great Northern Divers, Great Crested and Slavonian Grebes, Eider, Long-tailed Duck, Common Scoter, Goldeneye and even occasionally Black Guillemots. Red-breasted Mergansers have nested in Anglesey since 1953, and Newborough is now one of their regular sites, so you may be fortunate to see an attentive female and her brood during midsummer. Anglesey is the headquarters in Wales for breeding terns: Common, Arctic, Roseate and Little Terns have all nested in the past at Llanddwyn, and further east at Abermenai, and are regularly seen fishing offshore.

Forest birds include Sparrowhawk, Goldcrest, various tits, Siskins, Redpolls and with luck, Crossbills. There is a large Raven roost in the conifers. As many as a thousand may be seen during the winter months, with birds coming from all over the island and Caernarfon to roost in the pines.

In the dunes you should be able to locate birds such as Whinchat, Stonechat,

Grasshopper and Sedge Warblers and Whitethroat, and in winter Hen Harrier and Merlin, even quartering Barn and Short-eared Owls, while Ospreys appear on passage. Llyn Rhos-ddu provides a further variation in habitat; this small freshwater pool attracts Shoveler, Pochard, Tufted Duck, Goldeneye in winter, while Coots and occasionally Little Grebes breed.

Timing

For both Llyn Rhos-ddu and Llanddwyn, the timing of visits is not so critical, except for the latter on spring tides.

Access

Access to the shore at Llanddwyn Bay is possible by footpath direct from Newborough, or alternatively leave the A4080 at SH424656 and follow the road right through to the shore car park, where facilities for wheelchair users include a nature trail and toilets. A brisk walk along the shore brings you to Ynys Llanddwyn where there is an exhibition in the Pilots' Cottages, or for the more energetic the long walk may be taken southeast towards Abermenai Point. Several other footpaths are open through the forest and dunes, one of which passes close to Llyn Rhos-ddu with its hide, another crosses the dunes to the Braint estuary.

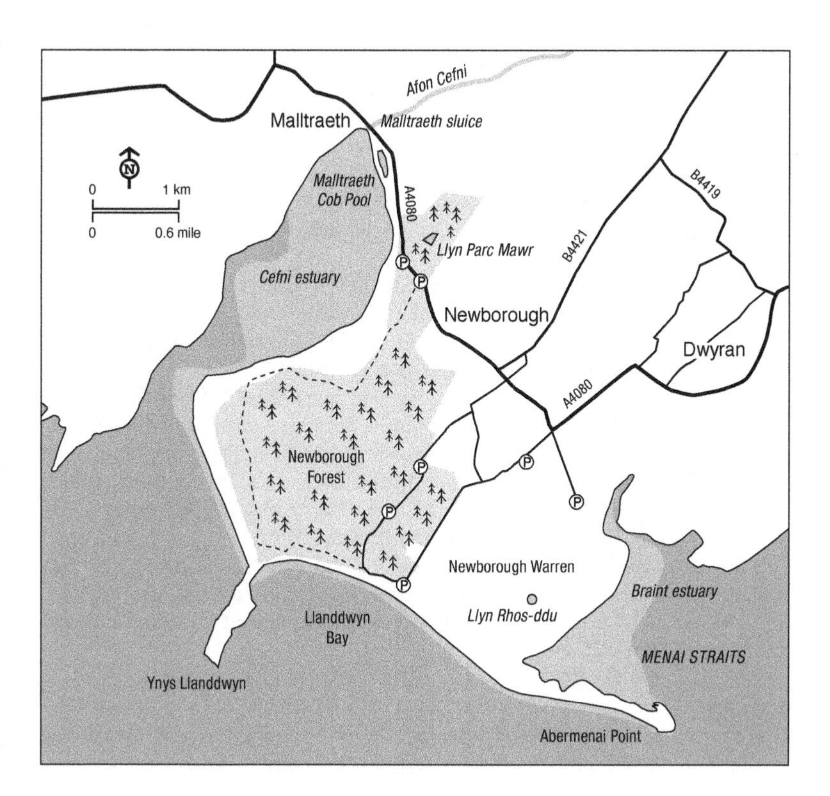

CALENDAR

Resident: Little Grebe, Cormorant, Shag, Coot, Sparrowhawk, Kestrel, Red-legged Partridge, Turnstone, a wide range of passerines of woodland edge and open country including Raven, Siskin, Lesser Redpoll and Crossbill.
December–February: Red-throated and Great Northern Divers, Great Crested and Slavonian Grebes, Shoveler, Pochard, Tufted Duck, Common Scoter, Goldeneye, Red-breasted Merganser, Hen Harrier, Merlin, Woodcock in the Forest, roosting Ravens and Snow Bunting.
March–May: Departure of winter and arrival of summer visitors, Sandwich Terns being one of the first to be seen.
June–July: Breeding season coming to an end, first wader passage noted in the estuaries. Terns may be seen passing offshore.
August–November: Continual passage of summer visitors until September, when there can be large movements of Swallows and martins. Winter visitors arrive in early October, large numbers of Fieldfares and Redwings from the middle of the month in some years, Woodcock in the forest.

15 MALLTRAETH COB AND LLYN PARC MAWR

OS Landranger Map 114/115
OS Explorer 262/263
Grid Ref: SH409683

Habitat

It was to Malltraeth that the noted bird artist Charles Tunnicliffe came in 1945, and here he sketched and painted many of the birds from the surrounding area. His studies of wildfowl and waders on Malltraeth Cob Pool are among his most outstanding work and many can be seen at Oriel Ynys Môn, the museum and gallery in Llangefni. Those unable to visit the gallery should search out his books such as *Shorelands Summer Diary* – Shorelands being the name of his house.

The mouth of the River Cefni at SH408687 is a recommended starting point, with a variety of habitats within a short distance. The large estuary and saltmarsh to the east is worth a scan for waders and raptors in winter. Hen Harrier, Merlin, Water Pipit are all possible. The freshwater pool and the marsh to the south of Cefni bridge can be viewed from the A4080 with care, but preferably from the coastal embankment. The pool holds good numbers of Pintail and waders through the winter, although it is hampered by high water levels. The fields inland of the A4080 hold flocks of Lapwing and Golden Plover in winter. They are also worth checking for egrets and raptors. The River Cefni inland can yield waders and Kingfisher too.

Look also for Llyn Parc Mawr to the east of the A4080, a relatively new lake having been established by Forest Enterprise as a visitor facility. It is a somewhat odd lake but holds good numbers of Gadwall in winter, and the woodland now has a healthy population of Red Squirrel.

Species

The saltmarshes of the Cefni Estuary extend for over 400 acres (160 ha), dotted with pools of brackish water and tiny creeks. Greylag Geese, which after introductions breed at several sites on Anglesey, regularly flight to the marshes, where they are joined by Canada Geese, several pairs of which breed close by. Shelducks breed in the forest and among the sandhills, and bring their broods rapidly to the estuary for safety.

The Cob Pool at the nearby village of Malltraeth holds good numbers of Pintail during winter, while passage periods attracts numerous waders. Black-tailed Godwit, Greenshank, Ruff and Spotted Redshank are possibilities, and the area has produced a number of rarities such as Broad-billed, Baird's and Pectoral Sandpipers. The saltmarsh is a traditional Hen Harrier roost, where Peregrine, Merlin and Barn Owl can also be seen at dusk during the winter months.

Timing

For ducks and waders it is worth visiting during the winter and passage periods, while the reedbeds contain many breeding species in the summer. Best visited in the morning when the sun is behind you and on a rising tide.

Access

There are two Forest Enterprise car parks; use these for visiting the Cefni estuary, Malltraeth Cob Pool and for Llyn Parc Mawr (SH414672) where two hides are available overlooking the lake and where, in addition to waterfowl, woodland birds such as Crossbills come to drink. From the same car parks, one can walk southwest for some distance along the estuary shore.The Malltraeth sluice is also a good spot to look for Kingfisher and Spotted Redshank.

CALENDAR

Resident: Canada Goose, Greylag Goose, Mallard, Tufted Duck, Teal, Kingfisher, Water Rail and Crossbill.

November–February: Winter influx of ducks and waders on the marsh and Cob pool, including Pintail, Wigeon, Gadwall, Shoveler, Oystercatcher, Redshank, Lapwing and Golden Plover.

March–May: Wader passage may include Black-tailed Godwits, Greenshank, Ruff and Spotted Redshank. Marsh Harrier and Garganey may also be seen. In the reedbed are Reed, Sedge and Grasshopper Warblers.

June–July: Reed, Sedge and Grasshopper Warblers in the Cob Pools reedbed, with the possibility of Water Rail.

August–October: Wader passage under way, including Black-tailed Godwit, Whimbrel, Spotted Redshank, Ruff and Greenshank.

16 RSPB CORS DDYGA

OS Landranger Map 114/115
OS Explorer 262/263
Grid Ref: SH463725
Website:
https://www.rspb.org.uk/reserves-and-events/reserves-a-z/cors-ddyga/

Habitat

RSPB Cors Ddyga comprises of grazing marshes and pools, with extensive reedbeds and willow scrub, making it one of the largest lowland wetlands in

Wales. The RSPB has created reedbeds and lowland wet grassland from what was sheep pasture only 25 years ago. Access has been made somewhat easier with the Cefni cycle track and footpaths into the reserve. The Cors Ddyga Site of Special Scientific Interest (SSSI) is one of just three in Wales designated for the richness of their aquatic invertebrates, such as dragonflies and water beetles. The reedbeds are home to Otters, Water Voles, wetland birds and more than 30 species of scarce wetland plants.

Species

Typical wetland birds can be seen including Reed, Cetti's, Sedge and Grasshopper Warblers, also Lapwing during summer, while Marsh Harrier, Bittern and even Savi's Warbler have bred in recent years. Savi's Warbler heard in mid-June 2019 was present until 25 July. Two weeks previous a male and female were observed and a breeding attempt suspected. Subsequently, one was seen carrying a faecal sac and on another day carrying food: the first evidence of a Savi's Warbler breeding in Wales. The bird is named after Paolo Savi, an Italian naturalist who in 1821 recognised that it was different from other Sylviidae warblers. Garganey is a possibility during passage periods. During winter the whole area holds many ducks, including Teal, Shoveler, Pochard and Tufted Duck. Great and Cattle Egrets and Glossy Ibis are also seen occasionally.

Timing

For ducks and waders, it is worth visiting during the winter and passage periods, while the reedbeds contain many breeding species in the summer. A circular walk from the Cors Ddyga car park is recommended or an out-and-back along the cycle path towards Pont Marquis.

Access

Access to the RSPB Cors Ddyga carpark at Tai'r-gors is from the A5 which can be found just north of Pentre Berw at SH465728. A footpath also leads north through the reserve from a layby at Bryn-y-fedwen SH464710.

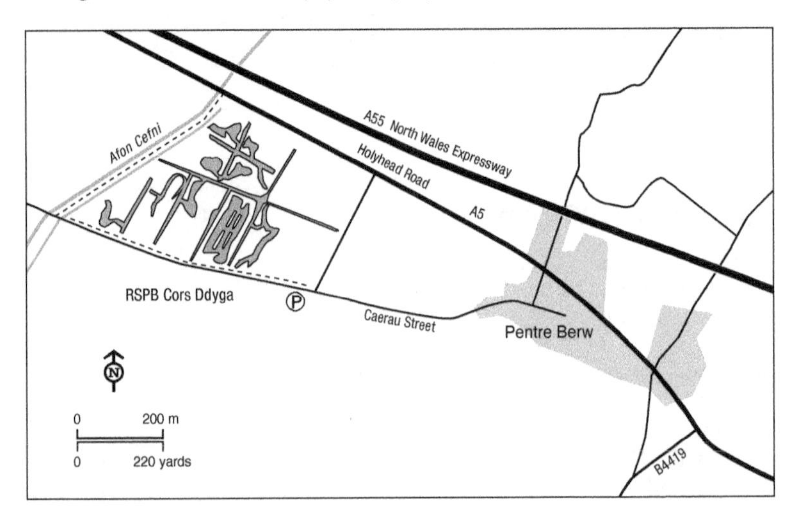

CALENDAR

Resident: Mallard, Teal, Kingfisher, Water Rail and Tufted Duck.

November–February: Winter influx of ducks and waders on the marsh and Cob, including Pintail, Wigeon, Gadwall, Shoveler and Golden Plover. There is also a large Starling roost.

March–May: Wader passage may include Black-tailed Godwits, Greenshank, Ruff and Spotted Redshank. Marsh Harrier and Garganey may also be seen. In the reedbed are Reed, Sedge and Grasshopper Warblers.

June–July: Reed, Sedge and Grasshopper Warblers in the marsh, with the possibility of Water Rail.

August–October: Wader passage under way, including Black-tailed Godwit, Whimbrel, Spotted Redshank, Ruff and Greenshank.

BRECONSHIRE (BRYCHEINIOG)

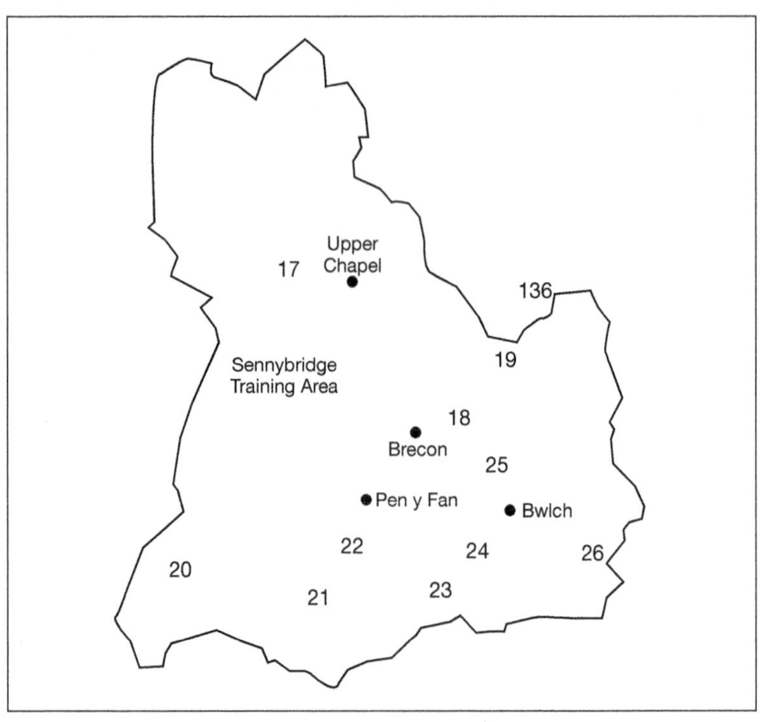

17 Nant Irfon
18 Priory Groves
19 Brechfa Pool
20 Usk Reservoir
21 Craig Cerrig Gleisiad
22 Mynydd Illtyd
23 Llwyn-on Reservoir
24 Talybont Reservoir
25 Llangorse Lake
26 Craig y Cilau and Darren Cilau

Breconshire is a large landlocked county, mainly of uplands, deep valleys and woodlands. As such the county is a haven for raptors, with large populations of Buzzards, Red Kites, Sparrowhawks – indeed, the Buzzard population is thought to have doubled in the last ten years. The mixture of woodland, deciduous and conifer, also makes ideal habitat for Goshawk. Other birds that can be found in many sites are breeding Redstarts, Wood Warblers and Pied Flycatchers. Up on the moors, there are both Red and Black Grouse, Hen Harriers and Merlins. The numbers of breeding waders have declined drastically in the last 50 years, although Golden Plover, Dunlin, Curlew and Snipe are still present but in much smaller numbers and in only a few sites. For such a landlocked county, there are few areas of open water, the main site being Llangorse Lake, where many species of waterbird breed. In passage periods, any area of open water is worth a look; ducks, waders and terns often turn up and stay a while to refuel. One of the most surprising finds was of a Paddyfield Warbler caught and ringed one morning by the side of Llangorse on 11 September 2004 – there are just three subsequent records for Wales, all on Bardsey Island.

17 NANT IRFON

OS Landranger Map 147
OS Explorer Map 200 (south part on 187).
Grid Ref: SN840550
Website:
https://www.first-nature.com/waleswildlife/e-nnr-nantirfon.php

Habitat

In the extreme northwest of Breconshire, close to the borders with Carmarthenshire and Ceredigion, is the Irfon Valley. The river is a major tributary of the Wye, which it joins at Builth Wells, not far from where Llywelyn ap Gruffydd, the last Prince of Wales, was killed in a skirmish with English soldiers in 1282. The upper reaches of the valley above Llanwrtyd Wells comprise attractive steep valley sides with stands of conifers and remnants of once extensive oak woodlands. Above the hamlet of Abergwesyn there are lines of rock outcrops and cliffs, the result of erosion in glacial times, quiet upland pastures, and beyond stretch the open hills.

Two small fields known as Vicarage Meadows are a nature reserve of the South & West Wales Wildlife Trust, with a rich flora comprising six species of orchid including Small White Orchid. Immediately adjoining is the Nant Irfon National Nature Reserve, some 353 acres (143 ha) owned by Natural Resources Wales. Situated on the steep western slopes of the valley, the reserve includes some of the highest Sessile Oak woodland in Wales, reaching to 1,300 feet (396 m). Bluebells carpet the woodland floor in late spring, while other interesting plants include the Alpine and Fir Clubmosses, Globeflower, Round-leaved Sundew, Common Butterwort, Bog Asphodel, Ivy-leaved Bellflower, Fragrant Orchid and, on damp rocks, the dainty Wilson's Filmy-fern. Extending north from Abergwesyn is an upland common rising to 2,115 feet (644 m) at the rock cairns of Drygarn Fawr, and forming one of the few extensive areas of open ground in the southern half of the mid-Wales plateau. Continue north for a further 3 miles (4.8 km) and you reach the Claerwen and Elan Valley Reservoirs (see Site 130). You will find when you reach the high ground that it is probably yours alone, for this is a lonely part of Wales with few visitors, though your solitude is likely to be shattered from time to time, at least on weekdays, by low-flying military aircraft on training flights.

Species

During the summer months, several pairs of Common Sandpiper breed along the river, and Grey Wagtails and Dippers are present throughout the year, except in the hardest weather. The oak woodland supports a range of species including Buzzard, Tawny Owl, Tree Pipit, Redstart, Wood Warbler and Pied Flycatcher. Where there are rocky areas, or in the steep-sided clefts along the escarpment, Ring Ouzels may be found, while Whinchat and Wheatear will be found on

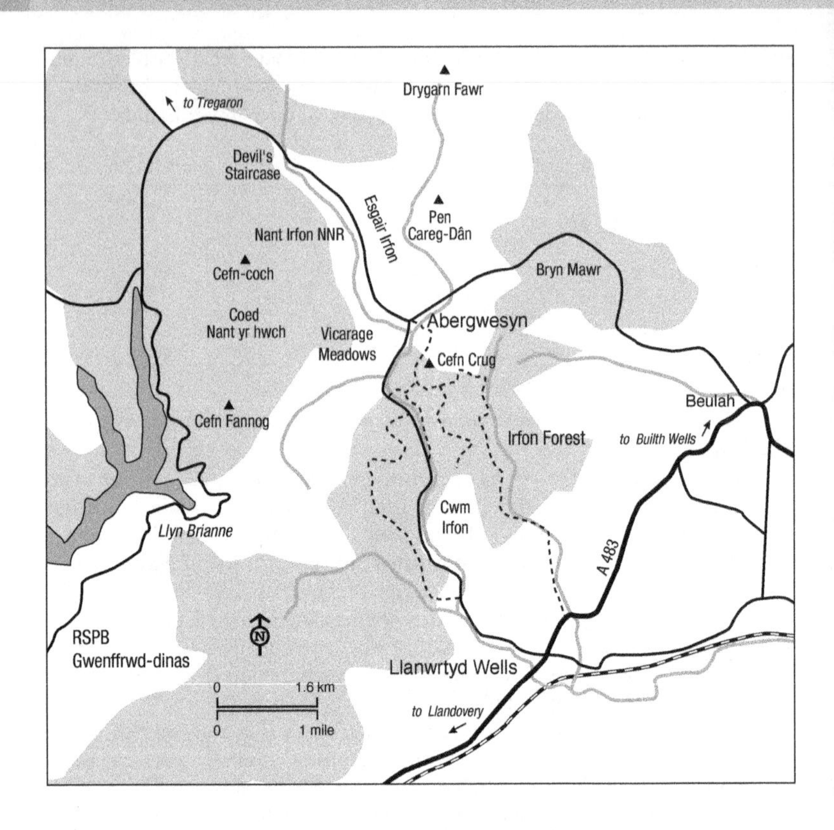

the more open ground. These days Red Kite and Buzzard vie with each other for the commonest raptor, with Peregrine and Kestrel often to be found along the ridges. Other birds of prey include Short-eared Owl, Merlin and in the large conifer plantations Goshawk.

Timing

This is not critical so far as the valley birds are concerned, but if you really mean to see some of the upland species such as Golden Plover and Dunlin then be prepared for an early start, or a late return. Both species can be extremely elusive during the main part of the day, and are heard and seen best about dawn, or in the evening towards dusk, when Snipe will also reveal their previously hidden presence.

Access

Leave the A483 in Llanwrtyd Wells by the unclassified road signed Abergwesyn. There are several Forestry Commission picnic areas close to the river, and these provide good vantage points or starting places for your visit. Beyond Abergwesyn the road is unfenced, and there are several good pull-off points for a car. From these the energetic can climb to high ground. The Nant Irfon (SN840550) and Vicarage Meadows (SN850526) nature reserves are always open.

CALENDAR

Resident: Sparrowhawk, Buzzard, Red Kite, Goshawk, Kestrel, Merlin, Peregrine, Red Grouse, Tawny Owl, Skylark, Meadow Pipit, Grey Wagtail, Dipper, Goldcrest, Raven, Siskin, Redpoll.
December–February: Resident species present except in hard weather.
March–May: Goosander, a chance of a trip of Dotterel on high ground, Golden Plover, Dunlin, Snipe, Curlew, Common Sandpiper, Cuckoo, Tree Pipit, Ring Ouzel, Redstart, Whinchat, Wheatear, Whitethroat, Wood Warbler, Chiffchaff, Pied Flycatcher.
June–July: All residents and summer visitors present.
August–November: Summer visitors depart during August and early September, tits and finches begin to flock, and autumn thrushes and finches move into the valley during October. Chance of Hen Harrier or Short-eared Owl passing through, possibly even lingering on the high ground.

18 PRIORY GROVES

OS Landranger Map 160
OS Explorer OL Map 12
Grid Ref: SO048301

Habitat

Broad-leaved woodland from close to the cathedral on the northern edge of Brecon extends north for just over a mile along the Afon Honddu, the river that has made its way southwards some ten miles from its source high on Mynydd Eppynt to join the Usk in the town itself. While popular with townspeople, Priory Groves offers a good range of woodland and waterside birds along easy laid paths and a chance of an Otter in the river.

Species

Although worth a visit at any time of the year, summer is best, when breeding birds include Tawny Owl, Great Spotted and Green Woodpeckers, Dipper, Grey Wagtail, Redstart, Chiffchaff, Wood Warbler, Garden Warbler, Blackcap, Spotted and Pied Flycatchers. Species during winter could include Woodcock, Kingfisher, Fieldfare, Redwing, Blackcap, Chiffchaff, Siskin, Lesser Redpoll and Brambling. The cathedral grounds should always be explored, if only to enjoy the Blackbirds scampering about as they hunt for worms.

Timing

All year.

CALENDAR

Resident: Tawny Owl, Kingfisher, Great Spotted and Green Woodpecker, Mistle Thrush, Song Thrush, Blackbird, Coal Tit, Goldcrest, Nuthatch, Treecreeper, Dipper, Grey Wagtail, Chaffinch.
November–February: Buzzard, Red Kite, Sparrowhawk, Woodcock, Kingfisher, Fieldfare, Redwing, Siskin, Lesser Redpoll and Brambling.
March–July: Redstart, Common Sandpiper, Blackcap, Chiffchaff, Willow, Garden and Wood Warblers, Spotted and Pied Flycatchers.

Access

It is recommended to use either of the town's pay-and-display car parks at SO047287 or the limited street-side parking closer to the Groves.

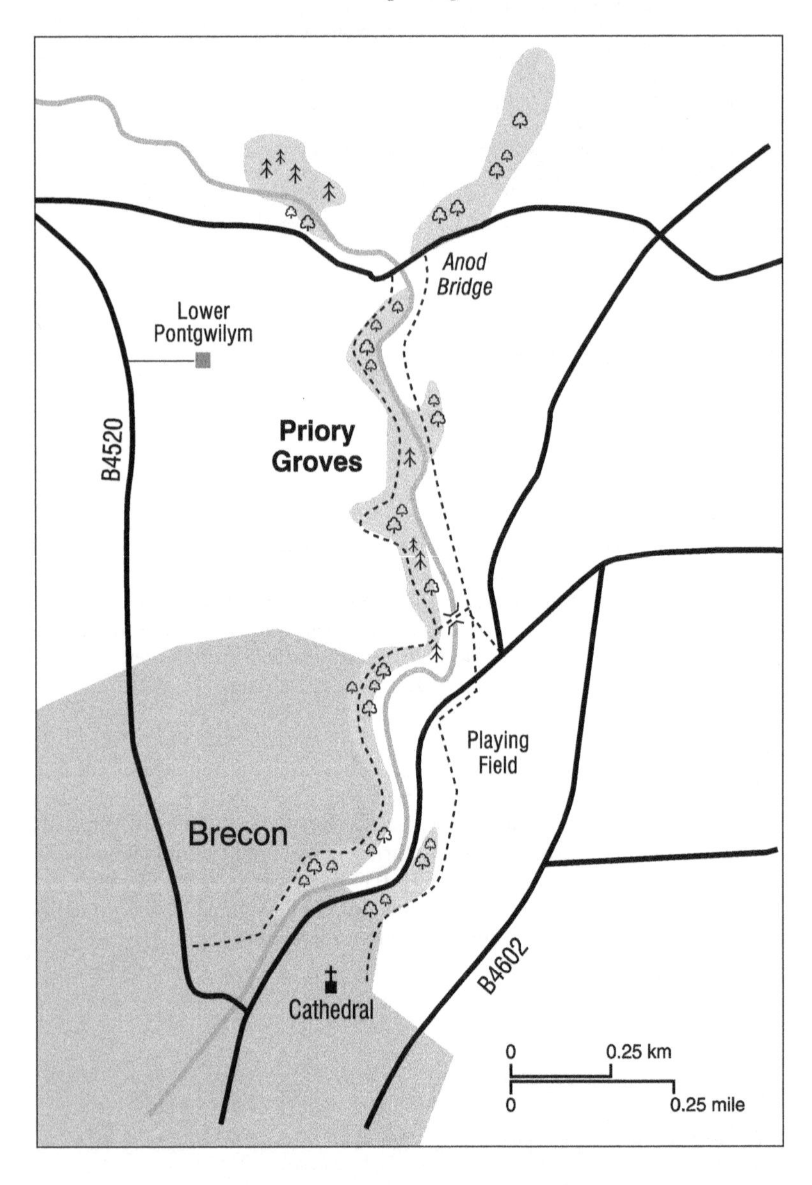

19 BRECHFA POOL

OS Landranger Map 161
OS Explorer map OL13
Grid Ref: SO118378

Habitat

Brechfa Pool is a small, shallow upland pool overlooking the Wye valley at a height of 900 feet (270 metres), with spectacular views eastwards to the Black Mountains. This SSSI and South & West Wales Wildlife Trust reserve since 1973 benefits from its lofty position and is noted for its rich aquatic flora including Pillwort – a Red Data Book species, mid-Wales being its European stronghold – Pennyroyal and Orange Foxtail Grass, while Water-crowfoots enhance the shallows. The muddy margins prove particularly attractive for waterfowl and waders on passage.

Species

In summer, there is much activity around the Black-headed Gull colony with up to 100 pairs in some years. Tufted Duck and Teal occasionally breed along with Canada Goose. Lapwing, Snipe and Curlew breed nearby and often feed at the Pool, with larger numbers gathering at the end of the breeding season.

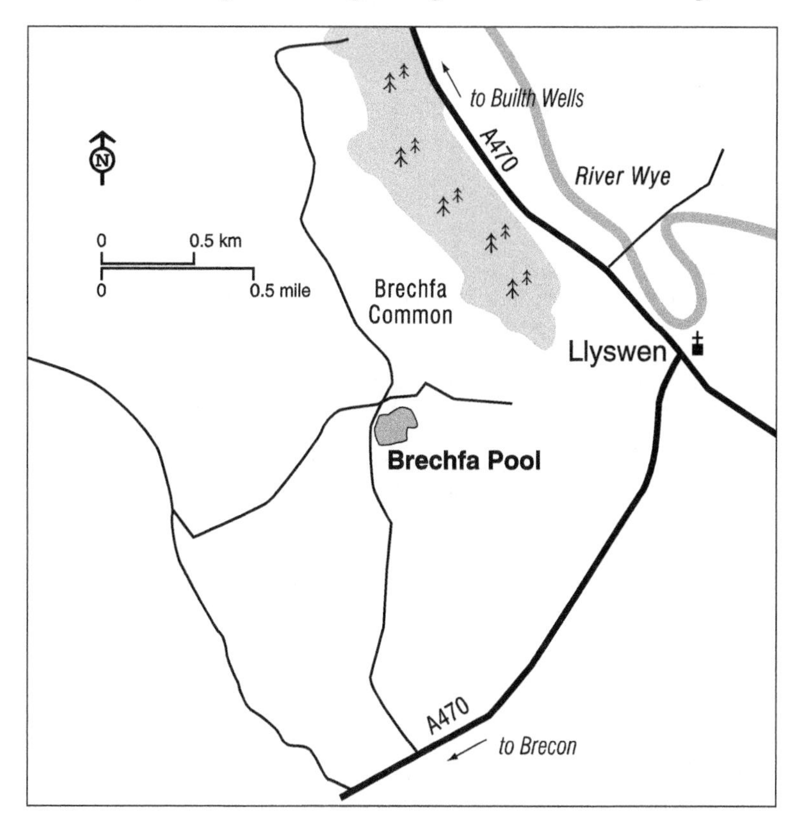

Rarities for inland Wales recorded include a Crane in September 2013, Garganey, Ruff was once seen feeding alongside a Pectoral Sandpiper, Black-tailed and Bar-tailed Godwits, Wood Sandpiper, Greenshank, Spotted Redshank.

Timing

The Pool is well worth a visit at any time of the year but especially during passage periods, March–May and July–October.

Access

Park at the T-junction by the phone box. There are no viewing facilities, but to limit disturbance it is recommended that the car is used as a hide.

CALENDAR

Resident: Mute Swan, Canada Goose, Mallard, Heron, Coot, Moorhen and Reed Bunting.
December–January: small numbers of Wigeon, Teal and Goosander.
March–July: Tufted Duck, Teal, Lapwing, Snipe, Curlew and Black-headed Gull colony. On the surrounding hills Skylark, Meadow and Tree Pipit, Wheatear and Stonechat.
August–November: Passage waders and ducks.

20 USK RESERVOIR

OS Landranger Map 160
OS Explorer Outdoor Leisure 12
Grid Ref: SN819284

Habitat

The source of the River Usk is located on the northern slopes of the Black Mountains, not far below the impressive crags of the Carmarthen Fans – the highest Fan Foel at 2,630 feet close to the border between Breconshire and Carmarthenshire. Stretching east is the high moorland of Fforest Fawr, or Great Forest, though the trees had long gone before it was designated a Royal Forest and subject to game laws in the Middle Ages. From much earlier times, there are standing stones, enigmatic reminders of when Bronze Age Beaker Folk passed this way, to be followed by Celtic settlers and then the Romans. On Mynydd Bach, the moorland stretching north from the Usk Reservoir, are several Bronze Age round barrows and the Roman fort of Y Pigwn, guardian of the now ancient road which ran across this high ground. The boundary between Breconshire and Carmarthenshire cuts across the reservoir completed in 1955.

Species

Great Crested Grebes nest from time to time, while both Black-necked and Slavonian Grebes have been recorded. Cormorants are usually present, though never in the numbers regularly experienced elsewhere in the county. Red Kites, Buzzards and Ravens never seem far away, while the woodlands can be productive for Goshawk, Sparrowhawk, Goldcrest, Crossbill, Siskin and Lesser

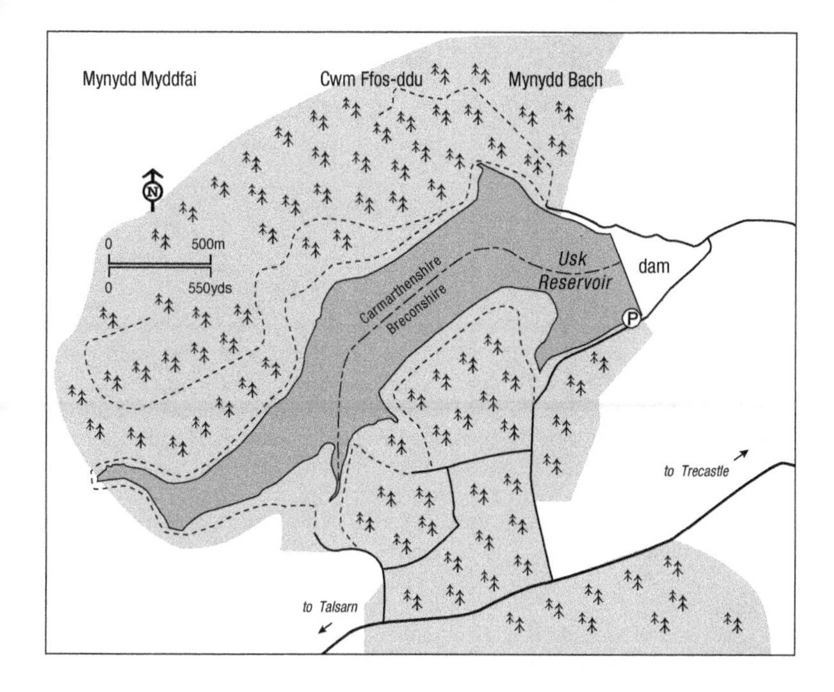

Redpoll, and Woodcock and Long-eared Owls have been recorded. Grey Wagtails at the water's edge where Common Sandpipers occur on passage. The high ground has Meadow Pipits, Skylarks, Wheatears and Linnets.

Timing

All year.

Access

A minor road from the A4069 south of Llangadog in the west to Trecastle in the east crosses the high ground from which a link passes close to a section of the reservoir edge, where there is parking not far from the dam.

CALENDAR

Resident: Great Crested Grebe, Red Kite, Goshawk, Sparrowhawk, Skylark, Meadow Pipit, Grey Wagtail, Stonechat, Goldcrest, Coal Tit, Great Tit, Raven, Chaffinch, Linnet.

December–February Hen Harrier, Merlin over the hill ground Woodcock in the plantations.

March–June Wheatears on the higher ground, Martins and Swallows foraging over the reservoir.

July–November A chance of passage waders at the reservoir, summer visitors depart and wintering thrushes occur from October onwards.

21 CRAIG CERRIG GLEISIAD

OS Explorer Outdoor Leisure 12
OS Landranger Map 160
Grid Ref: SN960218
Website:
https://naturalresources.wales/days-out/places-to-visit/south-east-wales/craig-cerrig-gleisiad-nnr/?lang=en

Habitat

Craig Cerrig Gleisiad National Nature Reserve, an impressive series of Old Red Sandstone crags and gullies which rise to 500 feet, occupying a north-facing position in the Brecon Beacons. These are spectacular remnants from when vast sheets of ice and huge glaciers gouged out the softer ground during the last Ice Age. Now they provide opportunities for alpine plants like Purple Saxifrage, here at its most southerly location in Great Britain. The woodland, just as you begin your climb, was planted in memory of Eric Bartlett, an indefatigable member of the then Brecknock Wildlife Trust and the Brecon Beacons National Park, who died in 1986.

Species

Breeding species include Red Grouse, Red Kite, Buzzard, Kestrel, Peregrine, Snipe, Curlew, Barn Owl, Redstart, Wheatear, Ring Ouzel, Whinchat, Stonechat, Tree Pipit, Grey Wagtail, Sedge Warbler, Raven and Linnet. Winter visitors may include Hen Harrier, Short-eared Owl, Kestrel, Merlin, Golden Plover and Brambling. Recent rarities recorded include Hobby, Dotterel, Long-eared Owl and Snow Bunting.

Timing

All year.

Access

For Craig Cerrig Gleisiad, park in the layby on the A470 at SN971222 from which a well-marked path climbs to the reserve. A Natural Resources Wales (NRW) display board and leaflets are available showing paths and species of interest. Care is required on some of the paths. Keep off the cliff face unless licensed by NRW.

CALENDAR

Resident: Red Grouse, Red Kite, Buzzard, Peregrine, Kestrel, Snipe, Skylark, Stonechat, Grey Wagtail, Meadow Pipit and Raven.
November–February: Hen Harrier, Kestrel, Merlin, Short-eared Owl, Golden Plover, Fieldfare, Redwing and Brambling.

March–June: Curlew, Barn Owl, Tree Pipit, Ring Ouzel, Wheatear, Grey Wagtail, Whinchat, Whitethroat, Sedge and Grasshopper Warbler.
July–October: The last Wheatears depart by end of period.

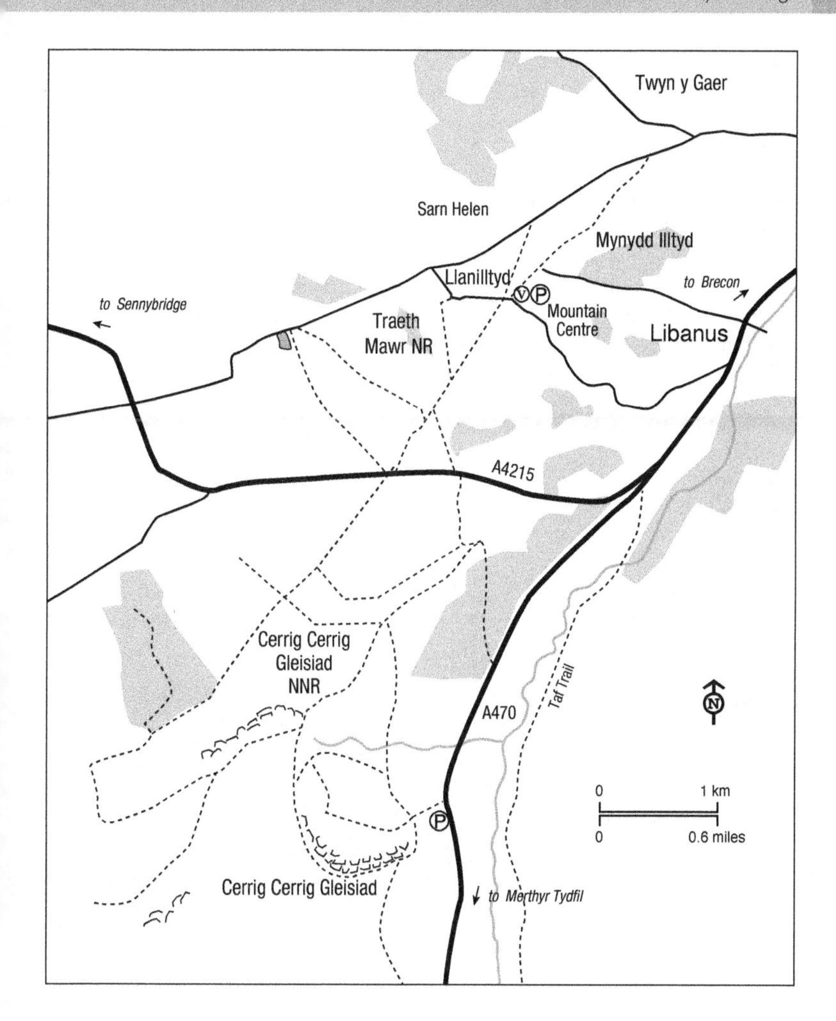

22 MYNYDD ILLTYD

OS Explorer Outdoor Leisure 12
OS Landranger Map 160
Grid Ref: SN979263

Habitat

A northern outlier of the main Brecon Beacons, these linear upland commons offer superb views and a good range of birds. The Brecon Beacons National Park Mountain Centre makes an ideal base, with walks extending across the area, somewhere in the vicinity of which is buried the Celtic monk Saint Illtud. Earlier still, Iron Age peoples chose to build a fort on Twyn y Gaer, the highest point of the common at 1,211 feet. The two most productive areas are Traeth

Mawr, a rain-fed raised mire with birchwood, and Traeth Bach, a basin mire. Both are South & West Wales Wildlife Trust nature reserves.

Species

Mynydd Illtyd is one of the best upland wetland habitats in Breconshire and a traditional site for breeding Spotted Crakes, Golden Plover, Skylark and Meadow Pipits, while winter thrushes are examples of species seen on migration. Traeth Bach at the western end of the wetland attracts passage waders such as Jack Snipe, Redshank, Green and Common Sandpipers. Breeding species include Red Kite, Buzzard, Kestrel, Peregrine, Snipe, Curlew, Barn Owl, Redstart, Wheatear, Whinchat, Stonechat, Tree Pipit, Grey Wagtail, Sedge Warbler, Raven and Linnet. Winter birds may include Hen Harrier, Short-eared Owl, Kestrel, Merlin, Golden Plover and Brambling. Recent rarities recorded here include Marsh Harrier, Hobby, Quail, Long-eared Owl.

Timing

All year.

Access

Traeth Mawr and Traeth Bach are located to the west of the Brecon Beacons National Park Mountain Centre. There is a small car-parking charge and facilities include restaurant, shop and interpretative centre.

CALENDAR

Resident: Red Grouse, Red Kite, Buzzard, Peregrine, Kestrel, Snipe, Skylark, Stonechat, Grey Wagtail, Meadow Pipit and Raven.
November–February: Hen Harrier, Kestrel, Merlin, Short-eared Owl, Golden Plover, Fieldfare, Redwing and Brambling.

March–June: Spotted Crake, Curlew, Barn Owl, Tree Pipit, Ring Ouzel, Wheatear, Grey Wagtail, Whinchat, Whitethroat, Sedge and Grasshopper Warbler.
July–October: Passage waders including Common and Green Sandpipers while the last Wheatears depart by end of period.

23 LLWYN-ON RESERVOIR

OS Explorer Outdoor Leisure 12
OS Landranger Map 160
Grid Ref: SO008118

Habitat

Most visitors heading north by way of the A470 into the Brecon Beacons National Park probably dash past Llwyn-on, the largest and most southerly of three reservoirs in the Taff Fawr valley, for it is well hidden from the road by large areas of mostly coniferous woodlands. Water draining from the southern slopes of the Brecon Beacons, with their spectacular peaks rising to almost 3,000 feet, passes through the reservoirs on its journey to Cardiff Bay 40 miles away. The valley is home to some deciduous gems: just to the south on the steep limestone crags is the entire world population in the wild, just sixteen trees, of Ley's Whitebeam.

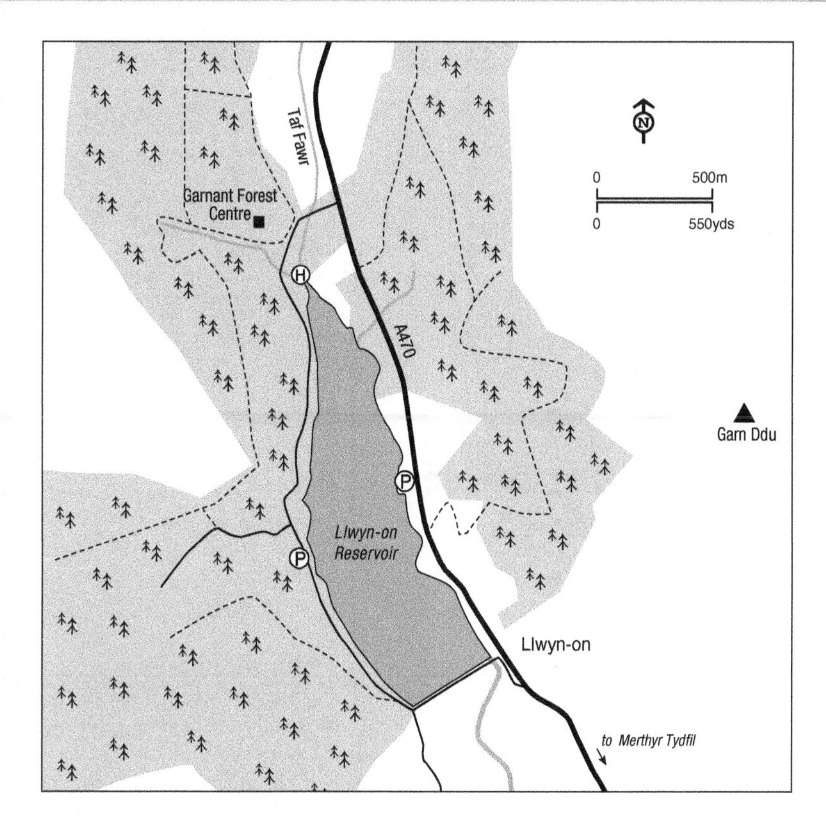

Species

The hide at the north end of the reservoir is a must, the access track passes a small heronry, where one or two pairs nest each year. Teal, Tufted Duck and Goldeneye, Great Crested and Little Grebes, Cormorants and Kingfisher. Grey and Pied Wagtails at the reservoir dam, Dippers on the river upstream from the reservoir. The mostly easy walks from the Garnant Forest Centre through the coniferous woodlands should produce Goldcrest, Jay, Lesser Redpoll, Siskin and Crossbill and, with some extra effort, Goshawk. The high ground on each side of the reservoir is home to Ravens and Red Kites and in winter Hen Harriers and Merlins.

Timing

All year

Access

At the Garnant Forest Centre, fee payable or at the picnic sites beside the A470. Limited parking on the western side of the reservoir, also beside the roadside leading to higher ground to the west.

CALENDAR

Resident: Mallard, Goosander, Little Grebe, Great Crested Grebe, Cormorant, Grey Heron, Red Kite, Sparrowhawk, Buzzard, Coot, Moorhen, Kingfisher, Grey Wagtail, Pied Wagtail, Wren, Dipper, Robin, Coal Tit, Jay, Raven, Siskin, Lesser Redpoll, Reed Bunting.

December–February: Teal, Tufted Duck, Goldeneye on the open water, check for Water Pipits on the spillway close to the main road.

March–May: Summer visitors arrive including Common Sandpipers on passage.

June–August: Swifts that have fed over the reservoir depart in early August, depending on water levels passage waders should occur

September–November: Summer visitors depart, winter visitors arrive both in the woodlands and on the open water.

24 TALYBONT RESERVOIR

OS Landranger Maps 160/161
OS Explorer map OL13
Grid Ref: SO101194

Habitat

One of the most beautiful valleys in all of South Wales, the Taf Fechan extends northwards from the outskirts of Merthyr Tydfil into the very heart of the Brecon Beacons. Follow the road through the valley and past the Pontsticill and Pentwyn reservoirs. Continue to rise to cross the watershed, and after several miles you come to Talybont Reservoir, built early in the 20th century to supply water for Newport. The reservoir is about 2 miles (3.2 km) long and has a maximum width of about 500 yards (457 m), the largest reservoir in the county and only marginally smaller than Llangorse Lake. Steep slopes with coniferous woodland surround much of the perimeter, though there are shallow areas along the reservoir margin, especially at the southern end.

If you have time, on leaving the valley at the northern end, turn east to Llangynidr, beyond which a road climbs south where, from near the highest point, one can travel east across the northern flank of Mynydd Llangatwg, a superb limestone escarpment with cliffs and screes.

Species

Among all the birds using Talybont, pride of place must go to Goosander. Although Goosanders may have nested in Wales since 1952, this was not confirmed until 1970. Since the early 1980s both the breeding population and the number wintering has increased. Breeding birds, of which there are now about 100 pairs in Wales, are still largely restricted to the river systems of mid-Wales. Although wintering birds occur on many waters, there are probably more at Talybont than anywhere else in Wales, with up to 90 having been recorded. Most are likely to be seen here from September through to April, particularly late in the day as birds arrive to roost.

Up to 600 Mallard have gathered here in late autumn and in early winter, a few remaining throughout the whole period. Some 100 Tufted Duck may be seen throughout the winter. Other winter wildfowl include Goldeneye,

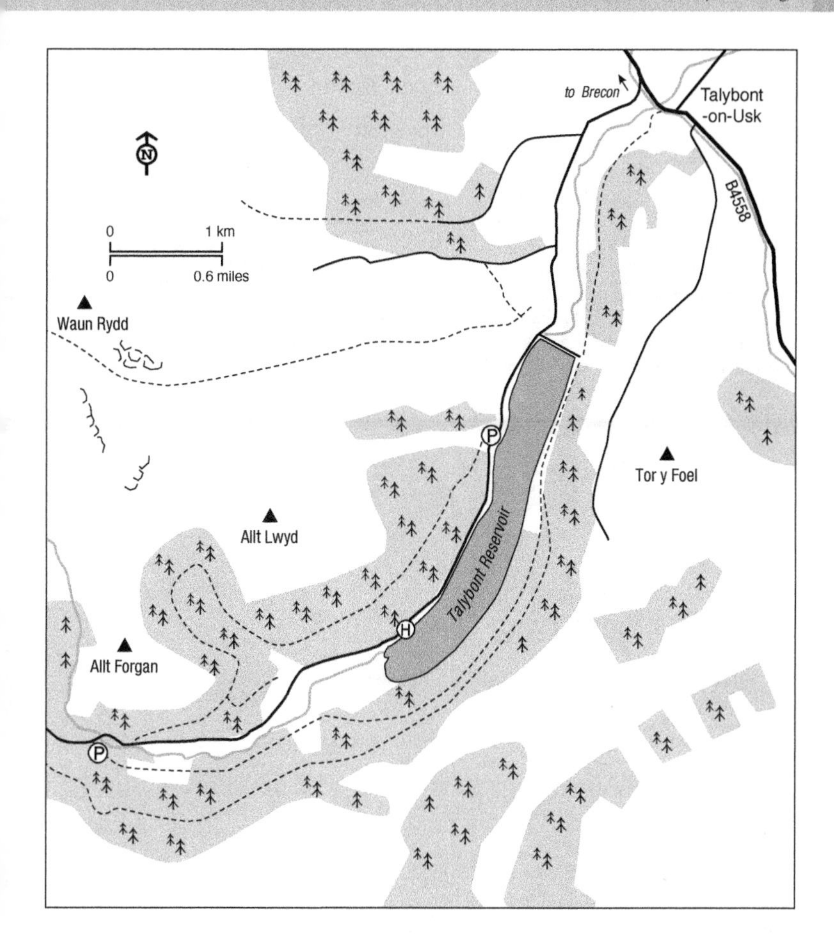

Wigeon, Teal and Pintail. Common Scoter and Red-breasted Merganser are occasionally recorded on passage. Wader passage in late summer can be good, though much depends on water levels, and several scarce species, at least for inland waters, have been recorded including Little Ringed Plover, Sanderling, Pectoral Sandpiper, Ruff, Green Sandpiper and Wood Sandpiper. In recent years, the shrubbery on the reservoir dam has attracted a winter roost of several hundred Greenfinch and Redwing. Other winter sightings have included Firecrest and Woodcock.

Timing

No particular time of day, though in the afternoon the light will be behind the observer.

Access

There is no access to the reservoir area, but this need not be a drawback as the unclassified road from Talybont-on-Usk (SO112227) skirts the western shore and provides excellent views before continuing south, eventually reaching the A4102 at SO067077 on the edge of Merthyr Tydfil.

CALENDAR

Resident: Grey Heron, Mute Swan, Great Crested Grebe, Little Grebe, Mallard, Goosander, Sparrowhawk, Buzzard, Goshawk, Kestrel, Grey Wagtail, Dipper.
December–February: Unless the weather is very hard most winter visitors present, including Great Crested and Little Grebes, Cormorant (numbers often reaching 100 at times), Wigeon, Teal, Pochard, Tufted Duck, Goldeneye.
March–May: Winter visitors depart, summer migrants arrive in surrounding woodlands. Occasional passage waders in May.
June–July: Passage waders if waters are low, including Little Ringed Plover, Ringed Plover, Dunlin, Ruff, Redshank, Greenshank, Green and Wood Sandpipers by end of July. Little Egrets have occurred while Common Scoter are occasionally seen on passage.
August–November: Passage waders and occasional Black Terns. Winter visitors including Pochard, Tufted Duck begin to arrive from early October while Mandarin, Gadwall and Goldeneye have been recorded.

25 LLANGORSE LAKE

OS Landranger Map 161
OS Explorer map OL13
Grid Ref: SO133264
Website:
https://www.breconbeacons.org/things-to-do/attractions/natural/llangorse-lake

Habitat

Llangorse Lake, some 370 acres (150 ha), about the size of 300 football pitches, is the largest natural lake in Wales. Long recognised as a gem by naturalists, it contains a range of wetland habitats, from the adjacent herb-rich pastures and damp meadows, through marshland, sedge and reed swamp, to the floating-plant communities, and finally the assemblage of submerged aquatic plants in the shallow waters. Even the mud and silt of the lake have their secrets: evidence of vegetation changes over many thousands of years has been recorded by analysing pollen grains. Close to the northern shore a tiny island was identified in 1876 as a 'crannog', an artificially constructed island, used for a settlement, perhaps dating from the 7th to the 9th century – more recently it has been suggested that it is of Iron Age origin, while there is no doubt this is the only crannog in England and Wales.

Aquatic plants include no fewer than 23 species that are rare in Wales and a further 15 rare in Breconshire, though to the layperson the reedbeds, the largest in inland Wales, and marginal aquatics such as Fringed, White and Yellow Water-lilies, Greater Spearwort and Flowering Rush, are probably the most striking. Giraldus Cambrensis, chronicler and ecclesiastic writing in the 12th century, referred to the lake, which he called Brecknock Mere, as being 'a broad expanse of water that supplies plenty of Pike, Perch, excellent Trout, Tench and mud-loving Eels'. Not surprisingly, Otters are present and also Water Voles.

Conservation of the privately owned lake and its high nature conservation interest has been a challenge for several decades. It is designated as a Site of Special Scientific Interest and a Special Area of Conservation under the European Habitats Directive. A management committee with representatives from the local community, lake users and conservationists has helped considerably in managing some of the pressures on the lake, caused by uses that conflict with conservation, such as water sports, angling, shooting and agricultural run-off.

Species

Great Crested Grebes have nested at Llangorse for over 100 years, one of their most important sites in South Wales and indeed the whole Principality. Numbers have been as high as 20 pairs in the past, but nowadays reduced to four or five pairs. Little Grebe occasionally nest and are a regular winter visitor in small numbers. Slavonian and Black-necked Grebes have been recorded at infrequent intervals, while Red-necked Grebe is much rarer in Wales but has been seen at Llangorse on a few occasions.

Coots have numbered up to 150 pairs but even these have declined, with only 15 or so young being reared in most years. Immigrants greatly swell numbers in midwinter when nearly 600 can be present. Breeding wildfowl are rather few, with Canada Goose most prolific. In winter, Tufted Duck are the most numerous ducks, with up to 500 being recorded, while smaller numbers, usually not exceeding 100 each of Wigeon, Teal and Mallard, are present together with smaller numbers of Pochard, Gadwall, Shoveler, Goldeneye and Goosander. Whooper Swans, Shelduck, Pintail, while Long-tailed Duck, Smew and Red-breasted Merganser are only occasional visitors.

Bitterns visit the Llangorse reedbeds in most winters, and there have been occasional sightings in spring and early summer. Whimbrels, not recorded in Breconshire until 1963, are now regularly seen on passage at Llangorse, and it is here that most passage waders are recorded in the county. These have included Oystercatcher, Ringed and Little Ringed Plovers, Dunlin, Black-tailed and Bar-tailed Godwits, Spotted Redshank, Greenshank, Green and Wood Sandpipers and Turnstone. Common Sandpipers breed as well as occurring on passage. Several species of tern have been recorded at the lake, most especially Common Tern and Black Tern, both annual visitors between May and September. Up to 2,000 Lesser Black-backed Gulls come here to roost in the autumn, with much smaller numbers at other times. Herring Gull, Common Gull, Mediterranean Gull and Yellow-legged Gull may be present in small numbers, while Little Gulls are seen almost annually during their spring and autumn passage. By contrast, Black-headed Gulls occur in large flocks, especially in late winter and early spring when up to 1,000 have been known to roost on the lake.

The reedbeds support a large colony of Reed Warblers, in recent years between 70 and 90 territories. Water Rails, Sedge Warbler, Cetti's Warbler and Reed Buntings also breed in the reedbeds, the latter noted as decreasing. The aquatic invertebrates flying low over the water provide a rich food source for Swifts and hirundines: many hundreds can be seen hawking low over the water, while in late summer enormous numbers of Swallows and Sand Martins can gather to roost in the reedbeds, up to 5,000 of each having been estimated in some years. Small numbers of Pied Wagtails also use the reedbeds as a roost, up to 150 being present in late summer, to be followed during the autumn by

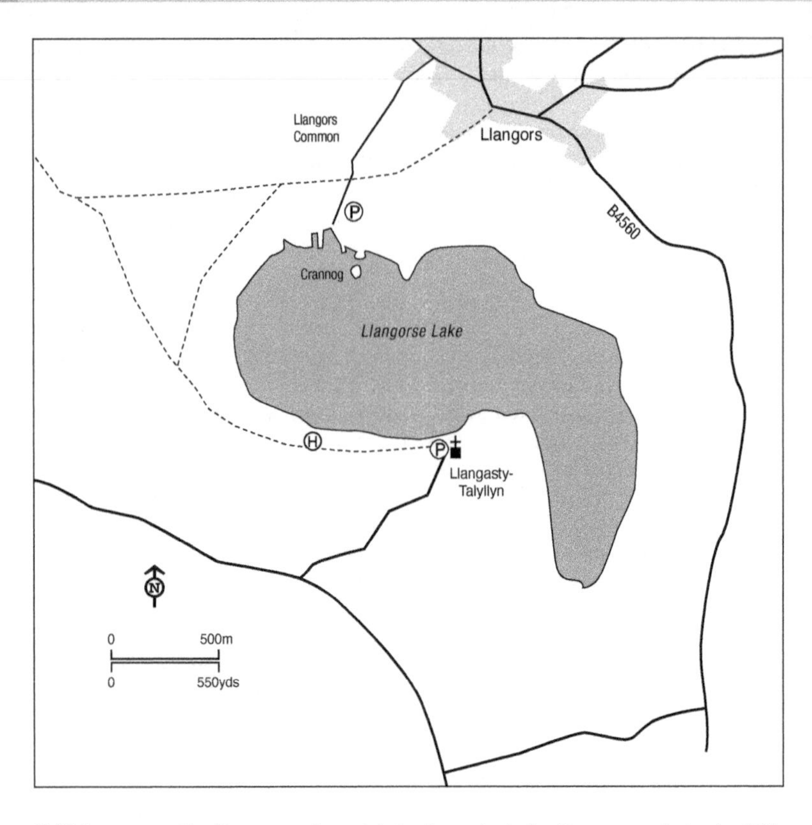

5,000 or more Starlings, coming nightly from their feeding grounds in the Usk Valley.

Scarce visitors in recent years have included Barnacle and Brent Geese, Lesser Scaup, Ferruginous Duck, Marsh Harrier, Osprey, Hobby, Spotted Crake, Temminck's Stint, Pectoral Sandpiper, Long-billed Dowitcher, Kittiwake, Roseate Tern and Bearded Tit.

The Llangorse Ringing Group is active throughout the year, concentrating both on residents and species that migrate through the reedbeds and marginal habitats. Over 2,500 birds of 50 species or more are ringed in most years, confirming the occasional presence of rarely seen skulkers such as Water Rail and Cetti's Warbler, as well as high numbers of more common species that use the reedbeds as an essential feeding site during migration. Extreme rarities have included Wales's first Paddyfield Warbler, which was trapped and ringed in September 2004, and several Aquatic Warblers.

Timing

In view of the numbers of other visitors, and in particular the boating activities, it is advisable to visit either early or late in the day.

Access

There are two main access points to the shore. The first is on the north shore at SO127271, near the village of Llangorse. The other is on the south bank where there is a superb hide beside the lakeside church of Llangasty-Talyllyn (SO132261), from where a footpath skirts much of the southern and western shore.

CALENDAR

Resident: Great Crested Grebe, Grey Heron, Mute Swan, Mallard, Buzzard, Moorhen, Coot, Water Rail, Lapwing, Black-headed, Herring and Lesser Black-backed Gulls, Reed Bunting.
December–February: Winter visitors include Teal, Shoveler, Gadwall, Pochard, Tufted Duck, Goldeneye, Goosander, Kingfisher.
March–May: Swallows and Martins arrive late March and early April, Reed and Sedge Warblers and Yellow Wagtails later in the month, Swifts in early May.
June–July: All breeding species present. Occasional passage waders and terns, hirundine numbers build up as July progresses.
August–November: Hirundine roost at peak during August and early September before main departure south. Passage waders and occasional terns until late September, when winter visitors start to arrive and include Little Grebe, Wigeon, Teal, Pintail, Shoveler, Pochard, Tufted Duck, and Goldeneye.

26 CRAIG Y CILAU AND DARREN CILAU

OS Landranger Map 160
OS Explorer Outdoor Leisure 13
Grid Ref: SO186160
Website:
https://www.first-nature.com/waleswildlife/e-nnr-craigycilau.php

Habitat

Hidden behind this impressive series of limestone crags which rise to 400 feet is a vast cave system, one of the largest in Europe, the discovered passages extending for at least 40 miles beneath the plateau of Mynydd Llangatwg. In the 18th and 19th centuries the cliffs were quarried, the limestone carried by tramway to the ironworks or to the nearby Monmouthshire and Brecon Canal for onward shipment. The quarries are long silent, the cliffs and screes now clothed in woodlands of Ash, Hazel, Large-leaved Lime and Yew, while the Beech trees are almost at their westernmost native limit in Great Britain. For botanists there are great treasures, especially rare Whitebeams, including one species, the Lesser Whitebeam, which is unique to Craig y Cilau. A National Nature Reserve mostly for its botanical and geological interest, nevertheless it produces a good range of bird species. Choice of heather/bilberry moorland above or broad-leaved woodland and farmland below the crags, each with their own characteristic birdlife.

Species

Summer birds include Peregrine, Red Kite, Kestrel, woodpeckers, Whinchat, Wheatear, Redstart, Ring Ouzel, Mistle Thrush, Spotted and Pied Flycatchers. Winter species include raptors, Little Owl, Short-eared Owl, Brambling and winter thrushes.

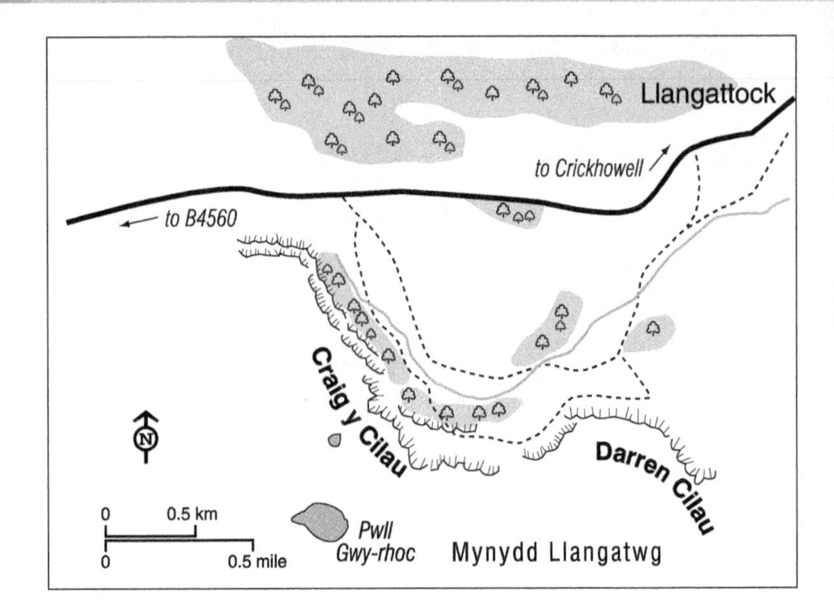

Timing

Craig y Cilau is worth a visit at any time in the year.

Access

For the west end, park off the road near the cattle grid at SO184169, and for the east in the car park at SO209154. Approach via several footpaths leading beneath the crags, open access common land, or tram-roads remaining from old quarrying activity.

CALENDAR

Resident: Red Grouse, Peregrine, Buzzard, Red Kite, Raven, Great Spotted and Green Woodpeckers.

November–February: Hen Harrier, Merlin, Kestrel, Short-eared Owl, Fieldfare, Redwing and Brambling.

March–July: Ring Ouzel, Wheatear, Whinchat, Redstart, Spotted and Pied Flycatchers.

August–October: Summer visitors and passage migrants depart while winter visitors arrive from early October.

CAERNARFONSHIRE

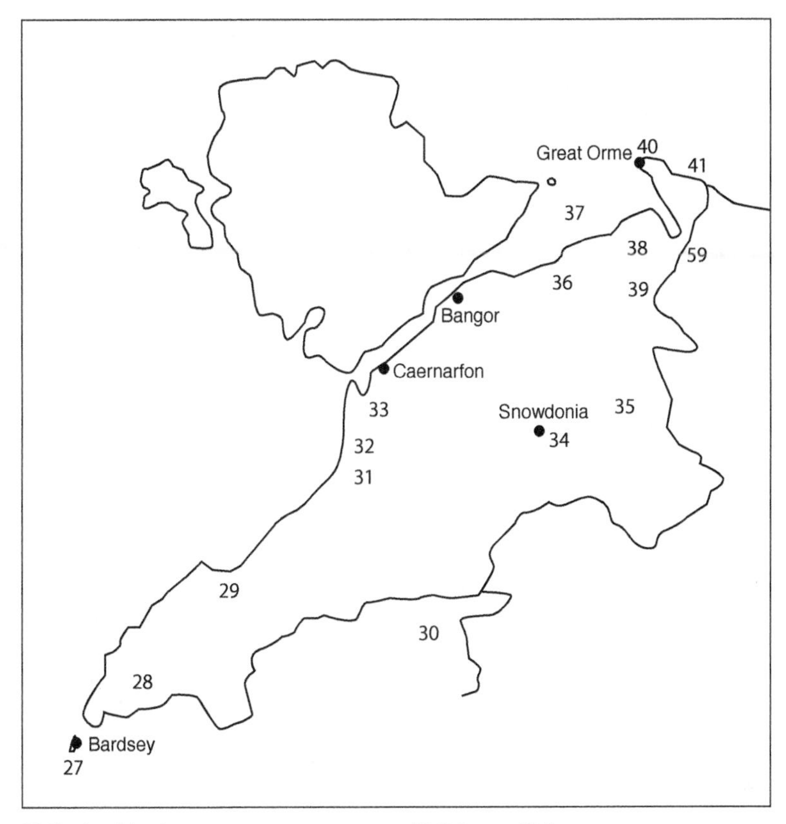

27 Bardsey Island
28 Aberdaron Headland
29 Porth Dinllaen
30 Porthmadog to Criccieth
31 Aberdesach and Pontllyfni
32 Dinas Dinlle
33 Foryd Bay
34 Snowdonia
35 Dolgarrog Station
36 Coedydd Aber
37 Traeth Lavan
38 Conwy Mountain and Pensychnant Nature Reserve
39 Caerhun Church
40 Great Orme
41 Little Orme

Caernarfonshire is a mountainous and picturesque county, with the splendour of the Snowdonia National Park in its centre, and the ruggedness of the Llŷn Peninsula jutting out into the Irish Sea. Birdwatching can be hard work in the mountains but with effort, the main upland species can be found. The numerous crags are ideal for Peregrines and Ravens, while the deciduous woods in the valleys are the summer homes for Redstarts, warblers and flycatchers.

Around the coast there are a multitude of interesting sites to visit in order to observe coastal specialties. Large numbers of sea-duck, divers and grebes winter off the north coast, by Llanfairfechan, and at the northern end of Cardigan Bay. Wintering ducks, waders and Brent Geese can be found along the long sandy beaches of Traeth Lavan and at the mouth of the Menai Straits.

There has been limited seawatching in the county, with most observers favouring the short trip to Anglesey, but dedicated observers have targeted the Great Orme, with some success. Indeed, many migrant land birds have also been found here. But when it comes to rarities, one site stands out from the crowd. At the end of the Llŷn is the island of Bardsey, one of only two bird observatories in Wales. There is nowhere that has turned up more first records for the country. Visiting is not all that easy, relying on weather and sea conditions, but anyone who has the chance should endeavour to stay on the island for a week or two.

27 BARDSEY ISLAND

OS Landranger Map 123
OS Explorer Map 253
Grid Ref: SH117213
Website:
http://bbfo.blogspot.com/

Habitat

Bardsey, legendary burial place of 20,000 saints, lies just over 2 miles (3.2 km) south of the tip of the Llŷn Peninsula. The sound that separates the island from the mainland provides its Welsh name, Ynys Enlli, 'the island in the tides'. The tide-rips and overfalls can make the journey, shall we say, interesting, and in stormy weather, impossible. Hilaire Belloc, in his classic *The Cruise of the Nona*, describes how he lost his dinghy there, and in a never-forgotten phrase said of the turbulence: 'The sea jumps up and glares at you.' Those who crossed in September 1997 to see the first Isabelline Wheatear for Wales, and whose boat broke down in Bardsey Sound on the return journey, know a little of what Belloc was talking about. Today crossings are much more reliable with a large twin-engined catamaran run by Mordaith Llŷn and skippered by Colin Evans.

When viewed from the mainland, Bardsey, which extends for 444 acres (180 ha), looks rather uninviting. The Mountain, as the highest point is called, rises to 548 feet (167 m), having steep rock and scree slopes tumbling to low cliffs and boulder beaches. This ridge of high ground effectively shields most of Bardsey from view. The far gentler western aspect of the island consists of lesser slopes richly covered in Western Gorse, Bracken and heathers, giving way to a network of small fields, 120 in all, divided by low stone and earth walls. Towards the northern end of the island stands the ruined tower of a 13th-century Augustinian abbey, while elsewhere there are traces of other earlier buildings. Spaced out across the field system and linked by a rough track, the Main Road as it is fittingly called, are no fewer than 11 cottages together with a school, built as farmhouses in the 1870s by Lord Newborough. The non-denominational chapel dates from this period, it being reputed that the islanders, who then numbered about 80, on given a choice between a jetty and a place of worship chose the latter. One of Lord Newborough's ancestors, John Wynn ap

Hugh, the standard-bearer at the Battle of Norwich (1549), had been rewarded with the gift of Bardsey by a grateful Edward VI. Island life changed his ways and he quickly became leader of a notorious band of pirates; indeed, the island remained a base for such activities for over 100 years. A descendant, the 7th Lord Newborough, was an equally doughty warrior, being awarded the Distinguished Service Cross as a Sub-Lieutenant for his services during the St Nazaire Raid in 1942. Badly wounded, he subsequently lost sight in one eye, when his motor torpedo boat was sunk; taken prisoner, he proved an inveterate escaper and so finished up in Colditz Castle. It was he who sold Bardsey to the newly formed Bardsey Island Trust in 1979.

Only one family farms Bardsey at the present time: mainly sheep, Welsh black cattle and a small area of arable crops. The walled gardens of the cottages, the four withy beds and a small conifer plantation provide excellent shelter for migrating birds. In 1998 Andy Clarke, a keen birdwatcher and organic gardener, noticed that the apples on a gnarled old tree in the garden were surprisingly free of disease so sent several to the National Fruit Collection which declared that the fruit and the tree were unique, thus the Bardsey Apple.

The garden at Cristin, now the Bird Observatory, appropriately harbours one of the largest trees on Bardsey, a sycamore. The observatory warden lives here, while there is also modern self-catering accommodation for those wishing to stay at the Bardsey Lodge, where they may wish to participate in the bird ringing and other activities or just have a relaxing break away from the hustle and bustle of mainland life. The cottages, renovated by the Bardsey Island Trust, are available for short-term letting.

The south of Bardsey is divided from the north by an isthmus, bounded on one side by the island's only sandy beach and on the other by shingle and pebbles. The landing jetty is situated here. Most of the southern extremity is an expanse of close-cropped maritime turf with Thrift, Sea Campion and Spring Squill.

On this low southern peninsula stands the lighthouse, constructed in 1821 by Joseph Nelson, one of the great lighthouse-builders. It is one of the few lights to have a square tower, but its chief claim to fame, or rather infamy, is that its light, giving five flashes every 15 seconds, can attract large numbers of nocturnal migrants, especially on very dark, cloudy nights. Many birds were killed by flying into the lantern and surrounding structure; the toll on bad nights could be hundreds of birds. From 1978 a 'false lighthouse', a tall mast with bright halogen lamps, has been in place close to the lighthouse. This draws birds away from the hazards of the tower and buildings, many settling on the ground around the mast and not moving away until daybreak. Not only are small passerines attracted, but Manx Shearwater, Storm Petrel, Water Rail, various waders, Kittiwake, terns, Swift and Cuckoo have been recorded. Some of the bigger attractions occurred in the early 2000s when up to 40,000 birds were drawn in, thankfully very few birds perished. To be at the Bardsey light when migration is at its height, when the beams are full of flighting birds, was an experience never to be forgotten. This all changed in 2014 when the five-beam lens was replaced with the much duller red LED; since then no further attractions have taken place.

Grey Seals breed infrequently on Bardsey, though they are always to be seen offshore, or hauled out on favoured low-lying rocks, with as many as 300 present at the late summer peak and shortly before dispersal to their breeding grounds; some 60 pups have been born in a single year on Bardsey, while other females are known to travel to the Pembrokeshire colonies 80 miles (128 km) to the south. Porpoises and dolphins of several species are regularly recorded, while Basking Sharks are occasionally seen. On land, there were Rabbits up to 1996 when they became extinct. Wood Mice and Common Shrews are still abundant, but House Mouse too became extinct in 1977. Palmate Newt is the sole representative of the amphibians, Slow Worm the only reptile.

John Ray, considered by many to be the greatest of all our field naturalists, visited Bardsey in the 17th century, while 200 years or so later the indefatigable William Eagle Clarke collected bird notes from lighthouse keepers. Ronald Lockley, doyen of pioneer island naturalists, came here in November 1934 to study bird migration and the problem of the lighthouse. Despite this early interest, and that of a scattering of other noted visitors during the first half of the 20th century, a writer in 1953 commented: 'It is remarkable that an island so well situated as Bardsey should be as little known to ornithologists.' Times were, however, changing. Birdwatchers in increasing numbers were taking up the challenge presented by small islands, and from the time the Bird Observatory opened in 1953, there has been a stream of visitors to help successive wardens assiduously add to our knowledge, not only of the birds but of all aspects of Bardsey's glorious natural history. Peter Hope Jones, who was closely involved with the island, provided in 1988 an eloquent account in *The Natural History of Bardsey* and ten years later he and the priest, poet and amateur ornithologist R.S. Thomas co-wrote *Between Sea & Sky: Images of Bardsey*.

Bardsey is one of only two observatories in Wales (the other Skokholm in Pembrokeshire), having been founded by the West Midland Bird Club, the West Wales Field Society and local residents. The first honorary secretary was the noted naturalist and author William Condry, and the first visitors arrived in late July 1953. A succession of wardens have manned the observatory ever since, save for the period 1971 to summer 1973, when transport difficulties meant a temporary closure. Over 294,000 birds of 196 species have been ringed on Bardsey, and the results of this dedication form a major contribution to our knowledge of bird populations and movements in western Britain. Bardsey Bird Observatory produces an annual report and has its own website, with details of recent sightings, membership and accommodation.

Species

Bardsey is a stronghold for Chough, and the seven to ten pairs that now breed make this an important site for the bird. They breed from about four years of age and tend to use the same sites year after year. A male that nested in Seal Cave was at least 17 years of age when it died. Young Choughs are much more mobile, wandering some distance along the Llŷn Peninsula; a few even reach into Snowdonia, while one moved south, 45 miles (72 km) across Cardigan Bay to Llangrannog, Ceredigion, and settled there to breed, and some venture even farther south: at least three have been seen at Strumble Head, Pembrokeshire. More remarkable, for it was way beyond normal Chough range, was the individual who travelled 88 miles (142 km) to reach south Lancashire. A key factor in maintaining so many breeding Choughs on Bardsey is the large area of close-cropped turf and the liberal quantities of droppings from large grazing animals. Particularly important in late summer are dung beetles, many hundreds of which may emerge from a single horse dropping. Ants are another important food, the mounds on the open sward being torn into by the Choughs searching for the yellow occupants. In winter, seaweed and other debris of the strandline is a source of food, mainly kelp flies and sandhoppers.

The other members of the crow family present include two pairs of Ravens, early nesters on secluded rocky outcrops. In most autumns there is an influx from the mainland, usually of up to 25 birds, though on several occasions double that number have been counted. With the wind in the east they soar along the face of the Mountain, hanging in the updraught, the air full of wild, joyful 'kronking' calls. Jackdaws nest in Rabbit burrows and crevices, the residents being joined in late summer by large noisy flocks which cross from the mainland. Magpies are gradually increasing, nesting in the withy beds and even in low bramble patches, a habit repeated on the Pembrokeshire islands. Hooded Crows are regular migrants – the coast of Co. Wicklow, Ireland, is after all only 60 miles (96 km) to the west. Not surprisingly, Jay, a resident of oak woodland, is a very rare visitor to offshore islands, but small flocks were seen in the late autumn of 1983, the year of a huge irruption of this bird into Britain.

After Chough, seabirds are the main attraction among breeding species on Bardsey, although numbers do not compare with those at some other sites in North Wales or on the Pembrokeshire islands. Lack of suitable nest sites restricts numbers of cliff-nesting birds. Fulmars first nested in 1957 and there are now around 40 pairs. For about 15 years from the early 1960s the number of breeding Kittiwakes was extremely low, although it has subsequently risen and roughly 125 pairs now nest, mainly at one site on the east coast. Up to 1,100 pairs of Guillemots share the same cliff, while Razorbills, numbering some 2,000 pairs, are more scattered; it seems surprising that more do not nest, as they take readily to sites among the lower scree slopes. Puffins were first confirmed breeding in 2000 and now number 150 pairs. It is on the cliffs on the east side that 50–60 or so pairs of Shags also nest. Although Cormorants are seen daily, none breed on the island; around 30 pairs nest in a nearby colony, on the Gwylan Islands, off Aberdaron Bay, 4 miles (6.4 km) to the northwest. Lesser Black-backed Gull has slowly increased since records began in 1956, now numbering about 200 pairs occupying a slope on the north of the island. About 500 pairs of Herring Gulls nest, numbers having shown a recent increase after a decline in the 1970s. Several pairs of Great Black-backed Gulls nest in most years.

The nocturnal Manx Shearwater is Bardsey's most numerous breeding bird, with between 25,000 and 35,000 pairs nesting in burrows all across the island. It is one of the events of a stay on Bardsey to hear after dark the calls of Manx Shearwaters flighting in from the sea, a signal to gather one's torch and head for the colony to view birds at close quarters – a magic world of calling shearwaters, of dew-soaked bracken, while away in the distance the Bardsey light stabs the night sky. Only recently were Storm Petrels proved to breed here too, with 11 chicks ringed in 2004, when the colony was estimated at about 111 pairs, increasing to 150 pairs at the last census in 2017.

Depending on weather conditions and the time of year, there can be spectacular movements of some species offshore, including 30,000 Manx Shearwaters, 15,000 Kittiwakes, 2,000 Guillemots and 15,000 Razorbills, with smaller numbers of Gannets (600), and terns (600 Sandwich and 700 Arctics). It is the scarcer species that attract attention, while others may be once-in-a-lifetime birds such as the Black-browed Albatross seen flying south one day in

2018 – what fortune for the lucky observers! Balearic Shearwater is a regular visitor in late summer, while Sooty Shearwaters are less so. Little Shearwaters have been recorded twice, while the first sighting of a Fea's Petrel for Wales took place on the evening of 10 September 1994, with others seen in 2013 and 2019. One of the great delights for seawatchers is when, on the windiest of days, the Leach's Petrel, with its distinctively buoyant and at times erratic flight, comes beating past; a few, usually singles, are seen each autumn. Arctic and Great Skuas are annual passage migrants, with over 100 of the former seen in some autumns. Pomarine Skuas are less frequent, while there are a few records of Long-tailed Skua each year. Other scarce species seen offshore have been Grey Phalarope, Mediterranean, Sabine's, Little and Bonaparte's Gulls, Black Tern, Black Guillemot, Little Auk. There have also been three records of Zino's/Fea's/Desertas Petrel.

North American passerines have provided some exquisite gems on Bardsey such as the first and so far only British record of a Summer Tanager in September 1957 and the first Yellow Warbler in August 1964 of which there have subsequently been just five occurrences elsewhere in Great Britain. How many others have set out to cross the North Atlantic aided by a swiftly moving depression, only to be lost at sea, or eventually arrive, and depart, quietly, unnoticed by birdwatchers? Other late-summer and autumn Americans have included a Sora Rail, three Grey-cheeked Thrushes, Blackpoll Warbler, Common Yellowthroat, a Red-eyed Vireo, White-throated Sparrow, Rose-breasted Grosbeak, a Buff-bellied Pipit and an American Robin in 2000. Visitors to Bardsey in the spring can also look forward with some anticipation to the chance of North American visitors, for a Dark-eyed Junco and a Song Sparrow have both been recorded then, the latter at the same time as a Baltimore Oriole in Pembrokeshire and a White-throated Sparrow in Caithness.

The list of rare migrants from Europe and Asia to Bardsey is equally impressive and includes the first Cretzschmar's Bunting for Wales in June 2015 and the only Crested Lark in Wales, June 1982. Other species have travelled much greater distances, among them Eye-browed Thrush, Pallas's, Dusky and Yellow-browed Warblers from Central Asia. Two scarce warblers from much nearer home are almost annual in their appearance: these are the Icterine and Melodious Warblers, both of which usually arrive during August and September. A River Warbler found dead at the lighthouse in September 1969 was only the third recorded in Britain. Great Reed Warbler has been seen on a couple of occasions in early summer, though few would have expected to hear its strident frog-like song on a remote Welsh island. As if there were not already enough to whet the appetite, other rare visitors include Night Heron, Cattle Egret, Great White Egret, White Stork, Honey Buzzard, Baillon's Crake, Great Snipe, Black-winged Stilt, Snowy Owl, Alpine Swift, Bee-eater, Wryneck, Short-toed Lark, Citrine Wagtail, Eastern Yellow Wagtail, Tawny Pipit, Nightingale, Thrush Nightingale, Isabelline Wheatear, Black-eared Wheatear, Lanceolated Warbler, Booted Warbler, Subalpine Warbler, Sardinian Warbler, Western Orphean Warbler, Greenish Warbler, Arctic Warbler, Radde's Warbler, Western Bonelli's Warbler, Red-flanked Bluetail, Red-breasted Flycatcher, which has been described as a Bardsey speciality, Collared Flycatcher, Penduline Tit, Woodchat, Red-backed Shrike, Isabelline Shrike, Rose-coloured Starling, Serin, Arctic Redpoll, Common Rosefinch, and Little, Yellow-breasted, Black-headed, Ortolan and Rock Buntings.

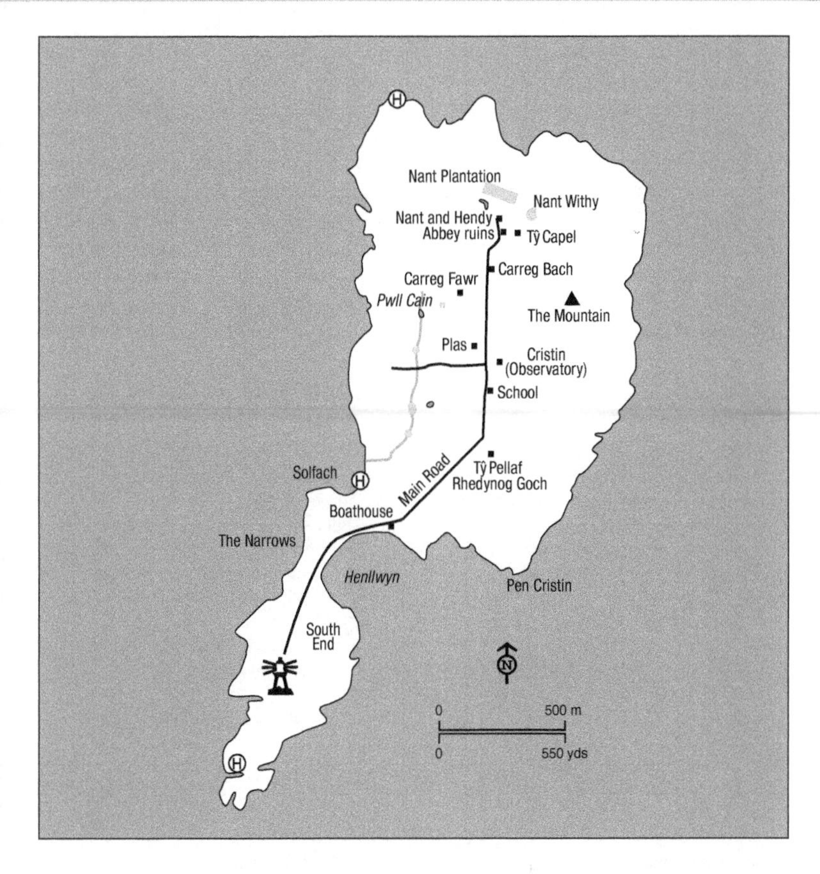

Rare birds yes, and from virtually all parts of the globe, but there is another, once widespread in Wales as a breeding bird but now alas not even seen annually: the Corncrake. It formerly nested on Bardsey, perhaps even as many as 10 pairs, but no longer can its strident calls be heard. Indeed, this bird must now be one of our rarest visitors, a salutary reminder of how a species' fortunes can change within just a few decades.

Timing

So much depends on your interests. If it is migrants, then the early autumn September/October is best; there is a good seabird passage and small migrants can be prolific at times, though this is heavily dependent on the weather situation over the whole of northern Europe. May must be the most attractive month, for the island flowers are then coming to their peak, spring migration is well under way, and the breeding seabirds have all returned to their colonies. The weather is often as good during May as at any other time of year.

Access

The Bird Observatory, the only one in North Wales, is open from late March to November and anyone over the age of 18 is welcome, beginners and experts alike. For enquiries and bookings it is best to check the observatory website.

There is room for up to 12 people self-catering. The observatory boat makes the journey from Aberdaron daily, weather permitting.

Bardsey is owned by the Bardsey Island Trust, details of which can be obtained from the Trust's website. The trust has a number of holiday properties for letting, while day visits are possible too. The boat crossing is daily, weather permitting, from Porth Meudwy (SH164255), between April and October. Bookings must be made in advance with the boat office 07971 769895. As with all islands, there are occasions when expectant visitors fail to cross because of inclement weather. If such a fate befalls you, then make the most of your time on the tip of the Llŷn Peninsula. It is well worth exploring the coastline south and west of Aberdaron (Site 28).

CALENDAR

Resident: Cormorant, Shag, Mallard, Kestrel, Peregrine, Moorhen, Oystercatcher, Herring and Great Black-backed Gulls, Little Owl, Rock Pipit, Chough, Raven.

December–February: Red-throated Diver, Fulmar, Guillemot and Razorbill return to the cliffs, the latter two making only sporadic visits, usually at dawn. Depending on weather conditions further east, there may be major influxes of thrushes, finches and Starlings. Merlin and Snipe usually resident.

March–May: Manx Shearwater, Lesser Black-backed Gull and Kittiwake return early in the period, winter visitors depart though there may be stragglers well into April. First summer visitors usually Ring Ouzel, Wheatear and Black Redstart, followed shortly after by Swallow, Sand Martin and Chiffchaff. Other warblers and more hirundines are a feature of April, together with smaller numbers of Whimbrel, Common Sandpiper, Redstart and Pied Flycatcher. Late summer migrants include Swift and Spotted Flycatcher. Puffins should be looked for offshore from early April (about 400 pairs nest on Ynys Gwylan off Aberdaron); Guillemots and Razorbills do not remain ashore permanently until the first eggs are laid, usually about the end of April/ early May. Vagrants likely to occur throughout whole period, Hoopoe and Woodchat Shrike being the two most regularly seen.

June–July: All breeding species present, though by end of the period Guillemots and Razorbills have left the cliffs. A few late migrants possible, though before July is out southward passage will have commenced. Purple Sandpipers from early July, while a few Turnstones usually summer on the island.

August–November: Kittiwakes complete their breeding season in late August and quickly forsake the cliffs, though many are seen offshore daily throughout the autumn. Fulmar, Manx Shearwater and Storm Petrel complete their breeding season during September. Seabird movement at its peak in September, with shearwaters, Gannets, skuas, gulls and terns the main species involved, joined in October by large passages of auk and occasional scarce species such as Sabine's Gull and Leach's Petrel. Given good conditions passerine passage can be spectacular, with the added attraction of scarce visitors. In late October and November usually large movements of larks, pipits, thrushes, finches and buntings, with occasional sightings of geese and swans.

28 ABERDARON HEADLAND

OS Landranger Map 123
OS Explorer Map 253
Grid Ref: SH159258

Habitat

Lying just offshore is Bardsey Island (Site 27), renowned for observing migration. The headland at Aberdaron can be just as productive as its famous neighbour and offers a great chance to find a scarcity. Seawatching can be rewarding, especially in autumn, while breeding birds include many Welsh coastal specialties such as Peregrine, Chough and Stonechat.

Species

During the summer months, Chough are present often feeding on the coastal grasslands and in the fields, while Yellowhammer, Stonechat and Linnets breed in the area. Offshore, large rafts of Manx Shearwaters sometimes in their thousands can be seen, with Gannets, Puffins, Razorbills and Guillemots observed from any vantage point.

Just to the west of Aberdaron is the Porth Meudwy valley, a small sheltered, vegetated valley which attracts any migrants during spring and autumn. A walk down the track to the secluded bay at the bottom of the valley can produce many warblers and Goldcrests passing through, while winter thrushes can number hundreds of birds if the conditions are right. Summer migrants such as Redstart, Whitethroat, Lesser Whitethroat, Sedge Warbler, Grasshopper Warbler, Pied and Spotted Flycatchers are frequently observed, while spring rarities have included Melodious, Ruppell's and Subalpine Warblers, Red-backed Shrike and Golden Oriole. During return migration, the valley can be full of Blackcaps, Goldcrests and winter thrushes. Yellow-browed Warbler and Firecrest are almost annual, while other autumn goodies have included Long-eared Owl, Waxwing and even a Red-eyed Vireo.

The village itself is worth exploring. Black Redstarts are regular visitors in late autumn, especially to the churchyard, while the beach here often holds a Mediterranean Gull among the resident gulls. Grey Wagtail use the stream which runs under the bridge and this can be a good place to survey the vegetation and rooftops for a migrant or two. Rarities seen in this area include Black Kite, Cattle Egret, Lesser Grey Shrike and Black-headed Bunting. You can take the coastal path from the village along to Porth Meudwy and then continue to Braich-y-Pwll, the very end of the headland. Alternatively, you can drive to the car park here which offers fantastic views across to Bardsey.

There are a number of footpaths on which you can wander around the headland. The walk up to the coastguard's hut offers an excellent chance to see Stonechat and Chough, while hundreds of pipits and finches pass over during migration periods. Autumn visible migration is especially good during the first few hours of light, but with the right conditions can continue all day. Snow Bunting and Lapland Bunting are regular, while Alpine Swift, Hoopoe, Subalpine Warbler and both Richard's and Tawny Pipit have also occurred. Seawatching is best from the area below the coastguard lookout in a north-westerly direction. Thousands of Manx Shearwaters pass here, while Balearic and

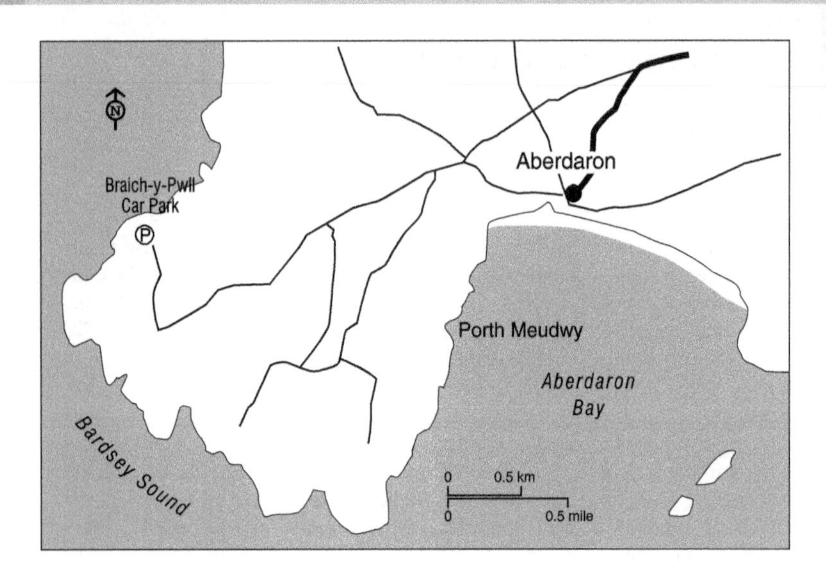

Sooty Shearwaters are annual. All four species of skua have been observed, as have both Storm and Leach's Petrels and Sabine's Gull.

The whole area is very under-watched, and anyone trying their luck here has a strong chance of finding some good birds and writing their name into the next bird report.

Timing

Spring and autumn migration periods, as well as the summer for breeding species.

Access

From Aberdaron Village, Porth Meudwy Valley (SH159258) and Braich-y-Pwll car park (SH143256).

CALENDAR

Resident: Cormorant, Shag, Peregrine, Rock Pipit, Stonechat, Chough, Linnet and Yellowhammer.

November–February: Red-throated and Great Northern Divers offshore, Skylarks, Meadow Pipits and finches along the coast. Goldcrests and numerous tits in the valley.

March–May: Early summer visitors, starting with Wheatear in early March. Swallow, Sand Martin, House Martin and Swifts pass through. Among the bushes various other migrants including Redstart, Chiffchaff, Whitethroat, Garden and Willow Warblers, Pied and Spotted Flycatchers.

June–July: Manx Shearwaters, Gannets, Guillemot, Razorbill and Puffin can be seen offshore. Rock Pipits, Stonechat and Whitethroat along the clifftop.

August–October: With the start of seabird passage, small numbers of Common Scoter pass the headland from late July onwards. Occasional Arctic and Great Skuas. Careful scrutiny of the passing Manx Shearwaters might produce Sooty and Balearic Shearwaters. The valley and coastal bushes may contain migrant passerines, Redstart, Whinchat, Ring Ouzel, Whitethroat, Lesser Whitethroat, Blackcap, Chiffchaff, Willow Warbler, Goldcrest, Pied and Spotted Flycatchers.

29 PORTH DINLLAEN

OS Landranger Map 123
OS Explorer Map 253
Grid Ref: SH276411

Habitat

The spectacular scenery of the Llŷn Peninsula welcomes those taking this circular walk around the headland of Porth Dinllaen with good views over the sweeping bay, the village owned by the National Trust since 1994, with the Ty Coch Inn arguably the best pub in Wales and in the top ten beach bars in the world. An Iron Age fort straddles the narrow headland which shelters the bay, once a ferry port for Dublin and formerly home to a prosperous Herring fishing industry, while pigs were sent by sea from here to Liverpool.

Species

Stonechat, Chough and Raven are present all year and can be seen from the National Trust car park. Take the footpath down to the bay and walk to the headland. In winter the bay plays host to Great Northern and Red-throated Divers, while Shags are common here. Continuing along to the headland, look out for Grey Seals which rest on the small islets offshore. In winter Snow Bunting can occur on the golf course fairways, while spring and autumn see good numbers of Wheatear, Skylark and pipits using the short grass to feed. A northwesterly blow produces skuas and Manx Shearwaters, while the migration periods are certainly worthy of more exploration at this site – the tip of the headland an ideal vantage point.

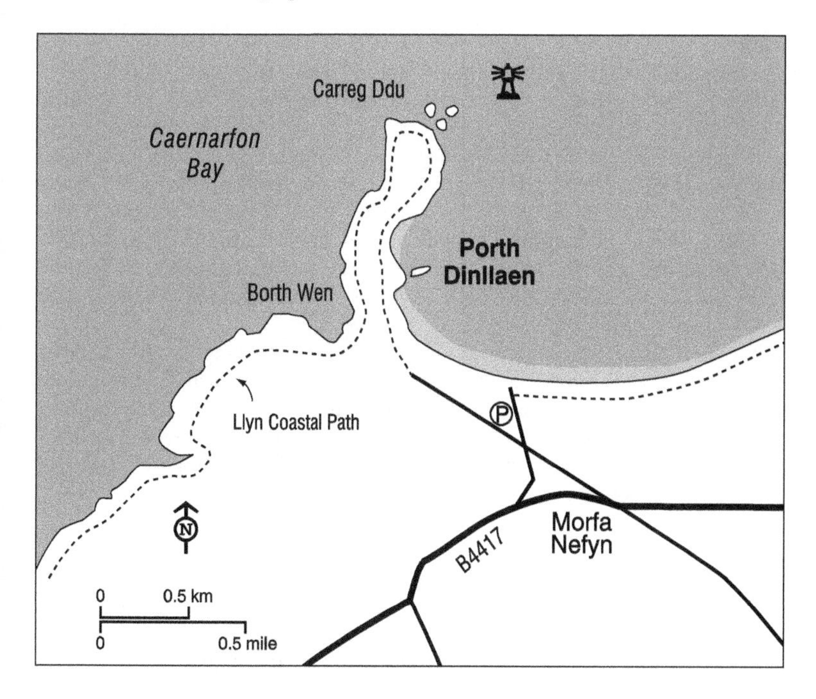

Timing
All year.

Access
From the National Trust car park at Morfa Nefyn (SH281406) – a charge payable. The footpath runs across a spectacular golf course and through coastal farmland. Birdwatchers can take a rest en route at the public house that nestles into the small bay on the headland, or at the golf clubhouse which stands above the bay.

CALENDAR

Resident: Cormorant, Shag, Skylark, Meadow and Rock Pipits, Stonechat, Chough and Raven.
November–February: Great Northern and Red-throated Divers may be offshore.
March–July: Spring migrants include Wheatear, Whitethroat, Swallow. Manx Shearwaters may be seen offshore during their feeding flights from distant colonies.
August–October: Manx Shearwater and westerly winds may bring in skuas and Petrels.

30 PORTHMADOG TO CRICCIETH

OS Landranger Map 124
OS Explorer Map OL18 and 254
Grid ref: SH543365

Habitat
Porthmadog is a small town at the mouth of Afon Glaslyn and Afon Dwyryd, in the top north-eastern corner of Cardigan Bay. The Glaslyn and Dwyryd Estuary is one of the most important sites in North Wales for wintering ducks and waders (see Site 91). Porthmadog Cob, a mile-long causeway, runs across this large estuary. On the seaward side is the estuary, while on the landward side is a lagoon and marshland, with spectacular views of Snowdonia. Ospreys have nested upstream of Porthmadog at Pont Croesor (Meirionnydd) since 2004. The adults can often be witnessed fishing in the bay, taking their catch back to the young in the nest. The Glaslyn Osprey Project has an information centre and Osprey viewing facilities (see Meirionnydd).

Cardigan Bay is internationally important for wintering sea-duck, grebes and divers. Southwesterly winds and oceanic currents push nutrients up into this top north-eastern corner and in doing so make this area the richest feeding area for birds in the bay, stretching from Criccieth in the north-west, past Black Rock Sands, Borth-y-Gest, Afon Glaslyn, Afon Dwyryd to Morfa Harlech (see Site 93) in the south-east.

Species
The lagoon by Porthmadog Cob holds good numbers of duck in winter, especially Wigeon, Teal and Pintail. Little Grebe, Goldeneye and Red-breasted Merganser are regular, while waders include good numbers of Redshank, Snipe, Curlew and Black-tailed Godwit. Whooper Swan are regular winter visitors.

During the summer months the Ospreys which breed upriver, fish the lagoons and can provide stunning views. On the seaward side, the large numbers of roosting gulls are joined in the summer by Sandwich and Common Terns. During winter and at high tide, Water Pipits feed in the Spartina grass, with the southern corner being best for this species.

Llyn Bach lies on the northern side of the lagoon and is worthy of investigation, with waders and ducks giving close views; a footpath runs around it and joins the causeway footpath. Many rarities have been found in this area over recent years including American Wigeon, Ring-necked Duck, Long-billed Dowitcher, Laughing Gull, Ross's Gull and Elegant Tern.

In winter, large numbers of sea-duck congregate in the bay, with Common Scoter numbering in the thousands. Among these flocks, small numbers of Velvet Scoter, Scaup and Long-tailed Duck can be found. Surf Scoter have also been seen over the years. Good numbers of Great Crested Grebe, Red-throated Divers and Red-breasted Mergansers are joined by small numbers of Slavonian Grebe and Great Northern Diver during the winter, while Common Eider are also present.

During passage times, especially after brisk southwesterly winds, seabirds may be pushed into this part of Cardigan Bay. Great and Arctic Skuas are seen annually, while Pomarine may also occur, particularly in spring. Little Gulls are regular offshore, sometimes in good numbers, with Iceland, Glaucous, Mediterranean and Sabine's a possibility. The beach at Black Rock Sands has held an Ivory Gull and a Ross's Gull in the past. Nearby at Criccieth and Borth-y-Gest, grebes, sea-duck and divers are also present during winter, while Criccieth holds a small flock of wintering Purple Sandpipers.

Timing

All year.

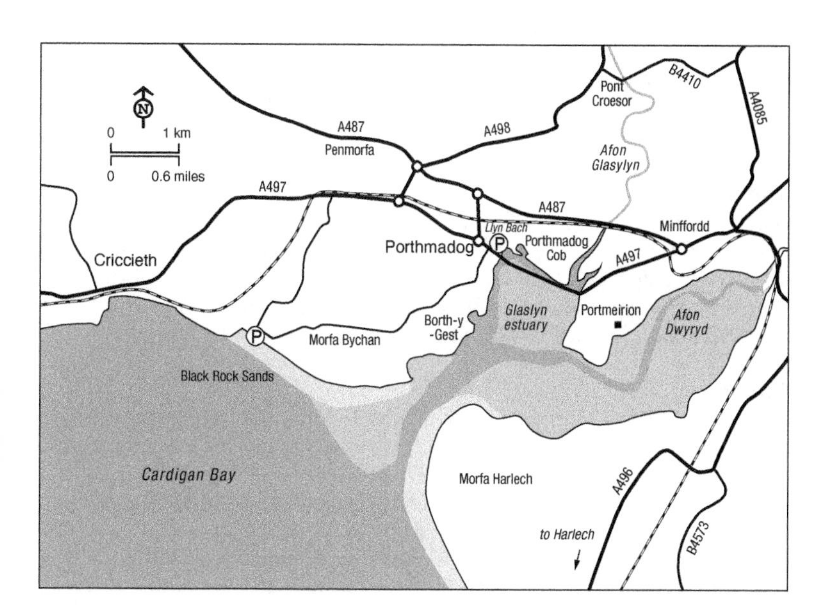

Access

Park at either end of the Porthmadog Cob, there is a free car park on the southern side or a pay-and-display near Llyn Bach, Porthmadog. There are two footpaths to choose from: the lower one runs below the road and gives sheltered views of the lagoon, while the other runs alongside the railway, which is exposed but gives views to the lagoon and estuary. For views of the bay there are car parks on Black Rock Sands (SH529372), in Borth-y-Gest (SH564375) and along the promenade in Criccieth.

CALENDAR

Resident: Cormorant, Grey Heron, Mute Swan, Shelduck, Mallard, Red-breasted Merganser, Kestrel, Oystercatcher, Ringed Plover, Lapwing, Curlew, Redshank, Black-headed, Herring and Great Black-backed Gulls, Skylark, Meadow Pipit, Pied Wagtail, Reed Bunting.

December–February: Red-throated and Great Northern Divers, Great Crested Grebe, Whooper Swan, Wigeon, Teal, Pintail, Tufted Duck, Scaup, Eider, Common Scoter, Goldeneye, Hen Harrier, Merlin, Peregrine, Short-eared Owl.

March–May: Wheatear, Swallow and Sand Martin from late March, Cuckoo and smaller migrants such as Grasshopper and Sedge Warblers, Whitethroat, Spotted Flycatcher during April, also wader passage at the same time including Whimbrel and Common Sandpiper.

June–July: Breeding species still present. Manx Shearwater occasionally seen offshore, also some terns. First autumn waders appear on estuary.

August–November: Wader passage throughout early part of period, with a chance of scarce species such as Little Stint, Curlew Sandpiper, Ruff, Black-tailed Godwit and Spotted Redshank. Winter ducks start arriving from late September, and a chance of seeing Whooper Swans from mid-November. Do not forget the sandhills for winter predators and possibility of birds such as Twite and Snow Bunting.

31 ABERDESACH AND PONTLLYFNI

OS Landranger Map 115
OS Explorer Map 254
Grid Ref: SH424514

Habitat

The wildly beautiful Llŷn Peninsula extends 30 miles south-west from the Menai Straits to reach Bardsey Sound, the west coast fronting onto Caernarfon Bay. As one guidebook (*Best Birdwatching sites in North Wales*) says, 'anyone who enjoys seeking out and identifying divers should head to Aberdesach and Pontllyfni, the two small seaside villages southwest of Caernarfon'.

Species

Both sites overlook Caernarfon Bay which plays host to divers and a variety of seabirds, including Gannet and Manx Shearwater. Divers are present during the winter and passage periods; March and April are best as numbers increase and Great Northern Divers regularly occur in double figures. This is also the best time for seeing Black-throated Diver if you are very lucky.

The boulder clay cliffs to the southwest of Aberdesach sometimes attract small numbers of Chough. Pontllyfni has a bit more to offer in the variety of

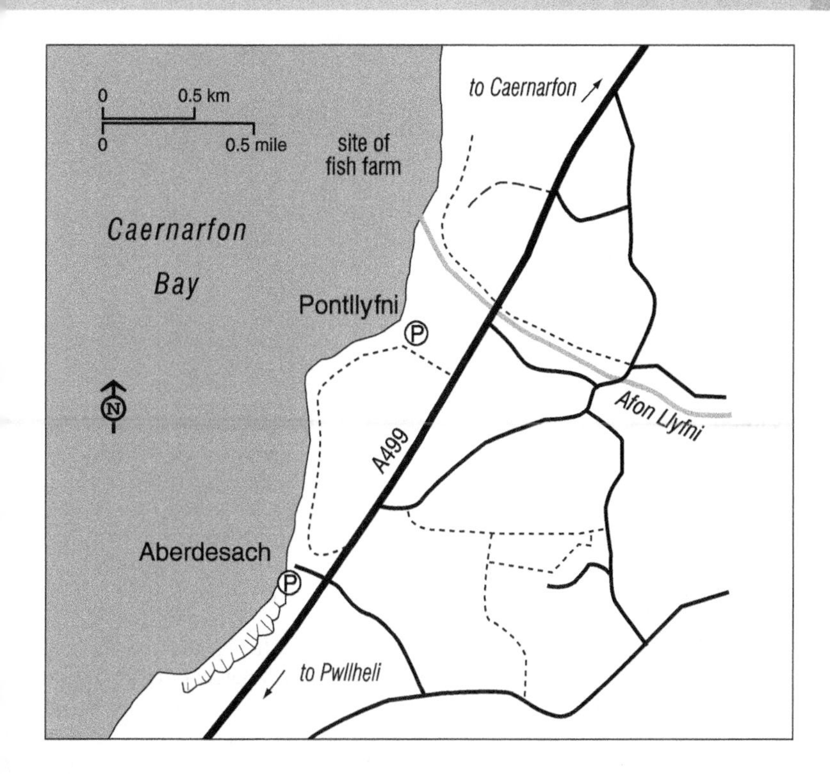

habitats and birds. Follow the public footpaths on either side of the Afon Llyfni leading to the beach, which can often produce Dipper and Grey Wagtail. The neighbouring coastal fields can attract flocks of gulls, including good numbers of Mediterranean Gulls with the occasional Iceland or Glaucous Gull.

There is a former fish farm on the eastern side of the Afon Llyfni. Much of the site is out of view but the settlement ponds can be viewed from the top of the beach. This is a good area for seeing early spring migrants such as Sand Martin and Wheatear. The ponds have attracted a wide variety of waders in the past, including Little Ringed Plover and Avocet. Unfortunately, they have since become too vegetated but are still worth a look.

Timing

Winter and passage periods.

Access

Aberdesach has ample parking on the beach at SH424514. Limited parking at SH434525, near Afon Llyfni.

CALENDAR

Resident: Cormorant, Shag, Peregrine, large gulls, Dipper, Grey Wagtail, Stonechat, Raven and Chough.

November–February: Red-throated and Great Northern Divers, Great Crested Grebe, Gannet, Turnstone and Merlin.

March–June: Gannet and Manx Shearwater, Sand Martin, Wheatear and other passage migrants.

July–October: Gannet and Manx Shearwaters offshore, with Red-throated and Great Northern Divers appearing in late autumn.

32 DINAS DINLLE

OS Landranger Map 115
OS Explorer Map OL17
Grid Ref: SH436566

Habitat

The small seaside resort of Dinas Dinlle is positioned on the south-western side of Morfa Dinlle, a peninsula that comprises a dune system in the north, a complex of low-lying fields in the interior, including Caernarfon Airport, and boulder clay cliffs to the southwest while the massive pebble upper-shoreline level contrasts with the lower sandy beach.

Species

Flocks of Lapwing and Curlew are a familiar sight during the autumn and winter months on the peninsula, while the airfield and neighbouring pastures are especially noted for sizable flocks of Golden Plover. This area can also attract a variety of migrants such as large numbers of White Wagtails and Wheatears. Raptors regularly occur, including Merlin and Hen Harrier during the autumn and winter.

The western shore overlooks Caernarfon Bay which plays host to good numbers of divers and a variety of seabirds. Both Great Northern and Red-throated Divers winter offshore in small numbers; however, during March and April, these numbers increase significantly. Black-throated Diver occasionally appears during this period. Late summer can produce skuas attracted by flocks of terns. The beach supports passage waders including Whimbrel and Sanderling. The road between the resort and the airfield is worth checking in winter for small numbers of Twite as well as the ubiquitous Rock Pipits. Snow Bunting occasionally pop up along the beach towards Belan Fort.

Rare visitors have included American Golden Plover, Osprey, Dotterel, Hoopoe, Richard's Pipit and Ortolan Bunting.

Timing

Winter and passage periods.

Access

Park at either SH431582 for the airfield and northern stretch of beach, or at SH436566 for the southern stretch.

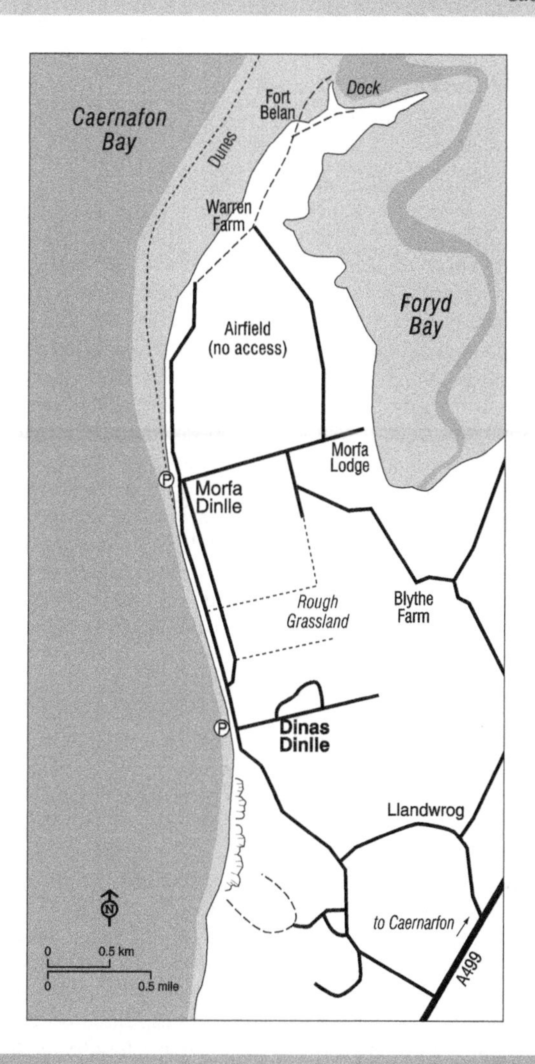

CALENDAR

Resident: Cormorant, Meadow Pipit, Rock Pipit, Skylark.
November–February: Red-throated and Great Northern Divers, Hen Harrier, Merlin, Peregrine, Lapwing, Golden Plover, Curlew.
March–May: Passage of Whimbrel, Sanderling, Wheatear and White Wagtails.
July–October: Return passage of Sanderling, Turnstone, Whimbrel, Lapwing, Golden Plover, Sandwich Terns and in late autumn divers.

33 FORYD BAY

OS Landranger Map 115
OS Explorer Map OL17
Grid Ref: SH452586

Habitat

Foryd Bay forms part of a network of estuaries at the southern end of the Menai Strait, including the Braint and Cefni Estuaries in Anglesey, that are used by large numbers of wintering wildfowl and waders. Birds can often be seen commuting between Foryd Bay and the Braint Estuary. The bay is a superb mosaic of intertidal habitats – mudflats and saltmarsh, river channels and a shingle spit which narrows the entrance to this wonderfully sheltered estuary. Here the Menai Strait is at its narrowest at just 350 yards to Abermenai Point on the south-western shores of Anglesey. At the northern extremity of the peninsula, Fort Belan built in 1775 advertises self-catering cottages where once Sir Ralph Payne-Gallwey (1848–1916), a colourful man best known to birdwatchers as the author of *British Decoys*, demonstrated in 1886 that ancient ballistas could hurl a stone ball across the Menai Straits, here just over a quarter of a mile wide, and that a crossbow could hit a target at the same range. His other interests included golf – it was he who designed the ball with its reticulated surface.

Species

Wildfowl and waders are the speciality of the bay, which holds large numbers of wintering Wigeon and relatively large numbers of Brent Geese – the Dwarf Eelgrass *Zostera noltei* growing on the intertidal zone is their major food plant, most of those wintering here being of the 'pale-bellied' race which breed in Arctic Canada, Greenland, Spitzbergen and Franz Josef Land. Wigeon generally start returning from late July onwards and tend to peak in November when numbers swell to over 3,000 birds. Numbers gradually decrease from January onwards, with small flocks remaining into April. This is one of the few sites in North Wales where both dark and light-bellied Brent Geese occur in reasonable numbers. Mixed flocks of over a hundred birds occur in midwinter. Time spent checking the wildfowl flocks can sometimes be rewarded by the odd Scaup. Greenshank is another site speciality and can often occur every month of the year. Most tend to appear on return passage July to October, when as many as 30 can be seen. Over the winter months up to 10 birds may be found. Spotted Redshank is a regular passage bird in small numbers and occasionally one or two may overwinter. Ruff, Curlew Sandpiper and Little Stint occur in small numbers from August to October. Little Egrets breed at a couple of locations in the Menai Straits, with numbers increasing to around 50 on the Foryd in autumn, with up to 20 wintering. Rarities have included American Wigeon, Spoonbill, Black-winged Stilt, Sharp-tailed Sandpiper, Pectoral Sandpiper, Ring-billed Gull, Iceland Gull, Glaucous Gull, Forster's Tern, Hoopoe and Shorelark.

Timing

Winter and passage periods.

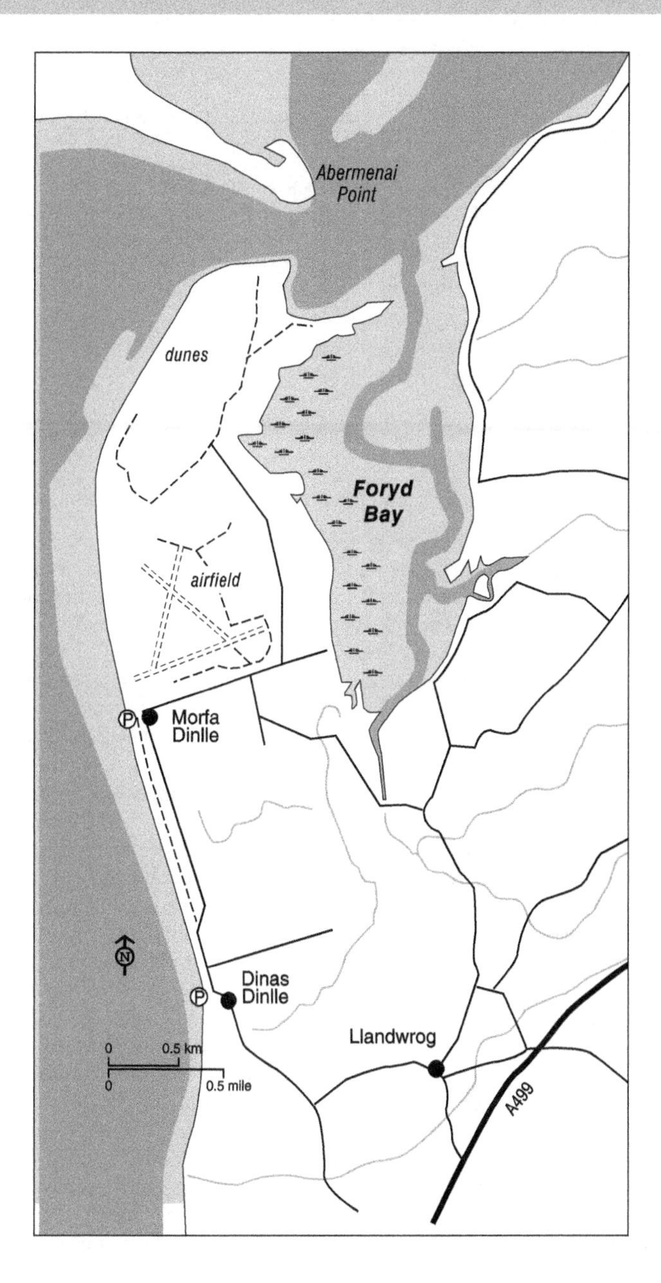

Access

Roadside parking along the foreshore road and picnic site at SH453603 on the eastern side of the bay.

Limited parking at the hide SH452586 and near the mouth of the Afon Gwyrfai SH453588 on the south-eastern side. Roadside parking near Caernarfon Airport at SH440585 to access the western side of the bay.

CALENDAR

Resident: Grey Heron, Little Egret, Oystercatcher, Curlew, Peregrine and large gulls.
November–February: Brent Geese, Wigeon, Teal, Pintail, Merlin, Ringed Plover, Dunlin, Redshank.

March–June: Passage waders including Whimbrel, Greenshank, Spotted Redshank, Knot and Turnstone.
July–October: Return wader passage, Knot, Whimbrel, Greenshank, Spotted Redshank, Ruff, Turnstone, terns may also be present, mainly Sandwich.

34 SNOWDONIA

OS Landranger Map 115
OS Explorer Map OL17

Habitat

Snowdonia, its Welsh name Eryri possibly derived from *eryr*, the word for eagle, suggests that Golden Eagles once soared over its narrow valleys, above its ridges and summits, nesting on remote crags, hoping no doubt to escape the attentions of shepherds. Alas, there are only tantalising hints in the literature. Edward Lhuyd himself, the greatest of Welsh naturalists, probably failed to see the bird, though the evidence he collected, and the reports of others, confirm that Golden Eagles had indeed been present here. Admittedly, the later records are sparse to say the least, but enough to lead us to believe they nested until the middle of the 18th century. The absence of Golden Eagle, and of other northern species, means that the mountains of Wales are somewhat impoverished ornithologically when compared with those of Scotland. Do not let this dissuade you from walking the hills, or going high into the mountains of Snowdonia, provided of course that you are properly equipped.

Natural Resources Wales manages a number of national nature reserves (NNR) in Snowdonia, including one of the most renowned, Cwm Idwal, at 984 acres (398 ha), the first NNR in Wales and aptly considered as holy ground for the history of early botany and geology. Many of the scarce and special plants can be found there, including the Snowdon Lily *Lloydia serotina*, named after Edward Lhuyd, who died far too young before he could write his planned *Natural History of Wales*, Purple Saxifrage, Alpine Meadow-rue and Moss Campion, while it also claims the highest tree in Wales. The glacial lake of Llyn Idwal, like a number of other unspoiled upland waters, has some interesting plants: Awlwort, Floating Bur-reed, Pillwort, Quillwort, Shoreweed and Water Lobelia, the latter fringing the shallows during July and early August with its pale lilac flowers. Another NNR, the largest in Wales, Yr Wyddfa (Snowdon), extends over 4,145 acres (1,677 ha) and includes the very summit of Snowdon itself.

Species

Ring Ouzel, one of the earliest returning summer migrants, is largely a bird of mountain and moorland except when on passage, its breeding range extending directly down the central spine of Wales, most occupying territories between 825 and 1,155 feet (250–350m) and often reaching the very summits at 2,500 feet (758 m) or so. Numbers have declined over the past 30 years, though Snowdonia remains a stronghold. Ring Ouzels require within their nesting territory a crag or gully, perhaps a stunted Rowan, even a disused mineshaft or quarryman's building, with plenty of good Heather and Bilberry. If searching for this bird, concentrate on areas where such features are evident, though usually the penetrating whistle and harsh 'tae-tac-tac tactac' notes will reveal the bird's presence even at long range. Ring Ouzels arrive in late March and early April and stay until late summer, before returning to winter quarters in north-west Africa. Another summer visitor to the high ground is Wheatear, arriving even before Ring Ouzel, and just as much at home in rocky areas as on the open grasslands, provided there are suitable holes for nesting. Wrens are encountered to the very tops of the mountains, and are often the only small bird present there during hard weather in midwinter. On the open mountain grasslands, Skylarks and Meadow Pipits are numerous, the latter a frequent host to the Cuckoo on slopes up to about 1,500 feet (457 m) in altitude.

Two members of the crow family are striking, at times noisy residents of the high ground. Choughs feed on the sheep-grazed turf of the mountains, having been recorded up to at least 2,900 feet (884 m) altitude. Pairs breed in the disused quarries and mineshafts, although numbers have declined by 72% since their peak of 27 pairs in the early 1980s to fewer than 10 pairs in recent years. Ravens frequent the high ground throughout the year, hard weather resulting in the death of sheep being to the birds' advantage.

Three species of falcon, Kestrel, Merlin and Peregrine, may be encountered in Snowdonia; Merlin is by far the rarest, for it needs large heather moors on which to breed. The lakes attract small numbers of winter wildfowl, groups of Whooper Swans being the most prominent. Goosanders usually winter here, with a small number remaining to breed. Pochard, Tufted Duck and Goldeneye, together with the Coot, may be seen on all the larger waters throughout the winter. A summer visitor to the shores of these lakes is Common Sandpiper, its shrill 'willy-wicket' song, certainly the sound of the lakes and rivers in early summer, soon attracting the watcher's attention; its flight is unlike that of any other wader, very low over the water with flickering wingbeats and short glides, and as soon as it lands, the characteristic bobbing action will be observed and the call heard. Other birds of the lakes and rivers are Dipper and Grey Wagtail, while Pied Wagtails seem to be everywhere, even up to the high ground.

The Snowdonia woodlands, especially the deciduous and mixed woodlands, a number of which have footpaths and nature trails to provide good access, support a rich range of species including Goshawk. With luck, and a little knowledge of its habits, a visitor to the Snowdonia woodlands may well be fortunate enough to see this magnificent bird of prey. Despite their size – females are almost as large as a Buzzard – they are rarely seen, preferring to hunt within or very close to woodland. Most noteworthy are the early morning aerial displays of the Goshawk high above woodlands from January to April, when the birds indulge in spectacular courtship display flights, often rising high above the nesting territory as they dive and swoop at one another.

Although breeding was not confirmed until 1972, it is thought that Siskins have been breeding in the Snowdonia woodlands since the late 1940s. Now they are widespread in the region, and indeed found in suitable habitats throughout much of Wales, maturing conifer plantations their favoured habitat. Woodcock winter in these same plantations but rather little seems to be known of their distribution or numbers.

Timing

Not material. It is more a question of the weather, which can change extremely rapidly on the mountaintops so that care must be taken, even in high summer, to acquaint oneself with the forecast before setting out. Make sure that you are properly equipped and do not proceed unless you are absolutely certain of both the conditions and your own capabilities, and that someone knows where you are going and your anticipated time of return. Do not neglect to inform them of any changes to your plans and ensure they know you have returned safely.

Access

While several of the Snowdonia lakes such as Llyn Gwynant, Llyn Dinas and Llyn Cwellyn are visible from the road, probably the best way of seeing upland birds is to follow one of the three trails.

The Miners' Track. Built during the 19th century for the copper mines near Glaslyn, you head south from the A4086 at Pen y Pass (SH647556). The track takes you close to Llyn Teyrn and eventually to Llyn Llydaw, one of the largest of the Snowdonia lakes. For those properly equipped, it continues, linking with the Pyg track above Glaslyn, to the summit of Snowdon. If you are road-bound, try the lay-by at Pont-y-Cromlech (SH628567), search the high ground through your binoculars for Peregrine and Raven, possibly Chough, while close by Skylarks will be singing and Wheatears nesting.

Cwm y Llan. Leave the A498 at Pont Bethania (SH627506) and follow the Watkin Path, the first part taking the watcher to the Snowdon Slate Works below Craig Ddu. For those properly equipped, it is possible to continue beyond to Bwlch Ciliau and Bwlch y Saethau and so on to the summit of Snowdon.

The following sites are on slightly lower ground but all amid spectacular scenery and worthy of exploration, for most have a rich variety of birds.

Aberglaslyn Pass (SH593467). This provides excellent views, while access to the footpaths on the east side of the river is from Beddgelert or from Pont Aber Glaslyn. Common Sandpipers on the river, Pied Flycatcher and Siskins in the woods, Whinchat and Wheatear on the higher ground.

Croesor Cnicht (SH632446). Those who revel in the tough walking towards Llyn Cwm-y-Foel may well encounter Red Grouse, Tree Pipit, Wheatear, Mistle Thrush and Raven.

Coed Hafod (SH807575). The oak woodlands here on the east side of the Conwy are accessible by footpaths from several parking areas on the A470 and should ensure an encounter with most birds of the Snowdonia woodlands. The nearby river should offer Red-breasted Merganser, Goosander, Grey Wagtail and Dipper.

Llyn Elsi (SH783554). A footpath leads on a circuitous walk from close to the church in Betws-y-Coed. The first woodlands you pass through should provide among others Crossbill, Wood Warbler, Marsh Tit, Nuthatch and Treecreeper. Common Sandpiper and sometimes Black-headed Gull nest on the lake. Winter wildfowl include Little Grebe, Pochard, Goldeneye and Goosander.

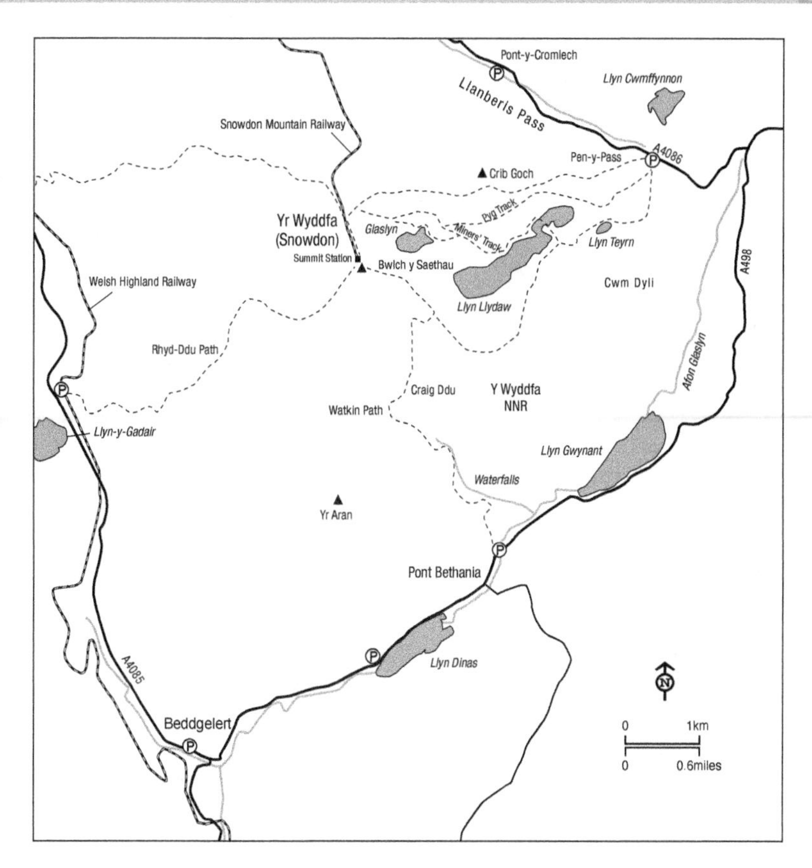

Llyn Conwy (SH780465). This lonely moorland lake needs to be approached on foot from Llyn Cottage on the B4470 at SH780446. The upland birds can well include Hen Harrier, Merlin and Red Grouse, while on the lake some duck and Common Sandpiper.

Llyn Glangors (SH773604). At the northern edge of the huge block of mixed woodlands to the east of the Conwy valley near Betws-y-Coed. A Forestry Commission trail leads north from near to Llyn Sarnau on the minor road which runs southwest through the woods from Llanrwst. A good area for birds of prey including Goshawk, while Goldcrest, Siskin and Crossbill can be located with care.

Llyn Idwal (SH646598). Leave the A5 at the Ogwen Cottage Mountain School (SH648603) and take the footpath south up the slope to the NNR, where a path leads right around Llyn Idwal. Common Sandpiper here, and up to the cliffs of the Devil's Kitchen with the possibility of Ring Ouzels.

Llyn-y-Gader (SH568520). One of the smaller Snowdonia lakes and immediately to the north of Beddgelert Forest. A footpath from the A4085 just to the north of the lake goes around the western shores and south to the forest. Whooper Swans winter here in all but the hardest weather.

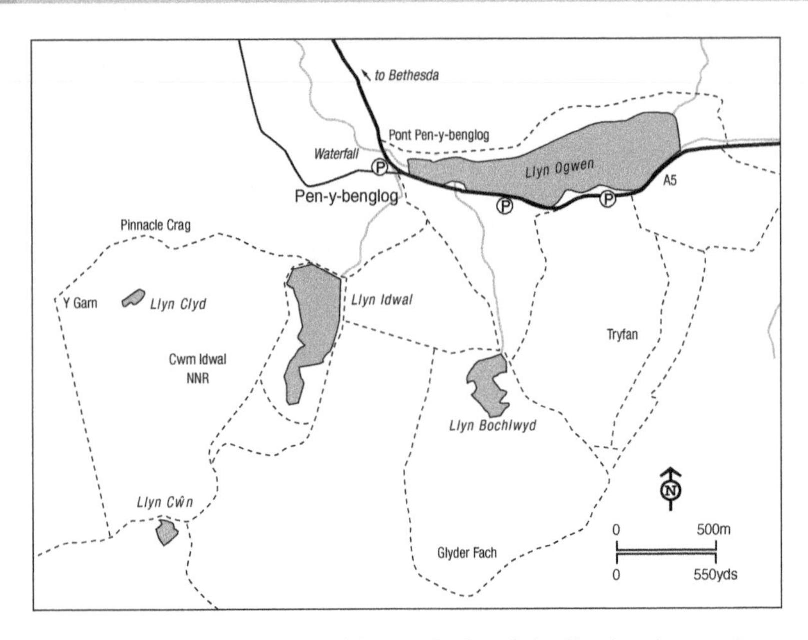

Llyn-y-Parc (SH792588). The lake may be largely birdless but the woods are excellent and easily traversed by a footpath running north from Betws-y-Coed.

Miners' Bridge Walk (SH792568). From the car park there is a fine riverside walk with Dippers and Grey Wagtails, while with good fortune you should encounter Redpolls, Siskins and other woodland birds.

Padarn Country Park (SH585604). A largely woodland area on the southeast shores of Llyn Padarn where there are Cormorants and Goosander all year and Common Sandpipers in summer. An unusual visitor was a Red-necked Grebe in February 1997. The woods contain most species to be expected including Redstart, Pied Flycatcher and Nuthatch, while the scrub areas support Wren, Whinchat, Stonechat and Whitethroat.

CALENDAR

Resident: Cormorant, Goosander, Grey Heron, Buzzard, Goshawk, Sparrowhawk, Kestrel, Merlin, Peregrine, Herring and Great Black-backed Gulls are regular visitors in all months, Little Owl, Great Spotted Woodpecker, Skylark, Meadow Pipit, Grey and Pied Wagtails, Dipper, Wren, Goldcrest, Nuthatch, Treecreeper, Chough, Jackdaw, Carrion Crow, Raven, Siskin, Lesser Redpoll, and Crossbill.

December–February: Depending on conditions, Whooper Swan, Teal, Pochard, Tufted Duck, Goldeneye, Goosander and Coot on the lakes.

March–May: Winter visitors depart. Wheatear and Ring Ouzel the first summer visitors to arrive, though by early April small numbers of Swallows over the lakes, soon to be joined by Common Sandpipers. Other visitors include Cuckoo, Tree Pipit, Whinchat, Garden Warbler, Blackcap, Wood Warbler, Chiffchaff and Redstart. Dotterel have been recorded in the Carneddau (particularly around the summit plateau of Foel Fras) while passing through in late April to early May.

June–July: Summer visitors still present, occasional passage waders at lakes in late

July. Chough family parties merge into small flocks.
August–November: Summer visitors leave the high ground by early September. Winter ducks, Woodcock from early October, other visitors shortly afterwards include Fieldfare and Redwing.

35 DOLGARROG STATION

OS Landranger Map 115
OS Explorer Map OL17
Grid Ref: SH779667

Habitat

An area of wet woodland situated either side of the Conwy River.

Species

Woodland birds such as Long-tailed Tit, Bullfinch, Lesser Redpoll and Siskin prefer the wet, swampy woodland which can be found between the railway line and the bridge. During the summer months, many warblers can be heard singing, including good numbers of Garden Warblers and Blackcap. Spring and summer produce numerous Reed and Sedge Warblers in the reedbeds, while Grasshopper Warblers can be heard reeling from the deep vegetation. During the winter months, large flocks of Redwing and Fieldfare feed in the surrounding fields. The river holds Goosander, Red-breasted Merganser and Goldeneye, while Common Sandpipers are a regular sight and sound in the summer months.

Timing

Spring and summer.

Access

Park at the Dolgarrog station car park which can be found along the A470 just north of Llanrwst. A bridge crosses the Conwy River at this point, allowing a good vantage point to observe the river and surrounding floodplains.

CALENDAR

Resident: Blue, Great, Long-tailed and Coal Tits, Goldcrest, Treecreeper, Bullfinch, Lesser Redpoll and Siskin in the woodland, Dipper and Grey Wagtail on the river, Reed Buntings in the reedbed.
November–February: Redwing, Fieldfare, Goldeneye, Goosander and Red-breasted Merganser on the river.
March–July: Common Sandpiper along the river, Blackcap, Garden and Willow Warblers, Chiffchaff in the woodland, Grasshopper, Reed and Sedge Warblers in the reedbeds.
August–October: Summer migrants depart and are replaced by Goosander, Goldeneye on the river and Redwings and Fieldfares in the woods.

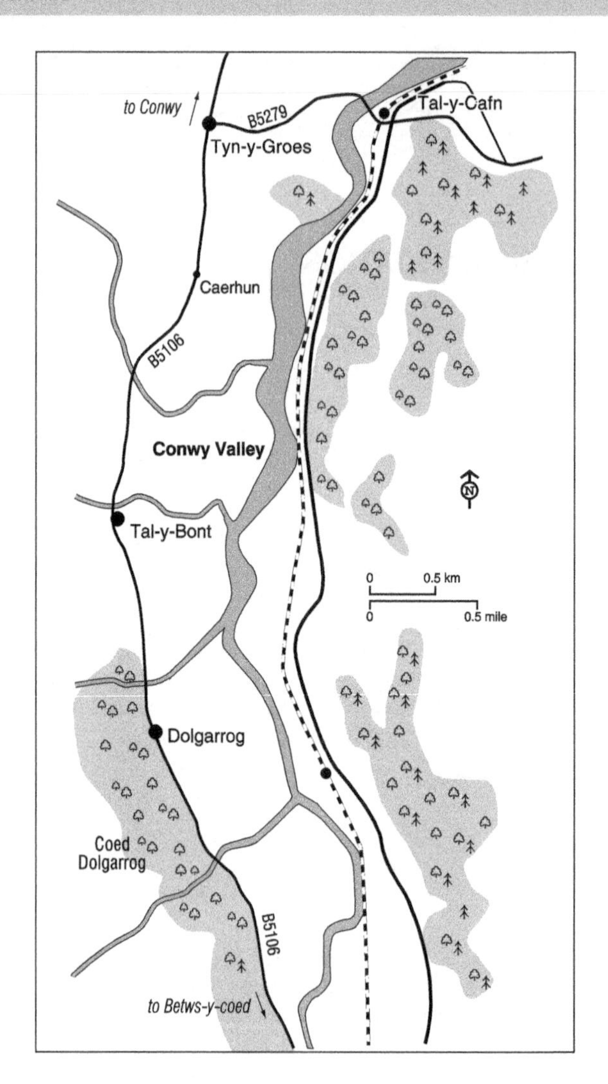

36 COEDYDD ABER

OS Landranger Map 115
OS Explorer Map OL17
Grid Ref: SH663719
Website:
https://naturalresources.wales/days-out/places-to-visit/north-west-wales/coedydd-aber-national-nature-reserve/?lang=en

Habitat

In a Sphagnum moss flush high on a shoulder of the Carneddau range, the Afon Goch commences its short, tumbling dramatic journey to Conwy Bay. Two miles (3.2 km) into its course it reaches one of Snowdonia's many geological highlights, a band of igneous granophyre forming a cliff some 120 feet (37 m) high. Over this the river plunges in a series of steps and terraces, to rush, swollen by several equally boisterous tributaries, a further two miles to the sea. The remains of long huts, round huts and hill forts offer striking evidence of occupation from the Iron Age; one ponders as to what the bird life was then. Much oak woodland remains on the steep sides of the valley, but even in this secluded area some has been lost, replaced by ranks of introduced conifers of several species. The Coedydd Aber National Nature Reserve extends for 420 acres (170 ha) and is one of the most popular woodlands in Wales, being visited by some 50,000 people each year. Close to the river, there are wet areas where Alder dominates, while as you climb out of the valley the woodland becomes more open and there are Hawthorn thickets, scattered Crab Apple and Rowan. The valley is famed for its mosses and lichens, an indication of minimal aerial pollution. A number of scarce plants have been recorded, while among the butterflies the elusive White-letter Hairstreak has been seen in the past. Alas, there are few recent records of this butterfly, which was formerly well distributed, if local, in Wales. Certainly, its food plant, Wych Elm, is found here. Pine Marten is even more difficult to locate and records are few, those reported are often road casualties, but sufficient, however, to indicate that it still occurs in Snowdonia – where there is a Castell y Bele, or The Marten's Castle – in dense woodland, or on open moorland where scree slopes and rocky tors provide shelter.

Species

Dippers and Grey Wagtails nest along the river and Kingfishers are occasionally seen. The deciduous woodland supplies the main bird interest, the summer visitors being Redstart, Garden Warbler, Blackcap, Wood Warbler, Chiffchaff, Willow Warbler and Pied Flycatcher. Do not neglect the conifers, for they support Goldcrests, Coal Tits and Siskins. A hundred years ago Siskin was virtually confined to the Scottish Highlands, yet now it breeds, at least locally, throughout Wales. Much of this expansion has occurred within the last 40 years, aided, at least partly, by a liking for peanuts at winter bird tables. In North Wales, Siskins probably first nested in the early 1960s, but such is their secretive nature at the nest that breeding was not proved until 1972. Now they frequent many of the larger conifer areas, their musical flight call, a short 'tsyzing', and their twittering song quickly attracting attention so that the bird is hardly likely to be overlooked. Look out also for Common Crossbill, which often frequents conifers; when its populations in Eastern Europe reaches high levels, large numbers move westwards and colonise new areas for several years, including Coedydd Aber. Firecrests are usually just winter visitors this far west in Wales, but don't neglect the possibility of breeding taking place, as it has on occasions in several other counties.

Resident woodland species in the valley include Sparrowhawk, Buzzard, Tawny Owl, Green and Great Spotted Woodpeckers, Robin, Mistle Thrush, Nuthatch, Treecreeper, Jay, Raven, Chaffinch and Bullfinch, with Dippers and Grey Wagtails on the river. Make the effort to climb southward out of the valley and you should encounter species such as the summer-visiting Tree Pipit and Whinchat, and on the open moor Wheatear and possibly Ring Ouzel, and if you are very lucky even a fleeting glimpse of a Merlin. The area around the

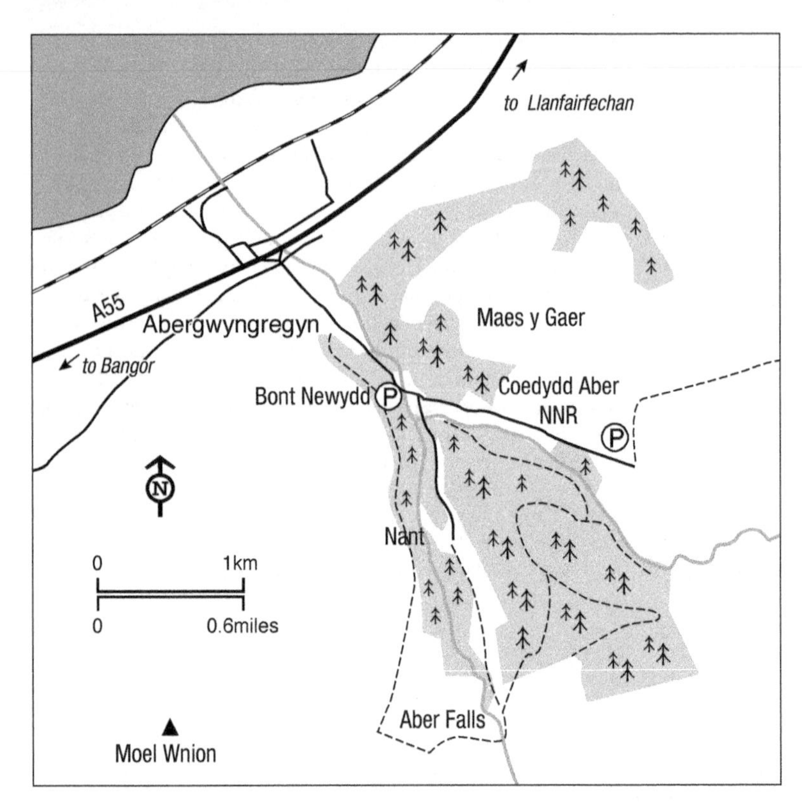

mountain gate in the Anarfon valley (SH675716) is excellent for Choughs, while Ravens are even more likely.

Timing

No particular time of day, though a visit in early morning during mid- to late May is probably the best.

Access

Turn south off the A55 Bangor to Llandudno road at SH656727 in Aber village; a minor road leads through the valley to Bont Newydd with a car park at SH663719. From here, a nature trail takes one on a circuitous route to Aber Falls and back over a hill and through some coniferous plantations.

CALENDAR

Resident: Goshawk, Sparrowhawk, Buzzard, Kestrel, Merlin, Pheasant, Woodpigeon, Tawny Owl, Little Owl, Kingfisher, Green and Great Spotted Woodpeckers, Skylark, Grey Wagtail, Dipper, Robin, Goldcrest, tit family, Nuthatch, Treecreeper, Jay, Raven, Chough, Siskin, Redpoll.

December–February: Woodcock, Fieldfare, Redwing and Brambling.

March–May: Arrival of summer visitors, including Cuckoo, Tree Pipit, Redstart, Whinchat, Wheatear, Ring Ouzel, Garden Warbler, Blackcap, Wood Warbler, Chiffchaff, Willow Warbler, Spotted and Pied Flycatchers.

June–July: All breeding species present, most now feeding young.
August–November: Summer visitors depart by end of August/early September, but the occasional late Blackcap and Chiffchaff may be located well into October. Winter thrushes appear during October, while resident tits, Treecreeper and Goldcrest form loose flocks which wander through the woodlands; check for Firecrest. During invasion years, search the conifer treetops for wandering Crossbills which may remain to nest.

37 TRAETH LAFAN

OS Landranger Map 115
OS Explorer Map OL17
Grid Ref: SH641746

Habitat

At low tide, the eastern end of the Menai Straits is almost blocked by the vast expanse of Traeth Lafan – the Lavan Sands; only a narrow deep-water channel remains, close under Beaumaris on the Anglesey shore. Traeth Lafan, one of the richest haunts for wildfowl and waders in the whole of Wales, extends for 26 square miles (42 square km) from Llanfairfechan westwards to Port Penrhyn and the Bangor Flats. At their maximum, the sands stretch some 3 miles (4.8 km) seawards, a seascape of sand, mudflats and mussel beds, with a low shingle ridge flanking the west side of the tiny Ogwen Estuary.

Species

Pride of place must go to the late-summer concentrations of Great Crested Grebes which use Traeth Lavan as a moult area, the numbers being of national importance. Present throughout the year, in single figures during early summer, increasing to around 150 in September, then falling away over the winter. The number of Red-breasted Mergansers gathering here rose towards the end of the last century to around 250 but has since declined dramatically, and now peak counts are of around 55. Small numbers of Slavonian and Black-necked Grebes occur annually, while the rarer Red-necked Grebe is only occasionally seen. Several species of diving duck regularly winter, including Scaup, Goldeneye and Long-tailed Duck. There is a large wintering Common Scoter flock in the bay, often numbering up to 1,500, among which there are a few Velvet Scoter and the very occasional Surf Scoter. Rarities have included Black Scoter, American Wigeon and Black Duck, the latter remained for six years until 1985, hybridising with the local Mallards.

Large numbers of Greylag Geese have been recorded, up to 500 gathering at Traeth Lafan, which is also important for several species of dabbling duck, total numbers of which may rise to 2,000 birds, mainly Shelduck, Wigeon, Teal and Mallard, with fewer (usually fewer than 100) Pintails. Large flocks of waders feed on Traeth Lafan, with up to 15,000 birds of ten species during midwinter. Oystercatcher number as many as 6,000, Dunlin 3,000, Redshank 1,300 and Curlew 2,000. Among which there may be up to 440 Knot, 250 Turnstone and smaller numbers of Lapwing, Greenshank and Ringed Plover. Avocet, Curlew Sandpiper, Ruff and Spotted Redshank have also been noted. Little Egrets have increased in number, with pairs breeding locally, autumn peaks are regularly over 100. Large numbers of gulls are also present, particularly in late summer and winter, mainly Black-headed, up to 2,000, and Herring, up to 1,500, with fewer than 200 Lesser Black-backed, Great Black-backed and Common. Mediterranean are regular, with Iceland and Glaucous less so.

The shoreline fields can in winter be quite productive, with flocks of thrushes and finches, including at times both Twite and Snow Bunting, while Lapland Buntings have also been reported. Rare visitors have included Richard's Pipit, Yellow-browed Warbler, Red-backed Shrike, Common Rosefinch and Wales's only Crag Martin in September 1989 with just eleven having occurred elsewhere in Great Britain. Wheatears are usually present on both spring and autumn passage, while in early autumn, migrant wagtails and pipits are prominent. Don't overlook the sewage farm, about a mile (1.6 km) west of Llanfairfechan, for wintering Chiffchaff, while Firecrest occur at The Spinnies, where there is a pool, reedbeds and woodland. Kingfishers are regular visitors here. Sedge Warblers breed in the reedbeds which are winter home for Water Rail and Snipe.

Timing

It is essential to check on the tide times and to arrange a visit for a couple of hours either side of high-water. At these times, ducks and grebes will move close inshore and the wader flocks will be concentrated on the restricted areas of mudflat.

Access

With a coastline and footpath some 6 miles (9.6 km) in length there is much to choose from and a number of access points to the shore, most notably at Port Penrhyn (SH592726), Aber Ogwen (SH614724), one of the best viewpoints for the shore and where for a contrast you can visit The Spinnies, a nature reserve

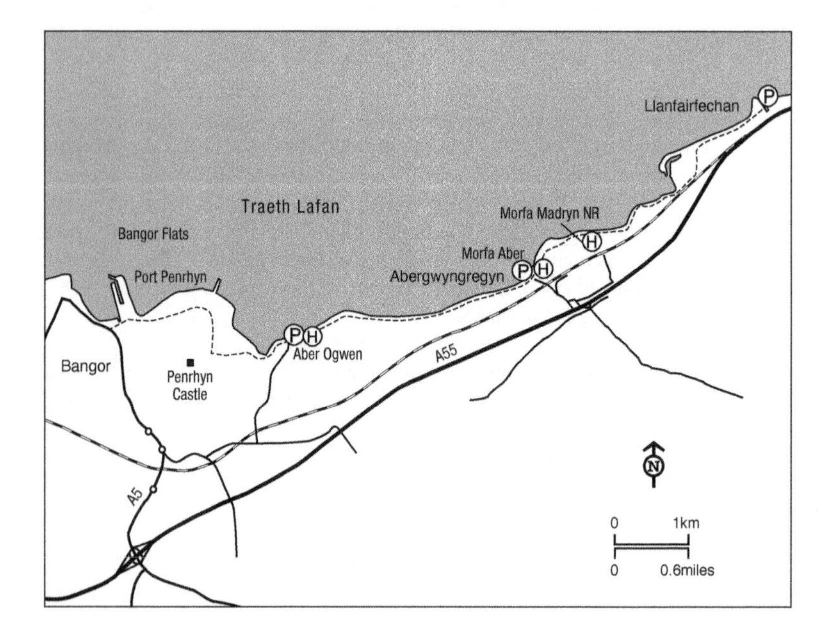

of the North Wales Wildlife Trust, with its brackish and freshwater lagoons and fine hide with wheelchair access. Morfa Aber (SH647731) also has a lagoon and hide and is best approached from Abergwyngregyn. Footpaths follow the coast in both directions, as they do from Llanfairfechan (SH678754) just east of which a series of pools with two hides, called Morfa Madryn, provide further excellent opportunities for birdwatchers.

CALENDAR

Resident: Great Crested Grebe, Cormorant, Grey Heron, Shelduck, Mallard, Red-breasted Merganser, Oystercatcher and large gulls.
December–February: Divers and small grebes, Wigeon, Teal, Pintail, Scaup, Long tailed Duck, Common Scoter, Goldeneye, Peregrine, Ringed, Golden and Grey Plovers, Lapwing, Knot, Dunlin, Black-tailed Godwit, Bar-tailed Godwit, Curlew, Redshank and Turnstone.
March–May: Departure of winter visitors, passage waders throughout much of period including Sanderling and Whimbrel, occasional terns, especially Sandwich Tern.
June–July: Numbers of Great Crested Grebes and Red-breasted Mergansers begin to increase during July and peak in August. Tern passage commences and wader numbers begin to increase.
August–November: Return passage of waders well under way throughout August, including small numbers of Little Stint, Curlew Sandpiper, Black-tailed Godwit, Spotted Redshank and Greenshank. Winter visitors begin to arrive during September and tern passage continues until October.

38 CONWY MOUNTAIN AND PENSYCHNANT NATURE RESERVE

OS Landranger Map 116
OS Explorer Map OL17
Grid Ref: SH765780
Website: http://pensychnant.co.uk/home.html

Habitat

Spectacular coastal upland with wonderful views over to the Great Orme and Anglesey. Pensychnant is a 120-acre Victorian estate within the Sychnant Pass where centuries of traditional farming have shaped the prehistoric landscape including Iron Age dwellings and Middle Age long huts when the whole upland was ploughed. Once abandoned Bracken, Gorse and Heather quickly took over until replaced two centuries ago by sheep grazing on a large scale. This is an eminently tranquil place, somehow set apart from the modern clamouring world, in which to enjoy and really appreciate nature. For moth enthusiasts, the mountain holds an impressive list of over two hundred species, including the very rare Ashworth's Rustic which inhabits the heathland here, one of only a handful of sites across Britain.

Species

A spring or summer walk through the grounds of the Pensychnant Nature Reserve will produce the sounds of singing Pied Flycatcher, Redstart, Wood

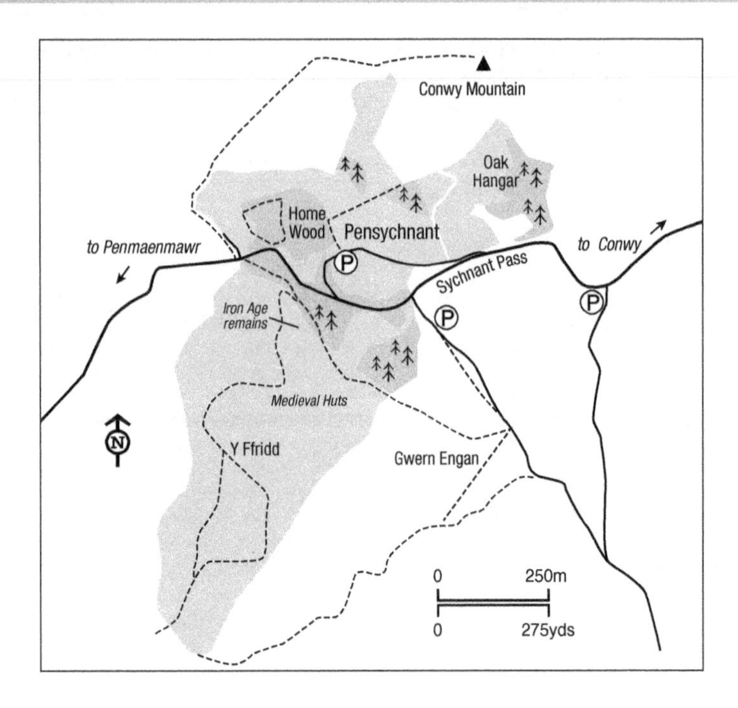

Warbler and a variety of common woodland species. The nearby mountain holds Peregrine, Buzzard, Stonechat and Linnet. Chough are regular, feeding in the nearby sheep and horse fields. In recent years there has been large autumn and winter counts of Chough, up to 50, on the mountain.

Timing

Summer

Access

Park in the Pensychnant Nature Reserve car park (fee payable).

CALENDAR

Resident: Buzzard, Peregrine, Chough, Stonechat, Meadow Pipit and Linnet on the uplands, Nuthatch, Treecreeper and Great, Blue and Coal Tits in the woodland.

March–July: Spring and summer visitors include Wheatear, Redstart, Blackcap, Garden, Willow and Wood Warblers and Pied Flycatcher in the woods.

August–October: Summer visitors depart. Wandering tit flocks in the woods. Numbers of Chough start to increase on the mountain.

November–February: A large wintering flock of Chough may number up to 50 on the mountain. Redwings and Fieldfares in the woodland.

39 CAERHUN CHURCH

OS Landranger Map 115
OS Explorer Map OL17
Grid Ref: SH776704

Habitat

The early medieval Caerhun Church with its magnificent Yew trees, one of the architectural and historical gems of the Conwy Valley, stands on the site of the Roman fort Canovium and overlooks the beautiful Conwy valley with its river and floodplains and westwards into the heart of Snowdonia.

Species

An excellent site for observing the secretive Hawfinch – the true status in North Wales of this our largest and most secretive finch remains somewhat obscure. The birds use the large Yew trees in the grounds of the churchyard and often drop onto the ground to feed. They seem to be especially active after rain when they come to drink the water in the church's guttering. If the birds are not present in the churchyard, they can often be located by scanning the tops of trees in the surrounding countryside; hopefully your patience will be rewarded.

During winter the river below the church often holds Goosander and Goldeneye, while the gull roost is always worth checking. The roost occurs on the sandbank at low tide. By scanning the surrounding valley you will increase your chances of sighting a bird of prey. Both Hen and Marsh Harrier as well as Barn Owl and Merlin have been seen hunting the reedbeds, while Peregrine, Buzzard, Red Kite and Sparrowhawk regularly break the skyline and there is a good chance of a Goshawk.

Timing

All year.

Access

Park in the church car park OS Ref: SH776704, from the A470 at Tal-y-Cafn.

CALENDAR

Resident: Lapwing, Curlew, Sparrowhawk, Red Kite, Buzzard, Peregrine, Great Spotted Woodpecker, Goldcrest, Treecreeper, Goldfinch, Nuthatch, Hawfinch and Reed Bunting.

November–February: Greylag Goose, Canada Goose, Wigeon, Teal, Goldeneye and Goosander on the river. Goshawk, Hen Harrier, Merlin, Redshank, large gulls, Barn Owl, Fieldfare, Redwing.

March–June: Common Sandpiper on the river, Whimbrels on passage, Willow Warbler, Garden Warbler, Blackcap.

July–October: Tit flocks. Winter visitors arrive from October onwards.

40 GREAT ORME

OS Landranger Map 115
OS Explorer Map OL17
Grid Ref: SH767833

Habitat

Some 2 square miles (3.2 square km) in extent and rising to 675 feet (206 m), the massive limestone headland of the Great Orme extends northwest forming the eastern arm of Conwy Bay. The holiday resort of Llandudno sprawls at its landward base, a continuation of human occupation extending back to Palaeolithic times, when some of Wales's earliest inhabitants huddled in the headland caves, and hunted across plains and marshes long submerged beneath the Irish Sea. Neolithic humans buried their dead here, in a cromlech near the halfway tram station. Almost an island, it is not surprising that early religious people sought solitude here, as did St Tudno in the 6th century. More recently but still nearly 700 summers gone, Bishop Anian, who had christened the first Prince of Wales, the future Edward II, built a palace here; perhaps not unexpectedly, it was burnt down by Owain Glyndwr. Despite such destruction and the passage of time, there are more than 100 known sites of archaeological and historical interest on the Great Orme.

The cliff and plateau vegetation are a rich mixture of limestone and heath with an area of limestone pavement, the fissures and knolls, grykes and clints supporting a number of plants rarely found elsewhere on the head. Despite its proximity to a large population and busy holiday resorts, the Great Orme is as much a botanist's paradise now as when the first naturalists passed this way in the late 18th century. One of the top five sites for rare plants in Great Britain, there are rich rewards for those with time to search: plants such as Goldilocks Aster, Dark-red Helleborine, Spotted Cat's-ear, Nottingham Catchfly and Spiked Speedwell. Cotoneasters have spread onto the head from nearby gardens, but one member of the family, native Wild Cotoneaster, first discovered on the Great Orme in 1783, and a Red Data Book species not known from anywhere else in Britain, still clings on here, despite past ravages by collectors, wild Kashmir goats descended from the royal herd at Windsor and scrub overgrowth. The five ancient bushes of a couple of decades ago have, with careful nurturing, been expanded to a population of 33, though much more needs to be achieved to ensure the future of this species. Butterflies abound on sunny days, the dwarf forms of Silver-studded Blue and Grayling here noted for being smaller and appearing earlier than their cousins just a few miles away.

Species

Seabirds are a speciality. The first record of Fulmar for Caernarvonshire was of two at the Great Orme in April 1937; breeding was confirmed in 1945, and there are now about 30 pairs occupying ledges above the road as well as on the sea cliffs below. Black Guillemot is a more recent addition, breeding in small numbers. Cormorants and a few pairs of Shags nest, as do Herring Gulls and a few pairs of Lesser Black-backed and Great Black-backed Gulls. Pride of place must go to a noisy colony of over 700 pairs of Kittiwakes (854 in 2019), this being their largest colony in North Wales. Sharing the same ledges are some

2,000 Guillemots and nearly 250 pairs of Razorbills, both species steadily increasing here as at a number of other colonies in Wales.

Peregrine, Kestrel and Chough breed on the cliffs. The varied habitats, especially on the more sheltered slopes, support a wide range of passerines, including Stonechat. Visits in winter should include a look at the beach near Llandudno pier: there are usually Oystercatchers, Curlews, Redshanks, Turnstones and with a bit of luck even Purple Sandpipers.

The Great Orme has benefitted from comprehensive observation at migration times from a small band of local birdwatchers. Their efforts have turned up many good birds, with near annual records of Dotterel, Wryneck, Red-backed Shrike, Great Grey Shrike, Firecrest, Yellow-browed Warbler, Ring Ouzel, Black Redstart, Lapland Bunting and Snow Bunting. National rarities have also featured, such as Bee-eater, Pallas's Warbler, Dusky Warbler, Booted Warbler, Isabelline Wheatear and Isabelline Shrike.

Timing

Not critical here, but beware: the marine drive and footpaths become very busy during the summer months, so it is well worth making an early start if you desire some degree of solitude. Certainly an early start will be essential for seawatching and for autumn migrants.

Access

The marine drive (fee payable) from Llandudno provides a complete circuit of the head, but parts can be closed due to rock falls. Another road crosses the high centre of the headland while a tramway, one of only three of its type in the world – the others being in Lisbon and San Francisco – provides an alternative means of reaching the high interior, where there is a visitor centre; a nature trail

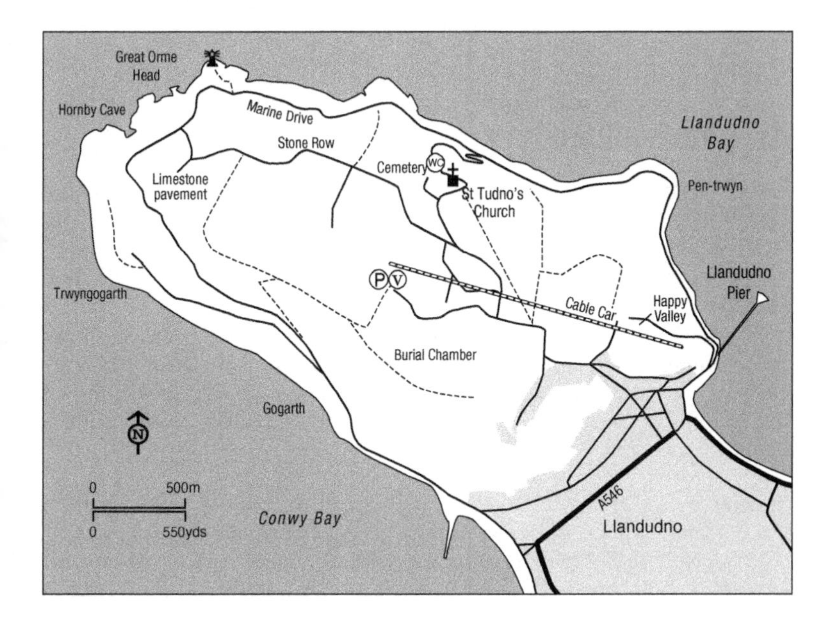

leads west to the seabird colonies and the south-west coast. Excellent views of these are obtained from the pleasure boats that sail out of Llandudno.

CALENDAR

Resident: Cormorant, Shag, Kestrel, Peregrine, Oystercatcher, Herring and Great Black-backed Gulls, Chough Skylark, Rock Pipit, Stonechat, Jackdaw, Carrion Crow, Raven.

December–February: Winter visitors such as divers and grebes offshore, Common Scoter and Eider. First Fulmars return during January, with Kittiwakes in February, while Guillemots and Razorbills make infrequent visits to the cliffs, always during the early morning and usually when conditions are settled.

March–May: Gannets offshore, Herring Gulls, Guillemots and Razorbills return to the colonies permanently about the end of April, by which time Kittiwakes are commencing nest-building. Summer visitors arrive, with Wheatear and Black Redstart in mid-March usually the first, quickly followed by Swallow, Sand Martin and Chiffchaff by end of month. Sandwich Terns possible offshore. Other migrants move through in April, with continuing Swallow passage, and small numbers of Redstarts, Sedge Warblers, Whitethroats and Willow Warblers. Dotterel is annual at this time of year.

June–July: Gannets regular offshore, all breeding species present with Guillemots and Razorbills completing their short stay ashore.

August–November: Seabird passage under way during first part of period, Manx Shearwater, Gannet, Arctic and Great Skuas, Kittiwake and terns being predominant species. Kittiwakes leave colonies by about third week of August but are usually seen offshore, Fulmars depart during mid-September. Soon afterwards, winter species such as Red-throated and Great Northern Divers, Great Crested Grebes appear offshore, while Common Scoters pass in small parties. Some passage waders may be seen but usually only Turnstone is present on the head itself. Passerine migration from late August onwards, with departing Swallows conspicuous in September, Skylark, Meadow Pipit, Ring Ouzel, Red-backed Shrike, Fieldfare, Redwing, Starling, Brambling and Chaffinch during late October, when possibility of Black Redstart, Lapland and Snow Buntings.

41 LITTLE ORME

OS Landranger Map 116
OS Explorer Map OL17
Grid Ref: SH813823

Habitat

A limestone headland just to the east of Llandudno though on a smaller scale than its larger cousin, rising to 463 feet (141 m) compared to 675 feet (206 m), it is thus easier to cover all areas in a shorter time. The headland partly comprises the Rhiwledyn Nature Reserve of the North Wales Wildlife Trust, including limestone grassland with a rich flora.

Species

Cormorants are the main seabird nesting on the cliffs which are easily viewed, with 500 nests in 2019; in the same year Kittiwakes numbered 324 nests, while Guillemots numbered 348 individuals, Razorbills just 24 and four Fulmars were reported. All can be observed easily from the beach at Craig-y-Don or from the small bay, known as Angel Bay on the eastern side of the headland. These quarries sometimes hold Black Redstarts from late autumn, with most years seeing at least a few records of the species among the limestone boulders. Peregrine, Stock Dove, Raven and Chough can be observed throughout the year.

The Little Orme has gained a reputation as one of the best seawatching sites in North Wales. If there is a strong northwesterly wind, birds can stream out of Liverpool Bay, passing below the headland. There is a sheltered spot to watch from on the eastern side of Angel Bay. Large numbers of Leach's Petrel can be seen some years, with Balearic and Sooty Shearwaters, Storm Petrel, all four skuas and Sabine's Gull almost annual amid the commoner seabirds. In winter Great Northern Divers and Red-throated Diver are regular, while sea-duck and occasional Little Gulls pass if the wind is from the north-west.

During migration times, the scrub and Hawthorns can hold good numbers of migrants. Pipits, warblers and Goldcrests move through in numbers, while scarcities have included Richard's Pipit and Pallas's Warbler.

Timing

All year, but particularly autumn and winter.

Access

Park in the Penrhyn Beach housing state on the eastern side of the Little Orme. The North Wales Wildlife Trust Rhiwledyn Nature Reserve at SH813821.

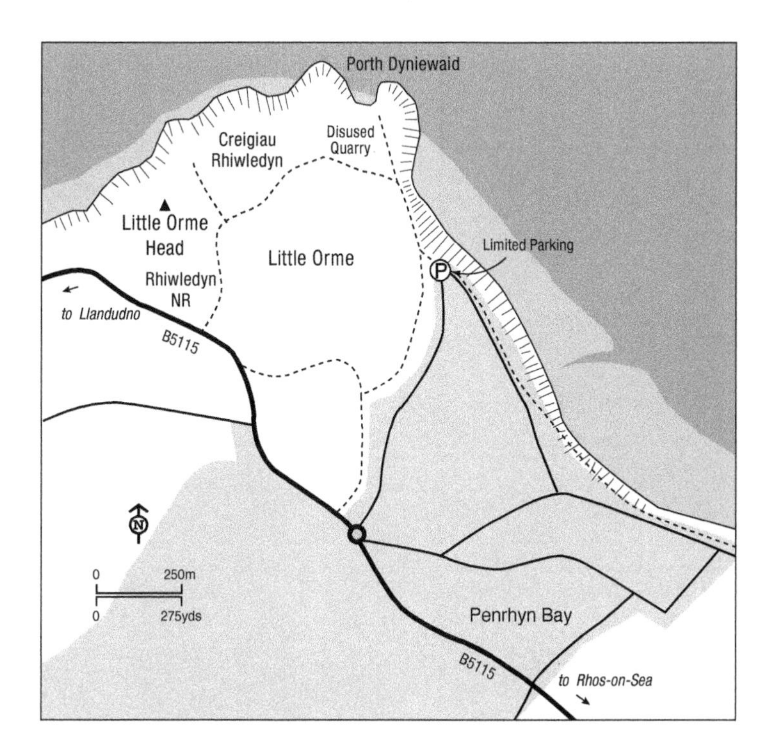

CALENDAR

Resident: Cormorant, Shag, Kestrel, Peregrine, Oystercatcher, Herring and Great Black-backed Gulls, Stock Dove, Feral Pigeon, Chough, Skylark, Rock Pipit, Stonechat, Jackdaw, Carrion Crow, Raven.

December–February: Winter visitors such as divers and grebes offshore, occasional sea-ducks including Common Scoter; Fulmars, Kittiwakes, Guillemots and Razorbills on the cliffs.

March–May: Gannets offshore, Herring Gulls, Guillemots, Razorbills make sporadic visits before taking up residence in late April, while Kittiwakes commence nest-building. Summer visitors arrive, with Wheatear and Black Redstart in mid-March usually the first, followed by Swallow, Sand Martin and Chiffchaff. Sandwich Terns possible offshore as their colony on Anglesey is not far away. Other migrants move through in April, with small numbers of Redstarts, Sedge Warblers, Whitethroats and Willow Warblers and greater numbers of Swallows and Martins.

June–July: Gannets regular offshore, all breeding species present, with Guillemots and Razorbills completing their short stay ashore as soon as the chicks leave the cliffs.

August–November: Seabird passage starts in late July, Manx Shearwater, Gannet, Arctic and Great Skuas, Kittiwake and terns being predominant species. Kittiwakes that have left their colonies are usually seen offshore as are Fulmars. Winter species such as Red-throated and Great Northern Divers, Great Crested and Slavonian Grebes, Goldeneye and Red-breasted Merganser appear offshore, while Common Scoters pass in small parties. Passerine migration from late August onwards, Swallows, Skylark, Meadow Pipit, Ring Ouzel, Fieldfare, Redwing, Starling and Chaffinch during late October, with possibility of Black Redstart taking up winter quarters.

CARMARTHENSHIRE (CAERFYRDDIN)

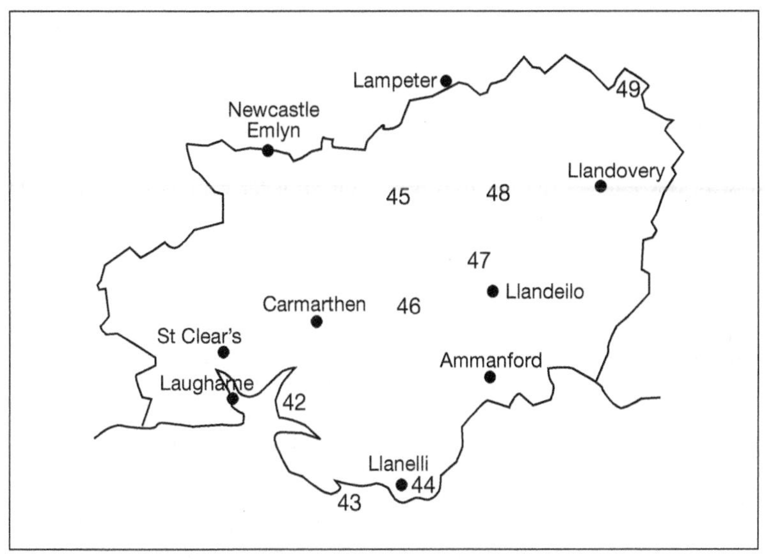

42 Taf, Tywi and Gwendraeth Estuaries
43 Burry Inlet North Shore and Loughor Estuary
44 Llanelli Wildlife Centre
45 Brechfa Forest
46 Crychan and Halfway Forests
47 Lower Tywi Valley
48 Talley Lakes
49 Upper Tywi Valley

Carmarthenshire is a county of deep wooded valleys and uplands in the north, dominated by slow meandering rivers, which flow down to the sea in one of the four estuaries, the largest and most significant of which, ornithologically, is the Burry Inlet. In terms of numbers of wintering waterfowl and waders, this site is the most important estuary totally within Wales. Several of the best places to view this estuary are in neighbouring Gower but hundreds of birds can be found all along its northern shore. One of the key developments over recent decades has been the Llanelli Millennium Coastal Park, a "tidying-up" of the once steel industry-orientated coastline. Now there are open parks, footpaths, a cycleway and new opportunities for birdwatchers to view the area. Within this zone is the Wildfowl and Wetland Trust Centre at Penclacwydd, which is a haven for birds, dragonflies and other wetland wildlife.

Inland, there are long river valleys and numerous woods and places to visit, too many to include in this book, where in summer the songs of Redstarts, Wood Warblers and Pied Flycatchers can be heard. Some of the best places to visit are in the Upper Tywi Valley, including the RSPB reserve at Dinas. It was in these areas that the Red Kite held on to its tentative existence in Wales for much of the of last century. Now Red Kites can be seen throughout Wales and far beyond, a real success story.

42 TAF, TYWI AND GWENDRAETH ESTUARIES

OS Landranger Map 159
OS Explorer Map 177
Grid Ref: SN342084

Habitat

From Tywyn Point to Ginst Point is just over 2 miles (3.2 km). To the north and east of this line are the estuaries of the Taf, tidal for 6 miles (9.6 km) to St Clears; the Tywi, tidal for 10 miles (16 km) to Carmarthen; and the Gwendraeth, a mere 3 miles (4.8 km) to Cydweli (Kidwelly). Here at this ancient town, Roger, Bishop of Salisbury and a minister for Henry I, built his castle in 1115. All three estuaries encompass thousands of acres of sand and mudbank, so exposed at low tide that the sea seems to disappear beyond the horizon. This was a hazardous spot to navigate in the days of coastal trading, as the numerous wrecks, some still visible, on Cefn Sidan Sands bear witness. One of these is the 834-ton iron barque *Craigwhinnie*, bound from Calcutta to Hull and stranded in December 1899, fortunately without loss of life. A vain effort was made to refloat the vessel, even to the extent of jettisoning her cargo of linseed, which washed ashore on the nearby beaches and provided a welcome food supply for thousands of geese, ducks and finches. Look in the churchyard at Pembrey where many of those lost on the sands were brought, and you will find against the church wall a notable stone recording three shipwrecks, in one of which Adeline Coquelin, a niece of 'Josephine, Consort of that renowned individual Napoleon Bonaparte' lost her life. A large tomb is that of George Bowser who 'was the first to discover the rare and most valuable properties of that part of Pembrey on which the harbours are situated'. I wonder what he would make of the situation today, when coastal trading vessels are but a distant memory even to the oldest inhabitants and birdwatchers and other naturalists who roam the shores.

A range of estuarine habitats will be encountered. At Pembrey the saltings contain much Rock Sea-lavender, a different habitat from its more normal rocky shores. There are flats dominated by *Spartina* Cord-grass, while there can be few finer open sandy beaches than those that run west from Ginst Point or southeast from Tywyn Point. There are deep narrow creeks on the south side of the Gwendraeth, acres of rocky shores at Salmon Point Scar, and scrub-crowned cliffs at Wharley Point, junction of the Taf and Tywi. Westwards from the Taf is more low-lying grazing marsh, here protected from the sea by the great sand dunes of Laugharne and Pendine Burrows. There is an impressive plant list, which includes the Welsh or Dune Gentian in a few dune slacks within the ranges at Pendine and at Pembrey. Fen Orchid, a nationally rare and endangered species, also occurs in the ranges.

Species

Cormorants frequent the shallow waters, which undoubtedly provide a rich feeding area for the birds, many coming from the colony on St Margaret's Island 15 miles (24 km) to the west. At times, the resting birds congregate in large flocks on the cliffs on the east side of the Taf or on favoured sandbanks, no fewer than 285 having been seen off Tywyn Point on one occasion. Grey Herons are equally frequent, and a colony was discovered in the cliffside scrub at Craig Ddu in 1945; numbers reached a peak of 34 pairs in 1953 and currently there are about 20 pairs present. There is another, smaller, colony close to the Tywi about 4 miles (6.4 km) downstream from Carmarthen. All the divers and grebes have been recorded in or close to the estuaries, and one can be fairly certain that the shallow waters of Carmarthen Bay support good numbers throughout the winter months, though systematic observations are difficult owing to the area and distances involved.

During the summer months, and especially in July and August, Manx Shearwaters can be observed off Cefn Sidan, sometimes in flocks totalling many hundreds, while very rarely Sooty Shearwaters occur and there are records of Cory's Shearwater, Mediterranean Shearwater and Storm Petrel. Gannets frequently fish in the same area, while Great, Arctic and Pomarine Skuas have also occurred. All five species of terns breeding in Great Britain have been recorded off Cefn Sidan, while Kittiwakes are also regular visitors. Usually keeping well offshore, though present in most years, are visiting Guillemots and Razorbills, even the occasional Puffin. Clearly this site is worth watching in late summer, especially during stormy weather. Other scarce passage seabirds may well occur.

Common Scoters are present in Carmarthen Bay, in a broad arc from Saundersfoot, Pembrokeshire to Gower, throughout the year. Maximum numbers are reached in late winter, when up to 40,000 birds have been present, the shallow waters and rich shellfish beds being the attraction for this handsome duck. At times, the flocks move close inshore, but the large area and difficulties of observation at sea make proper assessment hard to come by. The importance of the area to sea-duck was further highlighted by the *Sea Empress* disaster of early 1996 when oil, swept by tides and winds from the stricken tanker off Milford Haven, entered Carmarthen Bay and is known to have affected about 4,500 Common Scoters of which 3,326 died and a further 1,100 were rehabilitated and later released. How many others were not recorded is unknown, but upwards of 10,000 is thought to be a reasonable estimate. Velvet Scoters have occasionally been recorded, but systematic observation will surely reveal this species as a regular visitor in small numbers. Regular watching has shown that Surf Scoters are virtually annual among this massive Scoter flock, although admittedly they are not always that easy to see. Other diving ducks seen offshore, or just inside the estuaries, include Scaup, Eider, Long-tailed Duck, Goldeneye and Red-breasted Merganser. In winter, the upper reaches of the estuaries support modest numbers of Wigeon, up to 700 in the Taf, and smaller numbers of Teal and Mallard. Several hundred Pintails are seen each winter, while Shelducks breed.

Ringed Plovers breed at a number of isolated spots around the three estuaries and westwards within the Pendine Range, these and the handful a little further east being almost their only breeding sites remaining in south-west Wales. A wide variety of waders has been recorded in all three estuaries. One of the highlights must be the midwinter gathering of Sanderlings on Cefn Sidan, where up to 1,800 have been seen. Golden Plovers are plentiful, flocks of up to 4,000 being present on the Gwendraeth and Taf where, mingling with Lapwings, they resort to the low-lying grasslands and flight out to the estuaries. Small numbers of Grey Plovers are seen throughout the winter; they reach a maximum of about 30 during the early autumn, with Tywyn and Ginst Points being the places to search. Ruffs, occasionally in small groups, are seen on the Gwendraeth Estuary each autumn. Black-tailed Godwits are sometimes observed in their hundreds. Bar-tailed Godwits were formerly more numerous but their numbers have decreased notably in recent years. Greenshanks regularly overwinter, as probably does the occasional Spotted Redshank, while Turnstones occur sporadically at Salmon Point Scar. Little Egrets are a regular feature of the estuaries and may even breed nearby, and Great White Egret numbers are increasing.

The many rare visitors of recent years have included Green-winged Teal, Spoonbill, Marsh Harrier, Avocet, Little Ringed Plover, American Golden Plover, Black-winged Stilt, Wood Sandpiper, Lesser Yellowlegs, Grey Phalarope, Sabine's Gull, Gull-billed and Black Terns and Little Auk. An adult and an immature Crane in early 1984, the first to be recorded in Carmarthenshire, were a superb highlight. The first Glaucous-winged Gull for the UK, first seen in Gloucestershire, spent a few days on the Tywi Estuary at Ferryside in March 2007.

Black Redstarts are sometimes observed about the castles at Laugharne and Llanstephan, while Wheatears are regular along the coast at passage times, and with luck, you may encounter Yellow and White Wagtails. Buzzards, Merlins and Peregrines and are the most regular predators, while Hen Harriers hunt over the land to the south of the Gwendraeth. If you have journeyed to the western side of the Taf, before you leave the area spend a while at Pendine admiring the offshore Common Scoter flock and, if you are fortunate, small numbers of other species such as Velvet Scoter.

Timing

For estuary-watchers, a tide table is an essential part of one's equipment. A visit can then be planned which ensures that the birds are close inshore as the tide rises, or as the first mudbanks are exposed after high-water.

Access

Carmarthen Bay sweeps right round from the Burry Inlet in the east to Tenby in the west. It is internationally important for wintering Common Scoter among which there may be Scaup, Long-tailed Ducks, Red-throated and Great Northern Divers, Velvet Scoter and very occasionally Surf Scoter and Black-throated Diver. The scoter flocks may stretch from one end of the bay to the other and numbers have reached up to 40,000 individuals. Observation of the flocks is not easy as the birds may be quite far out, especially off Cefn Cidan (Pembrey). Other sites worth checking are off Pendine (SN234079), Telpyn Point (SN187073) and Amroth (SN170072) which is just inside Pembrokeshire.

Gwendraeth Estuary: One can walk eastwards from Salmon Point Scar into the Gwendraeth Estuary. Alternatively, you can go by car through Kidwelly to Kidwelly Quay; you do not have to move far from the car, if at all, to obtain good views of the head of the estuary from here. By contrast, the south shore is much more difficult of access, something accentuated by the presence of a RAF bombing range towards Tywyn Point. Access to the point itself is by way of Pembrey Forest to Cefn Sidan Sands.

Kidwelly Quay and Marsh: The area around the small town on Kidwelly has extensive saltmarsh, viewable from the Quay and wet-grazing meadows. For the Quay (at SN398063), turn down the small road opposite the Fisherman's Arms and continue over the railway crossing to the car park. High tides flood the marsh and drive the waders and other birds closer to the waiting observers. The best times to visit are during the autumn or winter when Redshank, Greenshank, Spotted Redshank, Curlew Sandpiper, Curlew and Whimbrel (on passage) may be seen along with the occasional Little Stint. Little Egrets, now numerous in south-west Wales, are always present, along with increasing numbers of Great White Egrets. In winter Hen Harriers, Merlins and Short-eared Owls may be seen patrolling the marshes and Peregrines are regular visitors.

Just before you reach the car park is Kidwelly Sewage Works. Although a very small works, the circular sprinklers are one of the best sites in South Wales to find wintering Chiffchaffs – as many as a dozen may be present.

The best way to watch Kidwelly Marsh is by access under the railway bridge alongside the main Llanelli to Carmarthen road, the A484, at SN413053. During spring tides, when the marsh is completely covered, good numbers of ducks and waders may be present. These include Greenshank, Ruff, Black-tailed Godwit and Green Sandpiper. Across the other side of the A484, Coedbach, though not as attractive to birds as it once was, is always worth a look. Great White Egret, Green-winged Teal, Spoonbill, Buff-breasted Sandpiper, Pectoral Sandpiper, Lesser Yellowlegs, Black-winged Stilt and Glossy Ibis to name but a few rarities have been seen here in recent years.

Taf Estuary: This is Dylan Thomas country. Following a visit in 1934 he described Laugharne as 'the strangest town in Wales'. Despite this he made regular visits and became a permanent resident four years before his death in

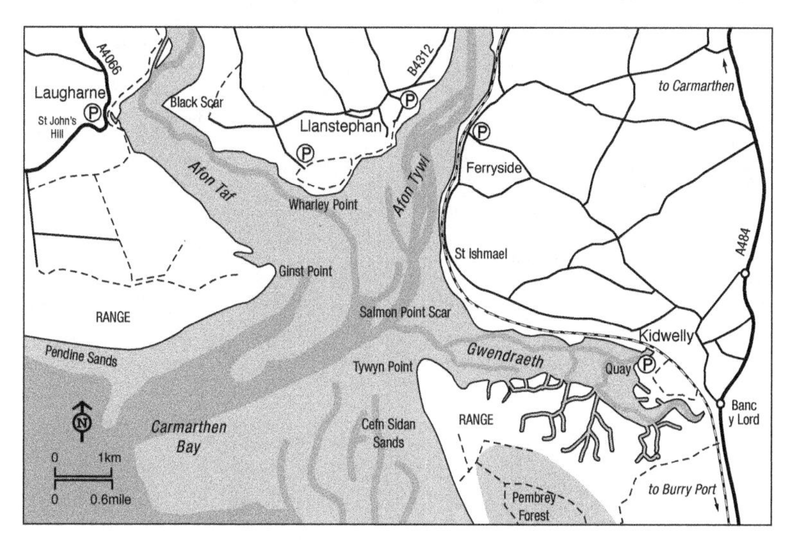

1953, his home by 'the heronpriested shore' and now a heritage centre he described as 'My seashaken house on a breakneck of rocks'. The town of Llareggub (try spelling it backwards) in *Under Milk Wood* was based on Laugharne from where Ginst Point is reached on foot by way of a path which skirts the estuary side of Sir John's Hill, or via the summit itself – Dylan's Birthday Walk – then south along the seawall or, better still, below it so as not to disturb birds in the fields or on the estuary itself.

An alternative route is by car, leaving the A4066 Laugharne to Pendine road at SN286096. This drops quickly to the coastal flat fields of Laugharne township, excellent for Lapwing and Golden Plover and for winter thrushes and finches. Follow the MOD boundary fence eastwards for nearly 2 miles (3.2 km) until the road reaches Ginst Point, though this last section may be closed on weekdays when the Proof and Experimental Station is testing weapons. This area can be excellent for birds of prey with regular Hen and Marsh Harriers, Merlin, Peregrine and Kestrel. Look out also for Barn and Short-eared Owls.

Tywi Estuary: Good general views may be obtained from Llanstephan, the finest view from any of the South Wales castles, where you may share the air with Buzzards, Red Kites and Ravens, and also from Ferryside where the first Glaucous-winged Gull for Wales was discovered in March 2007. Beyond Ferryside, an unclassified road runs south through St Ishmael's and provides access to the rocky shore of Salmon Point Scar and hopefully Purple Sandpipers. Access to the upper reaches of the estuary is not so easy, though it is possible to walk a little way north along the shore from Ferryside.

CALENDAR

Resident: Cormorant, Grey Heron, Little Egret, Shelduck, Mallard, Eider, Common Scoter, Buzzard, Oystercatcher, Ringed Plover.

December–February: Divers, grebes, Wigeon, Teal, Pintail, Goldeneye, Red-breasted Merganser, Knot, Sanderling, Dunlin, Black-tailed and Bar-tailed Godwits, Curlew, Redshank, Greenshank.

March–May: Main departure of wildfowl and waders during first part of period, but some passage until May, when summer visitors such as Whimbrel, Green and Common Sandpiper can be expected, the latter usually on the upper reaches of the estuaries rather than on the open flats. Occasional parties of terns seen.

June–July: Ringed Plover broods fledging, Shelduck broods in evidence on the estuaries, start of return wader passage, Common Scoter numbers build up offshore. Manx Shearwaters and Gannets offshore, first terns arrive on passage.

August–November: Peak wader passage, with greatest variety of species during first part of period. During October numbers of ducks increase both in the estuaries and offshore, divers and grebes may be seen, especially from November onwards. Seabirds including auks and skuas offshore, especially during bad weather during August and early September.

43 BURRY INLET NORTH SHORE AND LOUGHOR ESTUARY

OS Landranger Map 159
Explorer Map 178
Grid Ref: SS474997

Habitat

Burry Inlet with its vast mudflats, the largest area of saltmarsh in Wales and fine sand dunes is of international importance for winter waders and wildfowl. From Pembrey Burrows to Llanelli and thence into the Loughor narrows there are a number of excellent birdwatching spots. A journey by train, for the railway closely follows part of the shore, provides an excellent general impression of the low coastline, and of the estuarine expanse of up to 4 miles (6.4 km) reaching to the north shore of Gower (see Site 56, Burry Inlet South Shore and Whiteford Burrows).

The only truly wild area of hinterland is in the extreme west beyond the country park, though here large conifer plantations now almost completely shroud the dunes. Edward Lhuyd, who died far too young in 1709, referred to a 'Swan Pool' where swans wintered, while little more than a century ago considerable flocks of White-fronted Geese wintered until the construction of the Royal Ordnance Factory in 1914. Even the country park posed a threat at one time, when ambitious plans were proposed to take over the remaining unspoilt dunes and estuary – an idea successfully opposed both by local naturalists and by residents. This area is now the Pembrey Burrows Local Nature Reserve.

Elsewhere, there has been much urban development, though in many instances only the crumbling remains of factories remind us of the great industrial days. Most important of all, for birds and for those who gain so much pleasure watching them, has been the opening of the Wildfowl and Wetlands Trust Centre at Penclacwydd (see Site 44). The Millennium Coastal Park, a 20 mile (32 km) coastal path from Loughor Bridge to Kidwelly, provides further birdwatching opportunities on the north shore of Burry Inlet and beyond to the Gwendraeth Estuary.

Species

Burry Inlet has always been noted for its wader flocks, which in midwinter can total up to 45,000 birds. The main species, accounting for almost half the number, is Oystercatcher, many of Icelandic origin. Another speciality, Turnstone, is probably best encountered in numbers off Burry Port Harbour where as many as 1,000 have been seen in autumn. Golden Plover and Lapwing flocks are a conspicuous feature of Burry Inlet and Loughor and of the coastal fields in winter, while up to 250 Grey Plovers, 2,000 Bar-tailed Godwits, 250 Black-tailed Godwits, 1,000 Curlew, 1,000 Redshank, Knot and Dunlin can be found on the estuary.

Shelduck nest sparingly along the estuary, while Wigeon is the most numerous wintering duck. Brent Geese have steadily increased in numbers in recent winters and are best seen from the Loughor Bridge where up to 800 can occur. Little Egret have increased in numbers and breed locally, joined occasionally by the increasing Great White Egret. Spoonbills are annual on the inlet, with Penclacwydd the best place to see one. Burry Port and Pembrey Harbours host large numbers of Mediterranean Gulls in early summer and Sandwich Terns on passage, with highest numbers in the autumn, along with smaller numbers of Common, Arctic and Little Terns. Roseate Terns have also been seen here. Mediterranean Gulls occur in large numbers, sometimes as many as 40, at Llanelli Beach also.

Merlins and Hen Harriers winter on the coastlands, mostly to the west of Burry Port, with Black Redstarts regularly in the harbour at Burry Port and beyond at Pembrey. These hinterlands and the Pembrey Country Park are ideal spots for looking for migrants, including Wheatears, warblers, wagtails and pipits. The conifer plantations also hold a breeding population of Crossbill.

The marshland scrub along the Millennium Coastal Path and at Frwd Fen provide excellent habitat for both Whitethroat and Lesser Whitethroat, Cetti's, Reed, Sedge and Willow Warblers, Chiffchaff and Reed Buntings.

Offshore from Pembrey and to the west is Carmarthen Bay which has a large wintering flock of Common Scoter, sometimes as many as 40,000, which move up and down the Bay, and among which there may also be Scaup, Long-tailed Duck, Velvet and Surf Scoter. Also offshore are smaller numbers of Red-throated Divers, Great Crested Grebes, Scaup and Eider with the occasional Black-throated and Great Northern Divers and Slavonian Grebe.

The Sandy Water Park with a 16-acre lake as its centrepiece is a fine example of how a derelict industrial site could be regenerated. Only the railway embankment separates it from the Burry Inlet. Reed and Sedge Warblers in summer while Cetti's Warblers and Reed Buntings are resident. A striking feature in late summer are the gatherings of Swallows, House and Sand Martins hawking insects over the lagoon in preparation for their departure on migration. Nearby at North Dock both Herring and Lesser Black-backed Gulls nest on rooftops and tall buildings in the town centre.

Another feature hard by the estuary are Machynys Ponds, former claypits with Reed and Sedge Warblers and Reed Buntings breeding while scarce visitors have included Red-necked, Slavonian and Black-necked Grebes, Purple Heron, Long-tailed Duck and Green Sandpiper.

With the increasing numbers of birdwatchers, many scarce and rare species have been seen in the area, including Baird's Sandpiper, Wilson's Phalarope, Little, Sabine's, Ring-billed, Iceland and Glaucous Gulls, Gull-billed Tern, Black Guillemot, Little Auk and Richard's Pipit. Perhaps the biggest surprise was an Ivory Gull, the first for Wales seen and photographed in Burry Port Harbour in October 1988. Amazingly, in early September 2009 not one Glossy Ibis but a flock of 25 made a short appearance in a field at Pembrey Harbour, the first in the county since 1910. Four were colour-ringed birds from Lucio Cerrado Garrido, Doñana, Spain, part of an influx of up to 50 birds in Great Britain with sightings elsewhere in Wales from Anglesey, Caernarfonshire, Ceredigion and Pembrokeshire. Another excellent record was of the first Welsh record of Killdeer at Pwll, February–March 1978.

Timing

It is essential that you check the tide table. If you do not and you arrive at low tide or thereabouts, then the birds will have dispersed over a wide area of the estuary, for the tidal range is up to 26 feet (7.9 m). Watch during the hour or

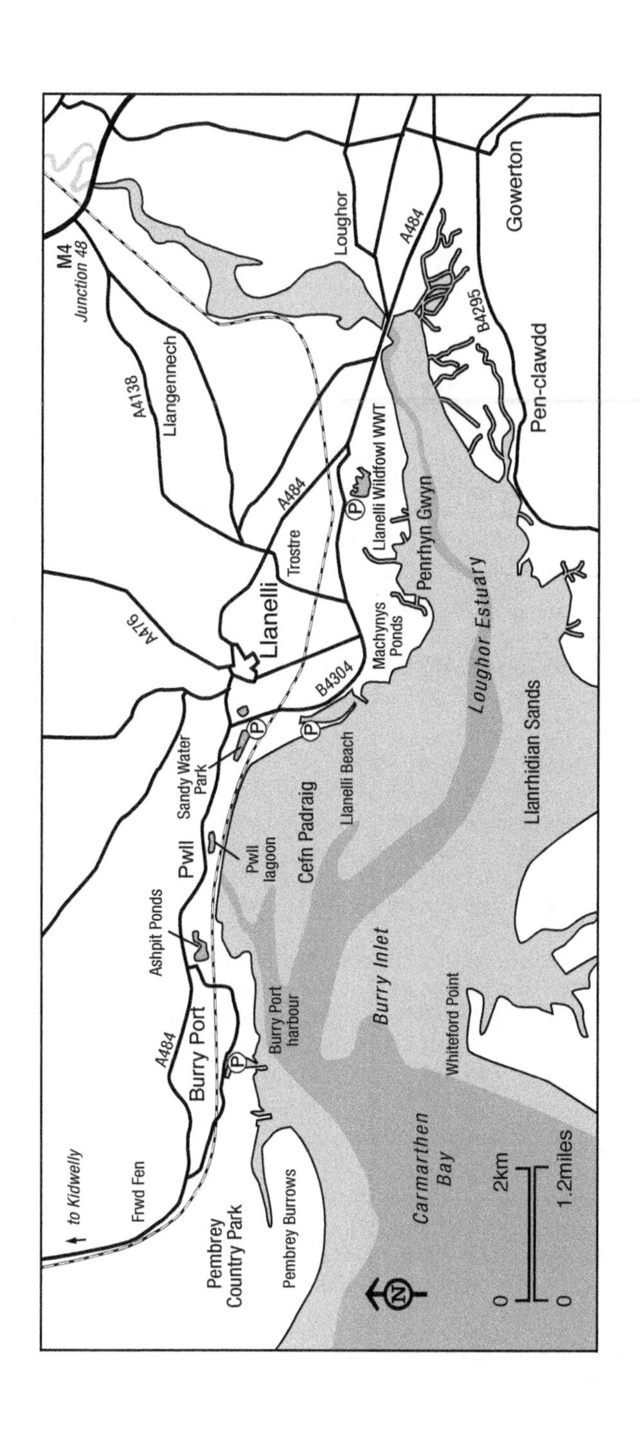
M4
Junction 48
A4138
Llangennech
Loughor
A484
Gowerton
B4295
Pen-clawdd
Llanelli Wildfowl WWT
Penrhyn Gwyn
A484
Trostre
Llanelli
A476
Machynys Ponds
Loughor Estuary
Llanrhidian Sands
B4304
Sandy Water Park
Llanelli Beach
Cefn Padraig
Pwll
Pwll lagoon
Ashpit Ponds
Burry Inlet
Whiteford Point
Burry Port harbour
Burry Port
A484
to Kidwelly
Frwd Fen
Pembrey Country Park
Pembrey Burrows
Carmarthen Bay
2km
1.2miles
0
0
N
P

two after highwater, or when the tide is coming in and the waders are forced off the lower sandbanks. If by any chance waders are not easily seen, do not forget to search such places as the Machynys area with its pools and reedbeds, Sandy Water Park and the harbour areas at Burry Port and Pembrey, not forgetting the Wildfowl and Wetlands Centre (see Site 44).

Access

Ashpit Ponds (SN462010): Formerly used as settling lagoons for pulverised fuel ash from the now demolished Carmarthen Bay Power Station, Ashpit Ponds is an important area for breeding wetland birds.

Ffrwd Fen (SN420025). A splendid, largely reedbed nature reserve of the Wildlife Trust of South & West Wales and the Llanelli Naturalists, the old towpath of the long-disused Ashburnham Canal is now a cycle/walking path but still provides good access along the north-eastern boundary from the B4317, or from the minor road at SN417028; limited parking at both.

Llanelli Beach (SS496995):There is a car park at the west-facing shore, while the river to the east is also worth checking for birds. Follow the signs to the beach from Queen Victoria Road in Llanelli or access of the B4304 which runs alongside Sandy Water Park.

Loughor Estuary (SS565985): There are several observation points for the estuary, which forms the boundary with Glamorgan. Loughor Bridge (SS561980), Pencoed (SS562998) and Llangennech (SN565010) with the reedbeds just to the north, are all easily accessible.

Machynys Peninsula (SS510978): Penrhyngwyn Pond which is accessed along the Millennium Coastal Path is good for wildfowl, while the tip of the peninsula provides a good vantage point both to view the Burry Inlet and search for passage migrants. Ring Ouzel, Wryneck and Firecrest have all been seen here. Head towards Morfa on the southern edge of Llanelli by way of New Dock Road then take either route south of the roundabout at SS512984 and so to the peninsula. This site can also be accessed off the B4304.

Pembrey and Burry Port Harbours (SN437002 and SN445004): Turn off the B4331 about half a mile west of Burry Port railway station; unclassified roads lead to each harbour where there is parking. There is also access to the east side of Burry Port Harbour from near the station and from here one can head east close to the shore and past the site of the now demolished power station.

Pembrey Country Park (SN405005): https://naturalresources:wales/days-out/places-to-visit/south-west-wales/pembrey-forest/?lang=en. A choice of parking sites, fee payable and includes a visitor centre from where there is a choice of routes through the dunes, some with boardwalks and steps to reach the shoreline, which for the energetic extends 5 miles (8 km) to Tywyn Point with views of the broad expanse of Carmarthen Bay, or alternatively head deep into the woodlands.

Pembrey Saltings LNR (SS425997): Head for Pembrey Country Park but turn left before the entrance and take the road/track to Cefn Sidan Beach. As an alternative you can walk the north shore from Pembrey Harbour.

Pwll (SN480007): Footpaths lead to the shore from the A484 at SN473011 and SN485007. The small fishing lake often holds Tufted Duck and Great Crested Grebe.

Sandy Water Park (SN495004): On the site of the former steelworks, the lake is a relatively new feature but already boasts 12 species of gull, including Laughing Gull in 2005, while Old Castle Pond just to the east is also worth a visit. The Water Park is well signed off the A484.

CALENDAR

Resident: Cormorant, Little Egret, Grey Heron, Shelduck, Eider, Kestrel, Water Rail, Lapwing, large gulls though none breed, Stonechat, Cetti's Warbler, Reed Bunting.

December–February: Great Northern Diver, Great Crested Grebe, Wigeon, Teal, Pintail, Hen Harrier, Merlin, Oystercatcher, Ringed and Grey Plovers, Knot, Sanderling, Dunlin, Black-tailed and Bar-tailed Godwits, Curlew, Redshank, Kingfisher, Black Redstart, Chiffchaff.

March–May: Winter visitors depart, but always at least a few individuals of most species remain throughout the period. Spring wader passage includes Whimbrel during April and May. Reed and Sedge Warblers take up residence at Ashpits Ponds and in the Sandy Water Park.

June–July: A chance of terns moving into the estuary late in July, build-up of Mediterranean Gulls, wader numbers begin to increase at about the same time, Whimbrel, Spotted Redshank and Greenshank being among the first to appear, and in quiet creeks possibly Green Sandpiper.

August–November: Seabirds such as Manx Shearwater and Gannet can move inshore during stormy weather. Sandwich Tern numbers build up and wader passage at peak during August and September as wintering species begin to arrive, winter ducks, divers and grebes usually appear from October onwards.

44 LLANELLI WILDLIFE CENTRE

OS Landranger Map 159
OS Explorer Map 178
Grid Ref: SS532985
Website: https://www.wwt.org.uk/wetland-centres/llanelli/

Habitat

The idea for a Wildfowl and Wetlands Trust (WWT) centre in Wales, broadly upon the lines of those developed so successfully by the WWT at Slimbridge in Gloucestershire, Martin Mere in Lancashire and elsewhere, was first proposed by Sir Peter Scott in 1984. With the close cooperation of the then Llanelli District Council, now Carmarthen County Council, the proposal was taken forward and an area of some 250 acres (101 ha) leased on the shore of Burry Inlet, just beyond the declining industrial area of Llanelli. Here a visitor centre was built on a convenient knoll, giving extensive views over the surrounding low-lying area and the estuary beyond. Wetland habitats were created, including brackish and freshwater scrapes, a 6-acre (2.4-ha) lagoon, and reedbeds, while footpaths and hides, one named the Sir Peter Scott Hide, were provided. The centre opened to the public in 1991.

With the acquisition of additional land, the centre now extends over 500 acres (202 ha), of which the 240 acres (100 ha) of marsh are within the Burry Inlet Site of Special Scientific Interest (SSSI) which is a Special Protection Area (SPA), a candidate Special Area of Conservation (SAC) and since 1992 has been designated a Ramsar site – a wetland of international importance. The reserve expanded further with the development of an additional 200 acres (83 ha) to the east, known as the Millennium Wetland (funded by Carmarthenshire County Council, Dŵr Cymru and the Welsh Development Agency). This formerly low-grade agricultural land has been transformed by exciting and in some cases pioneering management into a series of habitats for wetland birds and dragonflies, while at the same time other facilities have been provided including education projects, tourist accommodation, craft opportunities, fishing and a cycle track.

The wetland has been designed with the aim of encouraging 14 species in particular to breed, species that in some cases no longer breed in Carmarthenshire, while for others it would represent a considerable extension of range or a boost against declining fortunes elsewhere. The target species are Bittern, Little Egret, Spoonbill, Marsh Harrier, Oystercatcher, Lapwing, Black-tailed Godwit, Redshank, Kingfisher, Skylark, Sand Martin, Cetti's Warbler, Reed Warbler and Bearded Tit.

As with other WWT reserves there is a large collection of wildfowl, ducks and geese as well as a small flock of flamingos. There is an excellent information centre, interpretation and education area and café. The reserve is open from 9.30am – 5pm daily, with an entry charge for non-members. Most paths are wheelchair friendly and there are numerous activities for children.

Species

Many of the birds that occur at the centre are part and parcel of the total population of the whole Burry Inlet, to which reference is made in Site 43 and Site 137, the birds recognising no such boundaries. However, the importance of the centre within the estuary complex cannot be overstated: it is at the very heart, and this combined with the continual day-by-day coverage means there is excellent information available. A glance at the *Carmarthenshire Bird Reports* will very quickly confirm this.

One notable resident is Little Egret, numbers having risen to as many as 392 in late summer, a Welsh record. Small wonder this bird is on the list of those the centre hopes to attract to breed in the new habitats. Grey Herons are always present, up to nine Spoonbills have been recorded annually since 1995, while Bitterns are rare winter visitors.

Mute Swans are resident in small numbers and breed, while Whooper Swans have occurred in the winter. Greylag Geese are resident and several pairs now nest, Brent Geese, which largely winter on the Loughor, regularly pass this way. Wigeon and Pintail are the most numerous wintering duck, numbering in their hundreds. The number of Shoveler increases year by year, with those on the Glamorgan side of the estuary the highest concentration in Wales. Pochard and Tufted Duck occur throughout the year with about 120 of the former and smaller numbers of the latter in midwinter. Gadwall numbers have increased considerably, and other wildfowl present include Teal, Mallard, occasional Eider and Goldeneye, with Garganey seen on spring and autumn passage at what has become the best site in the county for this species.

Up to 10 pairs of Lapwings attempt to nest on the damp fields, though their success rate is not terribly high; Redshank; Common Sandpiper and Little Ringed Plover have nested here in the past. Wader numbers vary as birds move to and from other parts of the Burry Inlet and Loughor, with Oystercatchers, Golden and Grey Plovers, Lapwings, Knot, Dunlin, Black-tailed and Bar-tailed Godwits, and Curlew all occurring in many hundreds, sometimes thousands, on the estuary. Other regular visitors are Little Stint, Curlew Sandpiper, Jack Snipe, Snipe, Whimbrel which pass through between April and September, up to 30 Spotted Redshank, up to 60 Greenshank and Green Sandpipers. Occasionally Avocet, Ruff and Wood Sandpipers occur.

Black-headed are the most numerous of the gulls with upwards of 400 pairs nesting at Penclacwydd annually. Smaller numbers of Herring, Lesser Black-backed, Great Black-backed and Common Gulls are always present, while an increasing number of Mediterranean Gulls are seen annually, a few pairs of which have attempted to breed in among the Black-headed Gulls. Sandwich, Common, Arctic and occasional Black Terns occur on passage. Other wetland birds that breed include Moorhen and Coot, the two latter species readily taking to the wealth of new habitats and feeding opportunities being created. Other breeding birds include Cuckoo, Barn Owl, Kingfisher, Sedge and Reed Warblers, Lesser Whitethroat and Reed Bunting. Good numbers of passerines occur on both spring and autumn passage, while wintering birds include Hen Harrier, Merlin, Peregrine, Water Rail and Short-eared Owl. In some winters Brambling can sometimes be seen at the feeders at the visitor centre.

In its short existence, Penclacwydd has established a reputation as a place to see rare visitors to the county, indeed to Wales. These have included Cattle Egret, American Wigeon, Blue-winged and Green-winged Teals, Ring-necked Duck, Collared Pratincole, Long-billed Dowitcher, Marsh Sandpiper, Wilson's and Red-necked Phalaropes, Bonaparte's Gull, Franklin's Gull, Ring-billed

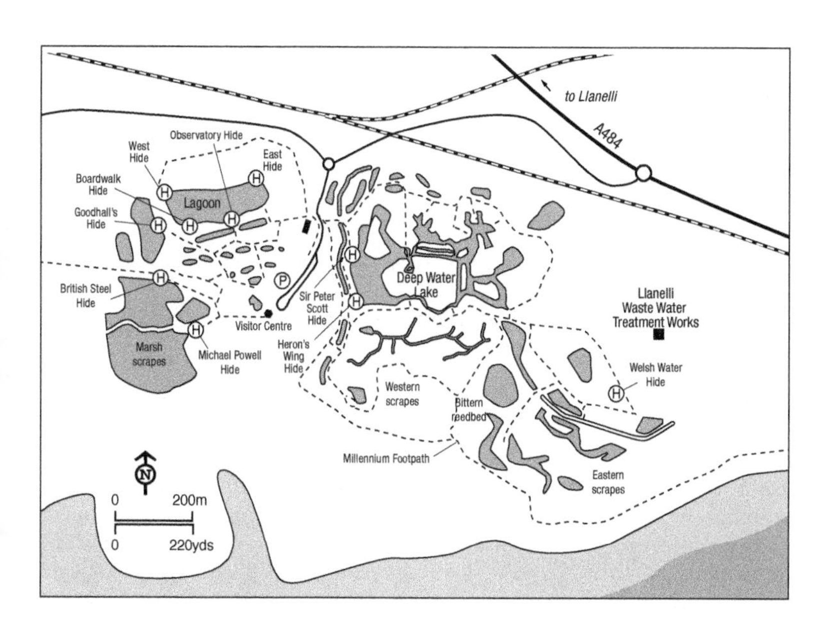

Gull, Caspian and Gull-billed Terns, Hoopoe, Wryneck, Savi's Warbler and Aquatic Warbler.

Timing

Open daily except Christmas Eve and Christmas Day, between 9am and 5pm (4:30pm in winter).

Access

Leave the A484 Swansea to Llanelli road about 3 miles (4.8 km) east of Llanelli and follow the signed road to the centre where there is ample parking. Entrance fee payable for WWT non-members. From the visitor centre, which has refreshment facilities, a series of tarmac paths provide easy walking to all parts of the site where there are plenty of hides to obtain excellent views of the birds.

CALENDAR

Resident: Little Egret, Mute Swan, Shelduck, Gadwall, Mallard, Pochard, Tufted Duck, Buzzard, Sparrowhawk, Kestrel, Water Rail, Oystercatcher, Black-tailed Godwit, Curlew, the large gulls, Kingfisher, Reed Bunting.

December–February: Wigeon, Pintail, Shoveler, Goldeneye, Hen Harrier, Merlin, Peregrine, Golden and Grey Plovers, Dunlin, Jack Snipe, Snipe, Greenshank, Green Sandpiper, Short-eared Owl, Fieldfare, Redwing, Brambling.

March–May: Wildfowl and wader numbers decline, summer visitors begin to arrive including Garganey, Whimbrel which on occasions number up to 200, Common Sandpiper, Sand Martins, Wheatear, being among the first arrivals. Breeding birds such as Lapwing and Redshank very noticeable by their aerial displays and loud calls.

June–July: Late summer passage waders begin to arrive and some species such as Curlew and Redshank rapidly build up numbers.

August–November: The best time of the year in many ways. A chance of pelagic species such as skuas and terns if the weather is stormy, but best of all the passage of waders followed by the arrival of winter visitors.

45 BRECHFA FOREST

OS Landranger Maps 146 & 147
OS Explorer Maps 186
Grid Ref: SN463285
Websites:
https://naturalresources.wales/days-out/places-to-visit/south-west-wales/brechfa-forest-byrgwm/?lang=en
https://naturalresources.wales/days-out/places-to-visit/south-west-wales/brechfa-forest-keepers-and-gwarallt/?lang=en

Habitat

At the southern outlier of the Cambrian Mountains is Brechfa forest, an extensive area of conifer plantations extending some 16,000 acres mixed in with small pockets of surviving oak woods north of where the Afon Cothi, noted for its fishing and its Otters, turns south to eventually merge with the Tywi. In the early Middle Ages this was the Forest of Glyncothi becoming, after Edward I defeated the Welsh in 1283, a Royal Forest with harsh Forest Laws that exacted severe penalties for those who transgressed. During the 16th century much of

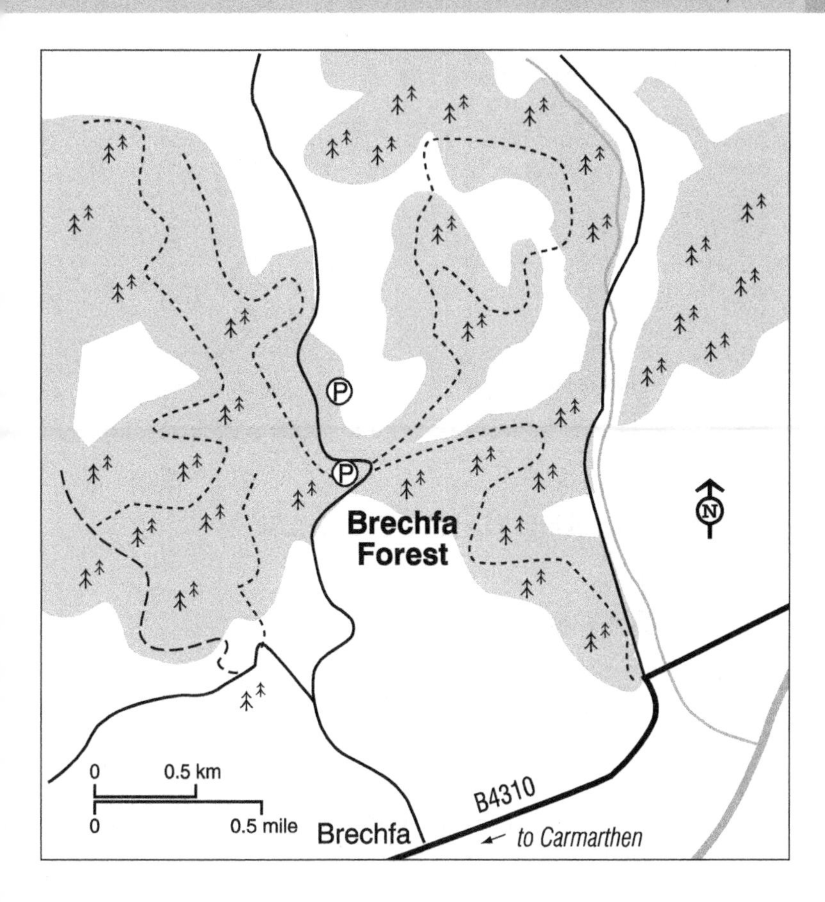

the woodlands, then deciduous, were felled and not replanted, now with mostly with conifers, until after the First World War.

Species

Siskins breed in the tall conifers while Lesser Redpolls frequent the younger plantations, and Crossbills are present most years. Since the early 1990s, Nightjars have bred in the clearfell areas and Great Grey Shrikes are irregular winter visitors. The surrounding deciduous woods are host in summer to Pied Flycatchers, Wood Warblers and Redstart. Kites, Buzzards and Great Spotted Woodpeckers may also be seen, while Goshawk are regular. Goosanders, Dippers, Grey Wagtails and Kingfishers may be seen along the Afon Cothi.

Timing

Spring and summer

Access

Parking available in Brechfa Forest at Abergorlech (SN586337) or along the minor road at Mynydd Llanfihangel Rhos-y-Corn (SN501354).

CALENDAR

Resident: Red Kite, Buzzard, Peregrine, Goshawk, Sparrowhawk, Great Spotted Woodpecker, Meadow Pipit, Dipper, Grey Wagtail, Goldcrest, Coal Tit, Siskin, Lesser Redpoll, Crossbill and Raven.
December–February Winter thrushes and finches unless weather is hard, in which case the forests at times seem almost birdless, save perhaps for the voices of a pair of Ravens high overhead.
March–May: Wheatears arrive in late March, early April followed by Tree Pipit, Redstart, Grasshopper and Willow Warblers, Whitethroat and Pied Flycatcher.
June–July: For most species the breeding season is completed by the end of July. River birds such as Goosander, Common Sandpiper, Sand Martin, Grey Wagtail and Dipper still obvious.
August–November: Summer visitors depart by October, about the time Fieldfares and Redwings return.

46 CRYCHAN AND HALFWAY FORESTS

OS Landranger 160
OS Explorer 187
Grid Ref: SN842413 & SN836331

Habitat

Crychan Forest, once part of the Glanbran estate with a mansion known as the fairest in the Tywi Valley, reached almost 30 miles from near Carmarthen to Builth Wells. The landscape from north-east and east of Llandovery changed massively from 1928, as the Forestry Commission began to buy sheep farms, and once-open high ground and the dividing valleys were transformed into vast coniferous plantations. The land of Dunlins and Golden Plovers, Meadow Pipits and Wheatears alas was no more. Halfway Forest, which is always worth exploring along two short walks and gentle trails, straddles the border between Carmarthenshire and Breconshire, the Nant y Dresglen stream, haunt of Dippers and Grey Wagtails, merging with Afon Gwydderig on its journey to Llandovery and the Twyi. Close by, beside the A40, which follows the old drovers route used to take cattle from mid-Wales to Smithfield Market, a small memorial, the Mail Box Pillar is worth pausing at to read the inscription, which tells how the driver, Edward Jenkins, drunk at the wheel of the Gloucester to Carmarthen coach in December 1841, caused the coach to plunge into the river. Jenkins survived, together with his five passengers and the guard, and was fined £5, about £310 today and costs.

Species

Siskins breed in the tall conifers, while Lesser Redpolls frequent the younger plantations and Crossbills are present most years. Since the early 1990s, Nightjars have bred in the clearfell areas and Great Grey Shrikes are irregular winter visitors. The surrounding deciduous woods are host in summer to Pied Flycatchers, Wood Warblers and Redstart. Kites, Buzzards and Great Spotted Woodpeckers may also be seen, while Goshawk are regular.

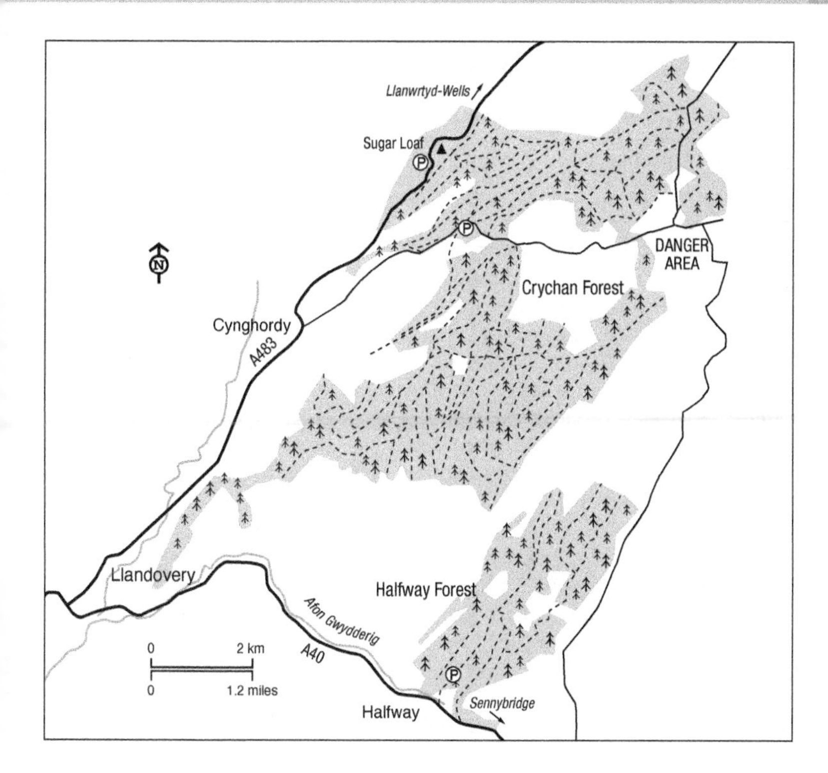

Timing

Spring and summer.

Access

There can be no better place to obtain a view than from the Sugar Loaf, the climb from the car park not too strenuous and well worth the effort for the fabulous views to the east and south across Crychan Forest. There are numerous forest tracks, including both cycle and horse riding trails, or take the minor road north from Cynghordy, which takes one right through the forest.

For the Halfway Forest, leave the A40 at Halfway, a car park is reached after a short distance from where there are two trails: the Nant y Dresglen Trail of just over a mile and never far from the waters of Nant y Dresglen, and the Melin y Glyn Trail of double the distance, being described as moderate.

CALENDAR

Resident: Red Kite, Buzzard, Peregrine, Goshawk, Sparrowhawk, Mistle Thrush, Dipper, Grey Wagtail, Goldcrest, Coal Tit, Siskin, Lesser Redpoll, Crossbill and Raven.
December–February Winter thrushes and finches, Buzzards, Ravens and Red Kites.
March–May: Wheatears arrive in late March, early April on the high ground outside the forest followed by Tree Pipit, and other summer visitors.
June–July: For most species the breeding season is completed by the end of July.
August–November: Summer visitors depart by October about the time Fieldfares and Redwings return.

47 LOWER TYWI VALLEY

OS Landranger Map 146
OS Explorer Map 186
Grid Ref: SN585208

Habitat

The lower Tywi from Llandovery to the tidal waters of Carmarthen, some 25 miles (40 km) to the west, meanders through a wide, flat valley. Picture the scene 1,000 years ago or more: thick carrs of willows and alder, and the river would have flooded frequently. Nowadays, the valley contains some of the richest grazing land in Wales, though even now in winter the floods appear. A climb to Paxton's Tower on the south side on a crisp January day will give a splendid, almost aerial, view of the scene, glistening water on green farmland and a touch of snow on the distant Carmarthen Vans.

There are several birdwatching sites in the valley. At Llanwrda, you can reach the riverbank while at Llandeilo, the Wildlife Trust of South & West Wales Castle Woods Nature Reserve occupies much of the woodland extending westwards from Penlan Park at the edge of the town past the summit on which ruined Dinefwr (Dynevor) Castle stands. At the foot of the steep escarpment is a wide expanse of water-meadows and remnant oxbows cradled by a great bend of the Tywi. Some areas of open water remain throughout the summer, but these are much enlarged in winter. North lies Newton House, a National Trust property, and Deer Park, 110 acres (45 ha) of pasture woodland complete with Fallow Deer and the ancient white cattle of Dinefwr, which wandered here from the time the castle was in occupation and before.

Four miles (6.4 km) downstream and high above the river with its adjacent water-meadows and flood pools there are the remnants of another castle, which stands on the great mound of Dryslwyn. Even in its ruinous state, the castle is a formidable structure, one of the most important to be built by a Welsh chieftain and long associated with the princes of Deheubarth, a kingdom in south-west Wales, and the great Lord Rhys who died in 1197. Foolishly, his sons fought over his inheritance and it fell to the English crown in 1287.

Another 6 miles (9.6 km) brings you to the hamlet of Abergwili, close by Carmarthen. We are told that on the night of Llanybyther Fair, 17 July 1802, in the episcopate of Bishop Murray, there was a great flood which caused the Tywi to switch its course to run at the foot of the southern slope, leaving in its turbulent wake a superb oxbow lake, the best example of its kind in South Wales and now known as the Bishop's Pond. In summer, the water levels drop and the surface is carpeted with Yellow Water-lilies, while the margins are fringed with Reed Sweet-grass, in Wales largely confined to the Tywi Valley and the coastal flats bordering the Bristol Channel.

The Tywi has a thriving population of Otters, while Grey Seals occasionally wander several miles upstream from the tidal limits just above Carmarthen. The most astounding occurrence, however, was a Sturgeon caught near Nantgaredig in July 1932 – the biggest fish ever to be caught on rod and line in fresh water in Great Britain; it measured nine feet two inches and weighed 388 lbs.

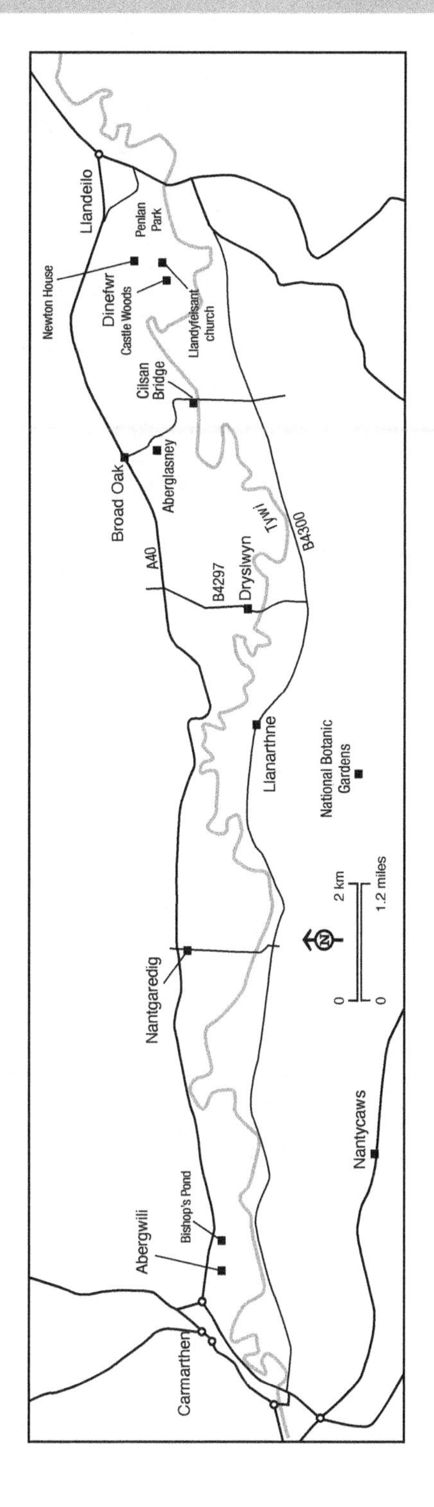
Llandeilo
Newton House
Penlan Park
Dinefwr
Castle Woods
Llandyfeisant church
Cilsan Bridge
Broad Oak
Aberglasney
A40
Tywi
B4300
B4297
Dryslwyn
Llanarthne
National Botanic Gardens
0
2 km
0
1.2 miles
Nantgaredig
Nantycaws
Bishop's Pond
Abergwili
Carmarthen

Species

Wildfowl and waders are the main attraction in the Tywi Valley, though the number of species remaining to breed is small. By late July, Mallard numbers begin to increase on the oxbows below Dinefwr as moulting birds take up temporary residence on these secluded waters, where up to 500 may be recorded well into the autumn. Passage waders are not infrequent, usually Green Sandpiper with occasional Dunlin, Whimbrel and Greenshank. Common Sandpipers breed on shingle banks along the river. Little Ringed Plover probably first nested on a shingle bank on the Tywi in 1984, while two years later the first confirmed breeding success was reported when three fledged juveniles were seen in early August. Since then, this attractive wader has colonised many riverside shingle banks and the population rose dramatically to about 76 pairs in 2007 and about 45 pairs in 2018. Breeding success at this bird's most important location in Wales is very variable and likely to be affected by sporadic shingle extraction, flooding and trampling by cattle.

Mallard breed, and there are usually several pairs of Mute Swans in the valley. A quite remarkable colonisation of the Tywi by Goosanders has taken place since breeding was first suspected in 1979. Most nest between Llandeilo and Carmarthen and along the Tywi tributaries, and up to 46 young have been seen in late July. Several pairs of Kingfishers nest along the river while the riverside banks are especially important for Sand Martins. Nearly 2,500 nest-holes were counted here and on the Cothi in the past. A former speciality in the valley, the Tree Sparrow, seems to have disappeared from what was its most westerly stronghold in South Wales, with very few records nowadays.

Dabbling ducks occur in modest numbers in Dinefwr, and at Dryslwyn. Whooper Swans are now regular winter visitors, mostly in the Dryslwyn and Cilsan areas, with up to 80 being seen. Bewick's Swan, which used to be a regular sight in the Tywi Valley, are now very rare visitors. It was here that White-fronted Geese once regularly grazed in the meadows, one of three main wintering grounds in Britain for the Siberian race of this goose. Numbers were as high as 2,500 in the late 1960s and early 1970s but since then have dropped steadily, and alas, no longer does the valley echo with their distinctive calls. Ones or twos still occur some winters along with occasional records of Pink-footed and Barnacle Geese, which can be found among the big flocks of Greylag and Canada Geese that are now the dominant geese present. Taiga Bean Goose has been seen here in December 1996, and in March 1971, a Lesser White-fronted Goose, the second record for Wales, was seen.

When flooded in winter the valley often has large flocks of feeding gulls. Mediterranean Gulls have been seen, but lucky observers found both Laughing and Franklin's in the same field during the 2005 November influx. Green Sandpipers occur most years along the river with birds occasionally overwintering, while other scarce visitors have included Ring-necked Duck, Osprey, Wood Sandpiper and Firecrest.

Cross the valley and visit Castle Woods, one of the most striking woods in south-west Wales, despite the decimation of the Wych Elms by Dutch elm disease in the 1980s. Immediately adjacent to this is Dinefwr Park, with more ancient trees contrasting with open grassland. Whatever the season, the woods always seem alive with birds. Three species of woodpecker are present, Lesser Spotted Woodpecker being noteworthy, although now very rare. Stock Doves, Marsh Tits and Ravens nest, the ringing calls of Nuthatch are a characteristic sound of these woods, while the shy Treecreeper is also a numerous resident. Among the summer visitors, Redstart and Pied Flycatcher are present in small numbers. Birds of prey include Sparrowhawk and Buzzard, and Peregrines regularly hunt this section of the valley in winter, no doubt attracted by the abundant waterfowl; Hobby is now also a regular summer sighting and breeds in the valley. From vantage points in the woods, it is usually possible to see Grey Herons feeding along the riverbank or around the oxbows; the largest colony in the county is at Alltygaer, Dryslwyn, and usually contains up to 55 pairs. Becoming more regular in the valley are Little and Great White Egrets, which are worth looking for from the river bridges at Dryslwyn and Cilsan and from Castle Woods.

Timing

When visiting Castle Woods, especially if it is in spring or early summer, try to make the early morning. You will then find birdsong at its peak, and the scent of Bluebells in the dew will be something else to remember. The timing of visits to Dryslwyn and the Bishop's Pond are not critical so far as the wildfowl are concerned.

Access

Bishop's Pond (SN443209): Not easily viewed other than from the bridge that crosses at one point, and which is reached from the A40 at SN445210. There is no general access along the banks of the pond. Do not neglect a visit to Carmarthen Museum in the former Bishop's Palace.

Castle Woods (SN615218): https://www.welshwildlife.org/nature-reserve/castle-woods-llandeilo/ The reserve is open at all times. A system of footpaths leads from Penlan Park, the town park of Llandeilo, through South Lodge Woods, past the disused Llandyfeisant church, and so to the main Castle Woods. Alternatively, drive to the National Trust property of Newton House where there is parking and views into the Deer Park and walk from there to the castle, though this route involves a steep climb, but well worth it when you reach the top: the view from the battlements is outstanding. Redstart, Spotted and Pied Flycatchers and Garden Warblers are summer visitors to the woods.

Cilsan Bridge (SN589214): The river and water-meadows here are usually the best place to see the Whooper Swans from October to early April. Check the river for Egrets, Goosander and Kingfisher. Otters are regularly spotted here.

Dinefwr Park (SN613225): The hundred-acre medieval deer park home to Fallow Deer and ancient white cattle is renowned for its ancient oak trees with the access point at Newton House, easily reached from Llandeilo, note this is only open 10am to 5pm.

Dryslwyn (SN554202): The water-meadows are best viewed from several points on the B4300, a quiet enough road for such activities. Excellent general views can be obtained of the whole area from Dryslwyn Castle, to which there is unrestricted access, and the river in particular from the bridge below. In November 2005, both Laughing and Franklin's Gulls were seen here.

CALENDAR

Resident: Cormorant, Grey Heron, Little Egret, Mallard, Goosander, Sparrowhawk, Buzzard, Red Kite, Coot, Tawny Owl, all common woodland species, Raven, Reed Bunting.
December–February: Little Grebe, Mute Swan, Wigeon, Teal, Shoveler, Pochard, Tufted Duck, Goldeneye, Peregrine, Snipe, Lapwing, Golden Plover and Curlew.
March–May: Great Crested Grebe, Little Ringed Plover, Common Sandpiper, Sand Martin and all other summer migrants, Spotted Flycatcher and Swift being among the last to arrive.

June–July: All breeding species present, Little Ringed Plover have chicks, first passage waders, usually Green Sandpiper begin to appear, Mallard numbers begin to increase.
August–November: Summer visitors depart usually by early September. Winter wildfowl start arriving, with large numbers of Lapwings and Golden Plovers and smaller numbers of Curlews by end of period. Winter thrushes move into the valley during early November.

48 TALLEY LAKES

OS Landranger Map 159
OS Explorer Map 186
Grid Ref: SN631336
Websites:
https://www.welshwildlife.org/nature-reserve/talley-lakes-llandeilo-carmarthenshire/
https://naturalresources.wales/days-out/places-to-visit/south-west-wales/talley-woodlands/?lang=en

Habitat

Talyllychau, 'the head of the lakes', or Talley it takes its name from its situation close by the two lakes which drain northwards to the Afon Cothi, a tributary of the Tywi. A short distance away is the shore of the upper lake, which together with its neighbour extends for some 44 acres (18 ha), one of the most important open freshwater sites in this part of Wales. The lakes are managed by the Wildlife Trust of South & West Wales. Hard by and overlooking the southern lake the great Rhys ap Gruffwd, Prince of Deheubarth, founded a house of the White Cannons in the late 12th century, the only one in Wales. Now all that remains is part of the tower and the remnants of the outer walls.

Species

The best place to watch diving duck in Carmarthenshire is at Talley where Pochard and Tufted Duck are the most frequent and have bred, and other visitors include Ring-necked Duck, Scaup, Goldeneye, Smew and Goosander. Mute Swans and both Great Crested and Little Grebes breed, while the Red-necked Grebe seen in October 1966 was the first recorded in Carmarthenshire.

Timing

The timing of visits to Talley are not critical so far as the wildfowl are concerned.

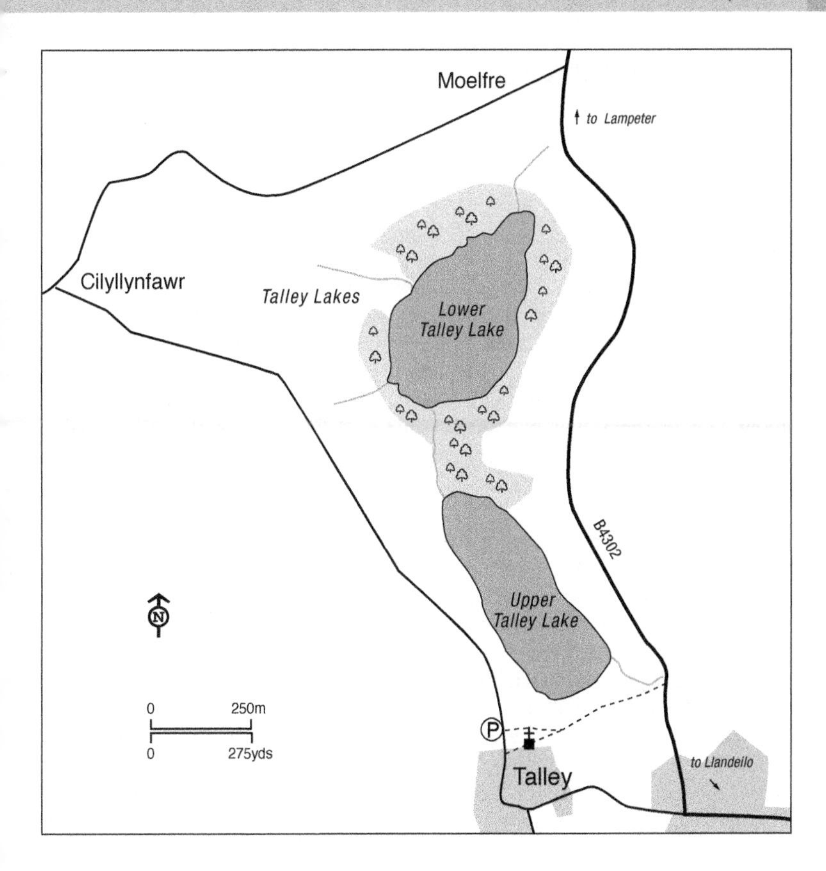

Access

Talley Lakes (SN632335). Access near the Abbey ruins, past the Upper Lake to the narrow neck of land, with a hide, which separates the Lower Lake. General views are possible from several points on the roads either side of the lakes.

CALENDAR

Resident: Cormorant, Grey Heron, Mallard, Goosander, Sparrowhawk, Buzzard, Red Kite, Coot, Reed Bunting.
December–February: Little Grebe, Whooper Swan, Wigeon, Teal, Shoveler, Pochard, Tufted Duck, Goldeneye.
March–May: Great Crested Grebe, Common Sandpiper, Sand Martin, Swallow, Reed and Sedge Warblers, while in the woodlands Blackcap, Chiffchaff and Willow Warblers.
June–July: All breeding species present, first passage waders, usually Green Sandpiper begin to appear, Mallard numbers begin to increase.
August–November: Summer visitors depart usually by early September, winter wildfowl start arriving, while Water Rails may be heard from the reedbeds.

49 UPPER TYWI VALLEY

OS Landranger Map 147
OS Explorer Map 187
Grid Ref: SN788472
Website: https://www.rspb.org.uk/reserves-and-events/reserves-a-z/gwenffrwd-dinas/

Habitat

In 1967, the RSPB acquired the Dinas where the Doethïe, emerging from its gorge, joins the Tywi at Junction Pool. This is the site of Twm Siôn Catti's cave, Twm being a somewhat legendary Robin Hood-type figure based largely on the early exploits in the life of Thomas Jones (1530–1609) of Tregaron, pardoned by Elizabeth I in the first year of her reign (for unknown offences). Elsewhere in the valley, Natural Resources Wales manage the Allt Rhyd y Groes National Nature Reserve, a superb hanging woodland with crags and scree slopes and a wealth of plant communities where the Afon Pysgotwr Fawr merges with the Afon Doethïe, in one of those remarkably well-hidden valleys of mid-Wales. Over the high ground to the south-west is the Nant Melyn nature reserve of the Wildlife Trust of South & West Wales, situated not far from where the Cothi starts a journey in one direction and the Melyn, a tributary of the Tywi, goes in another; they will not merge until they meet near the head of the estuary at Carmarthen.

Both the Tywi and its tributaries vary considerably within quite short distances. There are torrents cascading over boulders from pool to pool, then tranquil slower-moving waters, the summer haunt of that most jewel-like of damselflies, the aptly named Beautiful Demoiselle. Small fields and patches of woodland line the riverbank, hemmed in by ground that rises steeply, broken in places by rock outcrops, even cliffs. There are extensive areas of hanging oak woods, above which lie vast tracts of heavily grazed uplands where, in the wet hollows, the botanist can find Round-leaved Sundew and Common Butterwort. For those prepared to search, there are other gems such as the Alpine, Fir and Stag's-horn Clubmosses to be discovered on these, at times, bleak moors.

Much of the woodland has remained unmanaged for the past century; sheep have grazed unchecked, so that the ground flora is impoverished and regeneration of oak almost unknown. In the nature reserves, fences have been erected to exclude or control sheep-grazing and some tree-planting has been carried out. Certain birds, however, most notably Wood Warbler, prefer the open woodlands with little secondary growth to mask the ground, so that a careful balance of management has to be maintained. The RSPB has established a woodland, Coed Vaughan, a fitting memorial to Captain H.R.H. Vaughan, who lived for many years just down the valley and did so much to ensure that the Red Kite received protection.

Species

Most birdwatchers come to the Upper Tywi Valley in the hope of seeing Red Kites, for this part of Wales and over the high hills and sheepwalks into Ceredigion was for much of the previous century their last remaining stronghold in Britain. Towards the end of the 19th century, the British population had been so reduced by nest-robbing and the killing of adults that only a handful of birds remained, all in this part of Wales. The first Kite Protection Committee was formed in 1903, and as a result it was just possible to ensure the birds' continued survival, though for many years there were no more than 10 pairs and at times only half that number, so close to extinction had the Kite become. The population slowly increased during the tail end of the last century as a result of greater protection and reduced persecution. Red Kites then spread into many new areas across mid-Wales and in 1998 numbered 195 territorial pairs, of which 163 nested rearing 171 young. Since then, there has been what can only be described as a population explosion, the most recent full attempt at a census by the Welsh Kite Trust, that in 2010 recorded 578 apparently occupied sites. While the Welsh Kite Trust estimated that the Welsh population could be as high as 2,500 pairs in 2019. The pressures remain, however: bad weather, habitat loss as open hill land is taken up by forestry plantations, hard winters, and the continuing unscrupulous activities of egg-collectors.

Ravens are often obvious; their 'kronking' calls in May as a family party flies along the escarpment provide one of the joys of the early summer in these hills. Wheatears can be found on the high ground, and as you begin to descend from the open moorland to where scattered bushes and small trees have become established, you should find two other summer visitors, Whinchat and Tree Pipit, the latter extending into the open woodlands. Merlins are occasional on the high ground, Sparrowhawks, Buzzards and Kestrels are more frequently seen, and with any luck a Peregrine may speed across the valley. Goshawks nest in some of the larger coniferous plantations.

The woodlands of the upper Tywi in spring and early summer are alive with birds. Nest boxes in all the nature reserves attract large numbers of Pied Flycatchers. This boost for the population, which ensures one of the highest densities of Pied Flycatchers in Britain, may well be part of the reason for the species' westward extension down the valley in the 1980s and its colonisation of woodlands in southeast Carmarthenshire and Pembrokeshire. Redstarts and Wood Warblers are common, and there is a whole range of other woodland species including Mistle Thrush, Garden Warbler, Blackcap, Marsh and Willow Tits and Nuthatch. Lesser Redpoll, Siskins and Crossbills can be found in the conifer plantations. Lesser Spotted Woodpeckers are now rare. Hawfinches have bred recently in the area. This shy finch, one of the most elusive of our breeding birds, can often be located only by those familiar with its characteristic 'tick' call note. A few pairs of Woodcock nest in the damp woodlands of the Upper Tywi Valley, and visitors at dawn or dusk may well be fortunate enough to observe the roding display flight of a bird otherwise infrequently seen – its status in Wales remains poorly understood.

The rivers have Grey Wagtails, Dippers and in late summer occasionally Kingfishers, as they disperse from breeding sites further down the valley. There are several colonies of Sand Martins in the riverbanks and Common Sandpipers also nest here, their cheery song quickly drawing the observer's attention. Goosanders are a highlight on the rivers; they first nested here in 1980 and now pairs use the Tywi and Cothi valleys.

Timing

Early morning is best, though for late risers there can be few more pleasant ways to spend a late spring/early summer evening than by a visit to the Dinas.

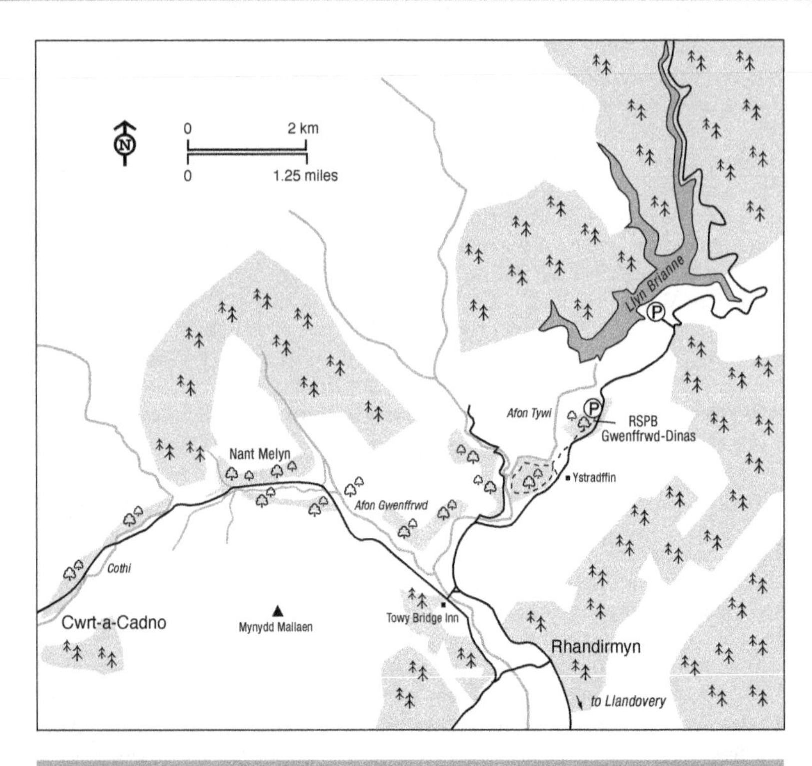

CALENDAR

Resident: Goosander, Red Kite, Sparrowhawk, Buzzard, Peregrine, Grey Wagtail, Dipper, Dunnock, Robin, Blackbird, Song and Mistle Thrushes, Goldcrest, Willow and Marsh Tits, Nuthatch, Treecreeper, Jay, Raven, Chaffinch, Greenfinch, Siskin, Linnet, Redpoll.

December–February: Winter thrushes and finches unless weather is hard, in which case the valley at times seems almost birdless.

March–May: Woodcock still present while summer visitors from about the beginning of April include Cuckoo, Sand Martin, Tree Pipit, Redstart, Whinchat, Wheatear, Garden Warbler, Blackcap, Wood Warbler, Chiffchaff, Willow Warbler, Pied Flycatcher.

June–July: Breeding season in full swing, but by end of period most woodland species have finished, and virtual cessation of song means that location of remaining birds is difficult. River birds such as Goosander, Common Sandpiper, Sand Martin, Grey Wagtail and Dipper still obvious.

August–November: Kingfisher likely on rivers. Summer visitors largely make an early departure, while from mid-October onwards winter thrushes and finches arrive. Resident species such as Goldcrest, tits, Treecreeper and Nuthatch form wandering bands which often leave the woods to work along hedgerows.

Access

Leave Llandovery on the unclassified road for Rhandirmwyn, beyond which you have a choice. Bear right up the valley to the Dinas reserve, which is open all the year and has an excellent woodland walk; there is a car park and information centre at SN788471. Beyond here, the road continues to the massive Llyn Brianne dam and reservoir, spectacular but ornithologically not very interesting, although Wheatears and Whinchats are usually present in the summer months. Alternatively, you can cross the Rhandirmwyn bridge and turn right for Cwrt-a-Cadno looking out for Redstarts, Spotted and Pied Flycatchers and Wood Warblers along the way. Further up the same road is the Nant Melyn reserve of the Wildlife Trust of South & West Wales, a footpath leads up through the woods from the bridge at SN730465 to the reserve itself. For those unable to leave the car, the roads do provide excellent vantage points for viewing large parts of this scenic area.

CEREDIGION

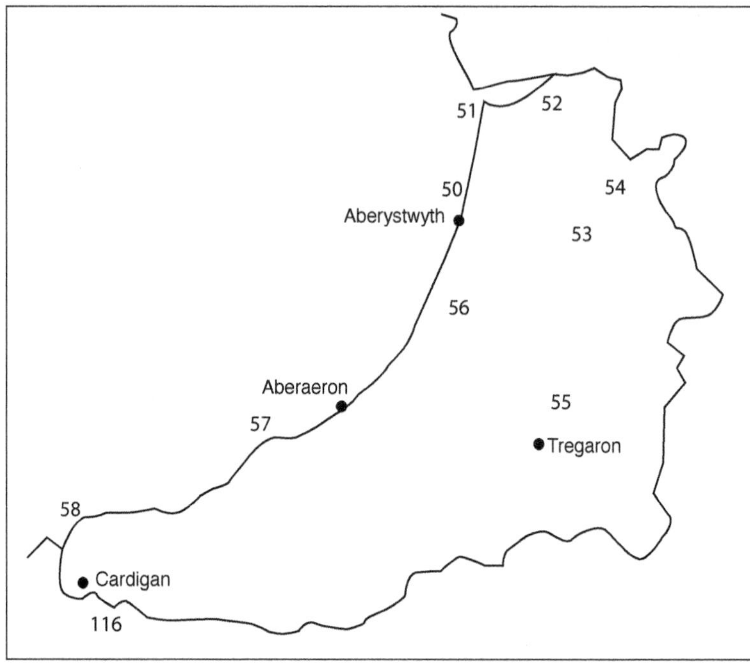

50 Aberystwyth and Tanybwlch
51 Dyfi Estuary
52 RSPB Ynys-hir
53 The Rheidol Valley
54 Bwlch Nant-yr-arian
55 Cors Caron
56 Llanrhystud–Llanon Coast
57 South Ceredigion Coast
58 Teifi Estuary

Ceredigion is a rather long and relatively thin Welsh county at the base of Cardigan Bay. The Bay itself is nationally important for wintering divers, grebes and sea-duck, but most of these tend to be at the north, from the mouth of the Dyfi Estuary (Borth) northwards to Criccieth (Caernarvonshire). Ornithological interest in the county is mainly at the extremes, the Dyfi Estuary (Ynys-hir, Cors Fochno and Ynyslas) in the north, the Teifi estuary (Teifi estuary and Marshes) in the south and Cors Caron in the east. In between, the rolling landscape can be quite wild in places and is generally overlooked by birdwatchers. Buzzards, Red Kites and Ravens suit this type of environment and can be seen virtually anywhere these days, although to get a true feeling of their splendour, with a backdrop of the majestic wild Welsh hills, observers should try to visit Cors Caron or Nant-yr-arian (Rheidol valley).

Many birdwatchers come to Ceredigion to see Redstarts, Wood Warblers and Pied Flycatchers, the characteristic species of the hanging oak woodlands of mid-Wales – as the county avifauna reminds us, 'one of the joys of spring is to listen to the trilling song of the Wood Warbler followed by its mournful call'. All three can be found in suitable habitat throughout the county, although some of the best sites are in the north, around Devil's Bridge, Cors Caron, Rheidol Valley and RSPB Ynys-hir.

The Dyfi Estuary attracts more wintering and passage birds than the Teifi, including a very small flock of wintering Greenland White-fronted Geese. At its mouth is Ynyslas and Cors Fochno (both National Nature Reserves), while just a few miles upstream is RSPB reserve Ynys-hir. Ceredigion's cliff coast is ideal for Chough, Peregrine and Raven, particularly in the south around Mwnt and Llangrannog (Ynys Lochtyn).

50 ABERYSTWYTH AND TANYBWLCH

OS Landranger Map 135
OS Explorer Map 213
Grid Ref: SN577816

Habitat

Aberystwyth is the largest town in the county and therefore has all the main amenities and transport links. The environs in and around the town are always worth exploring – the seafront and harbour for gulls, waders and wintering Black Redstarts, the woods around Constitution Hill and Pendinas for warblers and migrants and the flooded meadows beside the Afon Ystwyth, at Tanybwlch, for waders, gulls and passage migrants, while offshore spectacular passages of Manx Shearwater during the summer months, divers in the winter at which time the Starling roost at the pier should never be missed.

Species

Since the conversion from a small fishing harbour into a marina, the harbour itself attracts far less than it used to, with most of the gulls now frequenting the College Rocks, by the pier and castle. The gull roost here attracts quite a range of common species, including Black-headed, Herring, Lesser and Great Black-backed Gulls. Mediterranean Gulls are quite often seen, with numbers building during the late summer, July onwards, peaking in February and March when up to two dozen can be counted. Other gulls that occur irregularly include Kittiwake, Iceland, Glaucous and very occasionally Ring-billed. There has been a single sighting of a Ross's Gull, at the mouth of the harbour during a westerly gale in December 1994. The rocks, although suffering from disturbance by rock-poolers, have a small wintering population of Turnstone, Ringed Plover and Purple Sandpipers. The latter usually roost on a ledge underneath the castle at high tide.

Offshore Red-throated Divers are quite common during the winter, as are Cormorants, Shags, Gannets and during the summer months, auks and Manx Shearwaters. There are small breeding populations of both Guillemot and Razorbill all along the coast, while the Manx Shearwaters nest on Skomer, Skokholm and Ramsey in Pembrokeshire but feed out in the Irish Sea. It is quite normal to observe several thousand passing the seafront at Aberystwyth on a late summer's evening.

Black Redstarts winter in the area, and can be seen virtually anywhere around the harbour, with perhaps the Old College by the pier or around the back of the harbour, towards Tanybwlch, where the Ystwyth comes in, being the best spots. The river has been canalised in the past, but the grassland behind the beach is often flooded and provides a good chance to see passage waders and migrants such as Sand Martins, Wheatear and White Wagtail.

The woods around Constitution Hill put on a good Bluebell show and in April and May are alive with Blackcap, Chiffchaff, Garden and Willow Warblers. Occasionally Redstart, Wood Warbler and Pied Flycatchers are seen. The scrub around the top of Constitution Hill and Pendinas to the south of the town, overlooking the harbour, are best watched in spring, with chance of finding passage migrants such as Grasshopper Warbler, Lesser Whitethroat and Whinchat.

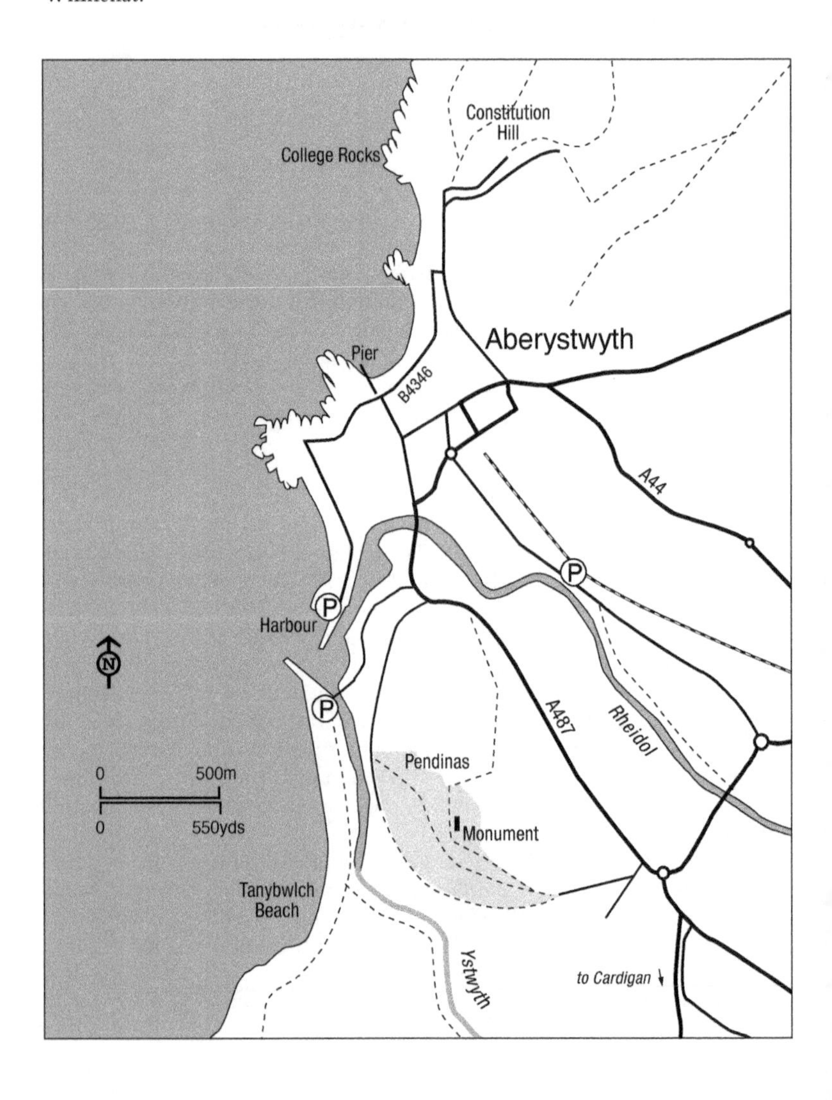

The most spectacular ornithological event in the town is the nightly Starling roost on the pier, from around 3pm onwards between November and February. Thousands upon thousands appear, maximum numbers reach 30,000 at times as the flocks swirl, twist and turn until they find their desired roosting spot. This in turn attracts avian predators, Sparrowhawks and Peregrines, for their nightly feast. The origin of the Starlings as revealed by ringing recoveries is Finland, Sweden and the Baltic states: birds on passage have been ringed in Belgium, Denmark and the Netherlands, while birds ringed at the roost have been recovered in all those countries as well as elsewhere in Great Britain, and in Belarus, France, Germany, Norway and Poland.

Timing

Low tide is best for checking the gull roost on Castle Rocks and for Turnstone and Purple Sandpipers (unless at roost). The winter Starling flocks start to gather from around 3pm onwards.

Access

The harbour is easily accessible on foot, while the promenade offers easy viewing points for the shore and enables one to watch from the comfort of a car if required; at least in winter there will be no parking problems, but in summer it is a different story.

CALENDAR

Resident: Grey Heron, Goosander, Red Kite, Sparrowhawk, Buzzard, Kestrel, Oystercatcher, Herring and Great Black-backed Gulls, Great Spotted Woodpeckers, Rock Pipit, Grey and Pied Wagtails, Wren, Dunnock, Robin, Blackbird, Song Thrush, Mistle Thrush, Goldcrest, tit family, Nuthatch, Treecreeper, Jay, Raven.

December–February: Red-throated Diver, Purple Sandpiper, Turnstone, Ringed Plover, Mediterranean Gull, Black Redstart and Starling roost.

March–May: Common Sandpiper, Sand Martin, Whitethroat, Garden Warbler, Blackcap, Chiffchaff, Willow Warbler.

June–July: Manx Shearwaters and first tern passage offshore by late July.

August–November: Manx Shearwaters offshore, Turnstones return, a chance of other passage waders, Mediterranean Gulls, tern passage continues until late September. Black Redstart, Fieldfare and Redwing during late October, while other winter visitors such as diving ducks and divers begin to arrive.

51 DYFI ESTUARY

OS Landranger Map 135
OS Explorer Map OL23
Grid Ref: SN609942
Website: https://naturalresources.wales/days-out/places-to-visit/mid-wales/dyfi-national-nature-reserve-ynyslas-visitor-centre/?lang=en

Habitat

The Dyfi Estuary is one of the most enchanting estuaries in Great Britain, being bounded to the north by Cadair Idris and its foothills, and to the south by the lesser slopes of north Ceredigion, the outliers of Plynlimon, which are intersected by hidden valleys such as the Clettwr, Einion and Leri. The Dyfi, which has flowed from its source below Aran Fawddwy, forms the boundary between North and South Wales at this point.

At the estuary mouth are one of the most popular holiday beaches in Wales and Ynyslas Dunes, which forms part of Dyfi National Nature Reserve, embracing the estuary, dunes, wet meadows and the raised bog and extending some 5,592 acres (2,263 ha), the second-largest NNR in Wales. In addition to other designations, the Dyfi is registered internationally as a Biosphere Reserve, only one of 11 such sites in Britain. There is a small visitor centre among the dunes at Ynyslas, open during the summer months, providing information and guidance. The dunes are protected by boardwalks at sensitive points to prevent erosion and their ultimate destruction. Remember as you look seawards from the height of the dunes that no less than 1,500 years ago the coastline would have been up to 20 miles further west. The legend of Cantre'r Gwaelod tells that this was a rich farmland protected from the sea by embankments with sluices. Alas, so the story goes, the sluice-keeper Seithennyn so enjoyed his ale one night that he failed in his duties: the land was flooded and the inhabitants fled. Are the 'sarnau' – the long stone ridges extending seawards and revealed at low tide – causeways used by these early people? There is no doubt that 7,000 years or so ago the land was forested, for at low water the fossilised remains of tree trunks are often visible.

On the landward side are several dune slacks. They hold shallow pools in winter, and even in midsummer, water is not far below the surface. Here thrive such plants as Common Bird's-foot-trefoil, Restharrow, Biting Stonecrop, Common Centaury, Early and Northern Marsh-orchids and Bee and Pyramidal Orchids. On occasions there may be upwards of 170,000 Marsh Helleborine plants: what a sight, a seemingly endless carpet of creamy yellow and pink flowers.

If you stand on the highest point of the boardwalk and look inland, the main part of the estuary extends some 5 miles (8 km) away to the east, while to the south-east lies Cors Fochno. Despite many inroads by reclamation for agriculture and peripheral peat-cutting, this remains the largest area of unmodified raised mire in the country, much of it owned by Natural Resources Wales. A wide range of peat-loving plants flourishes here, including Bog Myrtle, Bog Rosemary, Cranberry and Royal Fern. Of special interest is the colony of the Rosy Marsh Moth, a Red Data Book species discovered here in 1965, over 100 years after it became extinct in the fens of East Anglia. Subsequently it was found at Tregaron (see Site 54) and in 2005 in Cumbria. Access to the bog is limited to views from the Leri embankment or a small section of boardwalk in the northern section.

Further upstream is RSPB Ynys-hir (Site 52) and just over the border in Montgomery is the Dyfi Osprey Project (Site 98).

Species

It is well worth searching the sea off the mouth of the estuary and southwards towards Borth. The depth is shallow and the resultant warm waters make the area attractive to a range of marine species. In the past there have been large numbers of wintering Red-throated Divers (over 1,000 in January 1997), Great Crested Grebes (up to 500) and Common Scoter (up to 2,000). Sadly, in recent years there have been far fewer. Careful observation may also pick out Great Northern and Black-throated Divers, Slavonian Grebe, Eider, Long-tailed Duck and Velvet Scoter. Rarities have included Surf Scoter and even a King Eider.

During spring and autumn passage periods, there are usually quite large numbers of terns – mainly Sandwich but also Common and Arctic, rarely Little and Roseate, and even Gull-billed has occurred – around the mouth of the Dyfi and by the 'tern posts' to the right of the car park. The terns often attract passing Arctic and Great Skuas, while occasionally Long-tailed and Pomarine have been recorded. During the summer months, there is a large daily movement of Manx Shearwaters along the coast, viewable from here, of up to several thousand on their way back from feeding in the Irish Sea to their burrows on Ramsey, Skomer and Skokholm in Pembrokeshire.

The Dyfi estuary contains the usual estuarine wintering species like Little Egret, Shelduck, Wigeon, Teal, Mallard, Canada Goose, Greylag, Barnacle Goose, Dunlin, Curlew, Oystercatcher, Ringed Plover, Lapwing, Bar-tailed and Black-tailed Godwits, Redshank, Greenshank, Grey and Golden Plover. Smaller numbers of other wildfowl including Goldeneye, Red-breasted Merganser, Goosander and an unfortunately diminishing Greenland White-fronted Goose flock, once over a thousand, now down to a dozen or so.

Numbers of waders increase during passage periods, and Ynyslas is well known for large numbers of Whimbrel and Ringed Plover during April and May. Among the latter there have been several records of Kentish Plover. Other passage waders include Spotted Redshank, Knot, Common, Green and Curlew Sandpipers, Little Stint and Ruff. There have been several notable rarities recorded here: Semi-palmated, Pectoral and Buff-breasted Sandpipers, Lesser Yellowlegs, while sightings of Great White Egret, once a major rarity in Wales, are now quite commonplace on the Dyfi.

Raptors are often to be seen, including Kestrel, Peregrine, Buzzard, Red Kite, wintering Merlin and Hen Harrier, Marsh Harrier on passage and Osprey in summer. The bog at Cors Fochno is a good place to see many of these, particularly Red Kites and roosting harriers. Other species here include a good selection of marshland warblers, Grasshopper, Reed and Sedge, alongside both Stonechat and Whinchat. There is also a small population of breeding Nightjars around the edge of the bog. Rarities have included Squacco Heron and Montagu's Harrier.

Timing

Visits to Ynyslas for seawatching should preferably be in the early morning, as soon after dawn as the light permits; for resident species choose high-water, at which time birds will often be close inshore, sometimes just off the beach. The tide is also the important factor on the estuary itself, when the best watching is usually from about an hour after high-water and before the tide recedes sufficiently for birds to disperse to feeding grounds at greater distances. Spring tides push birds further up the estuary, and this is when the pools, and the river at Ynys-hir (Site 52), attract most species. For the woodland and the hill, timing is immaterial, though an evening walk as the shadows lengthen over the distant estuary is very pleasant.

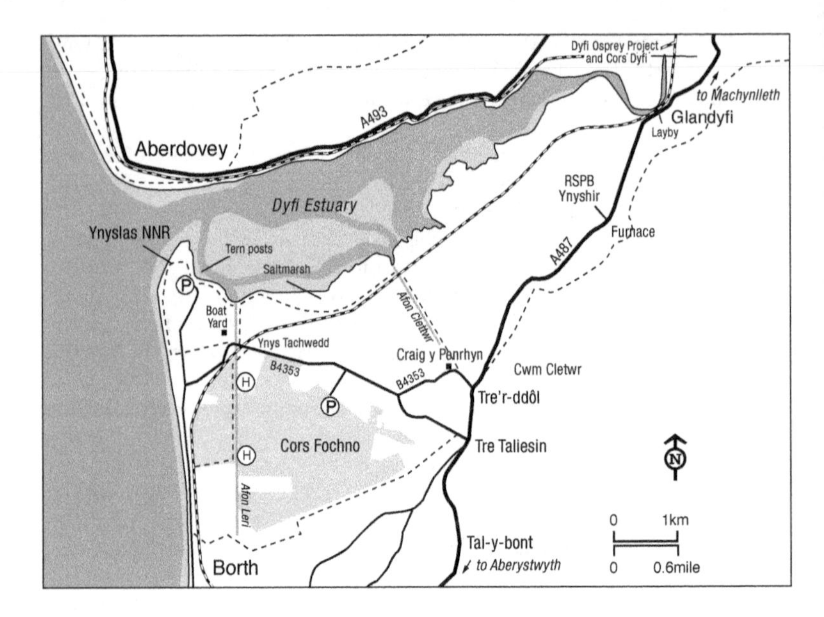

Access

The estuary itself is a mixture of saltmarsh, mud and sandflats. Birdwatchers need to be aware that birds move up and down the estuary on a daily basis, so to see the majority of what is present, visitors need to access as many sites as possible.

Ynyslas: Take the B4353 where it leaves the A487 Machynlleth to Aberystwyth road, at Tre'r-ddôl if coming from the north, or near Bow Street if coming from the south. At Ynyslas, leave the B4353 and continue straight on to the beach car park SN610941: this gives access to the dune system and the lower estuary.

Ynys Tachwedd/Leri Boat Yard: parking is limited by the side of the road, where the B4353 Borth to Tre'r-ddôl road crosses the Afon Leri by the Leri Boat yard. Walk down the track to the side of the boat yard out to the estuary.

Clettwr: The Clettwr is a 1 km ditch from Craig-y-penrhyn, SN654927, that flows into the estuary by the railway line at SN645943. Park by the ford (beware of high tides as this floods) and walk along the path to the estuary. Traditionally a good spot for wintering Water Pipits.

Cors Fochno: Small car park with access to boardwalks into the bog at SN633921. The bog can also be viewed from the coastal Borth to Ynyslas road. Better views can be obtained by parking on the side of the road, by the golf club and crossing the golf course at SN608911 to the Afon Leri and then walking along the embankment. There is also access at the other end of the Leri, with an area of small pools which often attracts Green Sandpipers. Limited parking on the roadside where the railway line crosses the B4353 at SN618931, with a path that goes on the land side of the railway and Leri.

Glandyfi Lay-by (SN691967): Along the A487 towards Machynlleth, road improvements have put in a lay-by on the eastbound side. This gives an excellent view up the upper Dyfi, from the edge of Ynys-hir reserve towards the Dyfi Osprey Project and Cors Dyfi (Site 98).

CALENDAR

Resident: Cormorant, Shag, Grey Heron, Little Egret, Shelduck, Mallard, Eider, Common Scoter, Red-breasted Merganser, Red Kite, Buzzard, Kestrel, Oystercatcher, Ringed Plover, Redshank, large gulls, Kingfisher, Skylark, Meadow Pipit, Raven, Linnet, Reed Bunting.
December–February: Red-throated Diver, Great Crested Grebe, Whooper Swans, Greenland White-fronted Goose, Wigeon, Teal, Tufted Duck, Goldeneye, Water Rail, Lapwing, Grey and Golden Plovers, Dunlin, Curlew, Black-tailed and Bar-tailed Godwits, Turnstone, Water Pipit, Fieldfare, Redwing, Snow Bunting.
March–May: Winter visitors depart, the last divers and grebes may be in summer plumage when they leave. First summer migrants arrive in late March, Sandwich Tern offshore, Wheatear on the dunes. Passage continues throughout period, with Whimbrel and Common Sandpiper among the waders.
June–July: All residents and summer visitors present. Manx Shearwater and Gannet begin to appear offshore and wader numbers begin to increase during mid-July.
August–November: Wader and tern passage well under way during August and September. Winter ducks begin to appear. Large movements of Skylarks, Meadow Pipits, thrushes and finches from mid-October, when first Greenland White-fronteds appear together with occasional Hen Harrier and Merlin.

52 RSPB YNYS-HIR

OS Landranger Map 135
OS Explorer Map OL23
Grid Ref: SN679964
Website: https://www.rspb.org.uk/reserves-and-events/reserves-a-z/ynys-hir/

Habitat

Ynys-hir is one of the Royal Society for the Protection of Birds' finest reserves in Wales and has played host to BBC *Springwatch* in some years. William 'Bill' Condry, author of 14 books who for 41 years contributed the 'Country Diary' column to the *Guardian*, once described Ceredigion as 'beginning in the north in sand, pebbles and dunes and an estuary ... which looks to a beautiful circle of the hills of mid-Wales.' This very circle of hills provides a magnificent backdrop to Ynys-hir, meaning 'Long Island', purchased by the RSPB in 1969 with Bill Condry as the first warden.

The Dyfi Estuary, on the last section of its narrow tortuous course, suddenly widens out at the western extremity of the nature reserve, with Ynyslas (Site 51) four miles to the west. The estuary at Ynys-hir is flanked by saltmarsh, reedbeds and remnant peat bogs, while on the higher ground is a rich mosaic of lowland grassland and oak woodlands where the spring flowers of Bluebells and Wood Anemones carpet the forest floor. With such a variety of habitats, there is always something to see.

Species

Some 70 species of bird breed on the reserve, including Redstart, Pied Flycatcher, Wood Warbler, Lesser Spotted Woodpecker, Nuthatch and Treecreeper. So, anyone wanting to see all classic Welsh woodland species in one go is well advised to visit this site in late April and May, although the best time to see Lesser Spotted Woodpecker is in February when they are most vocal and visible.

The Saltmarsh Trail, 1.5 miles (2.4 km), takes you through open woodland and meadows, with plenty of chances to see Redstarts and butterflies in the spring and summer, down to Marian Mawr hide giving stunning views over the estuary. This is the best spot to observe wildfowl on the estuary, where up to 2,000 Canada Geese can be present, being joined in winter by 500 Barnacle Geese. Shelduck nest, with numbers on the estuary reaching a peak in December and January, while Shoveler, Wigeon and Teal are winter visitors, and both Goosanders and Red-breasted Mergansers are present throughout the year. Little Egrets have become established in the heronry at the Domen Las Hide (closed during the breeding season) and are always to be seen along the estuary. Ospreys are regularly observed on their fishing journeys to and from the nearby Cors Dyfi nature reserve of the Montgomeryshire Wildlife Trust (see Site 98). The trail then drops down to the Saltings Hide, best visited at high tide when waders are more obvious, before making its way back to the reserve centre.

The Wetland trail, 2.5 miles (4 km), takes visitors out via a boardwalk, through reedbeds onto the saltmarsh and the Ynys Feurig Hide, again best used at high tide, and then on the Ynys Eidiol viewing screen, overlooking wet meadows and a scrape. In spring and summer this is ideal for observing breeding Lapwing and Redshank and passage waders, not to mention the many dragonflies. Water Rail, Reed Buntings, and in the spring and summer Grasshopper, Reed and Sedge Warblers may be seen or more often heard here.

The Woodland Trail, 1.5 (2.4 km) miles, takes visitors into the heart of the oak woodland, through a carpet of Bluebells in May, an area alive with summer visitors such as Pied Flycatcher, Redstart, Wood, Willow and Garden Warblers, Blackcap and Chiffchaff. It is also the best spot in the county to see Lesser Spotted Woodpeckers.

Scarce visitors include Great White Egrets, Spoonbill, Smew, Marsh Harrier, Spotted Crake, Avocet, Temminck's Stint, Pectoral, Buff-breasted and Spotted Sandpipers, Lesser Yellowlegs, Caspian Tern, Rose-ringed Parakeet and Serin, the first two records of the latter for Ceredigion being here, though pride of place must surely go to the Grey-tailed Tatler, a visitor from north-east Siberia which took up residence for over a month in the autumn of 1981, the first record for the Western Palaearctic; there has only been one subsequent British record, in Moray in 1994.

Timing

For the woodland and the hill, timing is immaterial, though an evening walk as the shadows lengthen over the distant estuary is very pleasant. The tide, however, does play its part, with high tide pushing waders up onto the saltmarsh and making them more viewable.

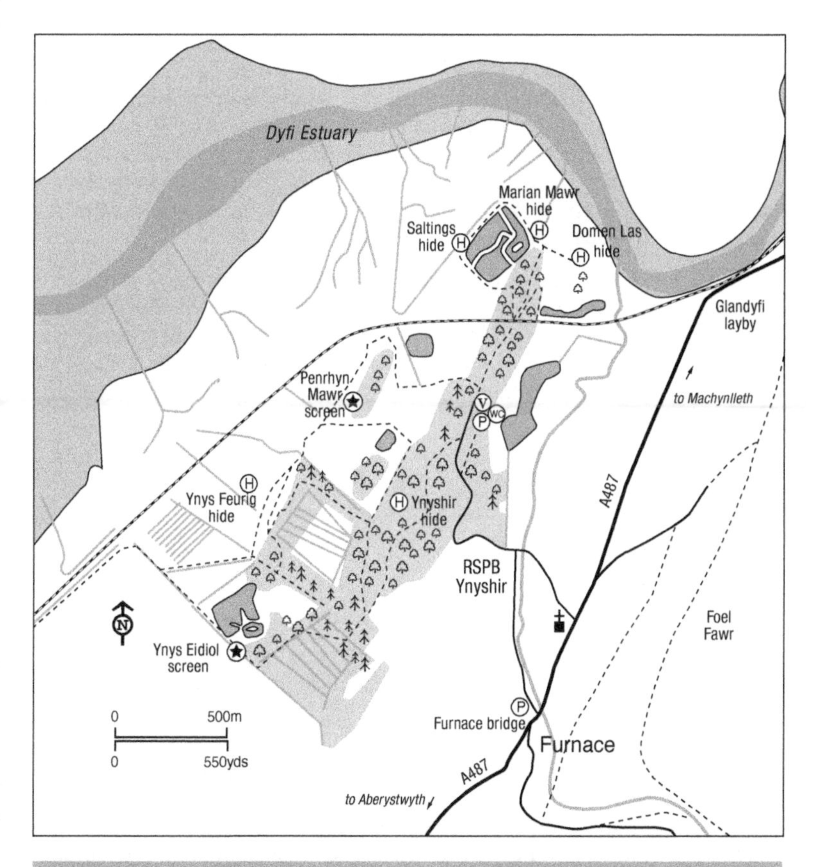

CALENDAR

Resident: Canada Goose, Barnacle Goose, Little Grebe, Cormorant, Shag, Grey Heron, Little Egret, Shelduck, Mallard, Red-breasted Merganser, Red Kite, Buzzard, Kestrel, Pheasant, Moorhen, Coot, Oystercatcher, Ringed Plover, Redshank, large gulls, Stock Dove, Barn and Tawny Owls, Kingfisher, woodpeckers, Skylark, Meadow Pipit, tit family, Raven, Chaffinch, Greenfinch, Goldfinch, Linnet, Bullfinch, Reed Bunting.

December–February: Great Crested Grebe, Whooper Swans, Greenland White-fronted Goose, Wigeon, Teal, Tufted Duck, Goldeneye, Water Rail, Lapwing, Grey and Golden Plovers, Dunlin, Curlew, Black-tailed and Bar-tailed Godwits, Turnstone, Water Pipit, Fieldfare, Redwing, Brambling, Siskin, Snow Bunting.

March–May: Winter visitors depart, the last divers and grebes may be in summer plumage when they leave. First summer migrants arrive in late March, Chiffchaff in the woods, quickly followed by Sand Martin and Swallow. Passage continues throughout period, with Whimbrel and Common Sandpiper among the waders, while Swift and Spotted Flycatcher are the last migrants to appear. Ospreys return around mid-April and are present until early September.

June–July: All residents and summer visitors present.

August–November: Wader and tern passage well underway during August and September, summer visitors largely slip away unnoticed from the woods though a late Chiffchaff may draw attention with a few final bursts of song. Winter ducks begin to appear. Large movements of Skylarks, Meadow Pipits, thrushes and finches from mid-October, when first Greenland White-fronteds appear together with occasional Hen Harrier and Merlin.

Access

RSPB Ynys-hir: Turn off the A487 at SN685952, close to Furnace bridge, and follow the signs to the reception centre. The reserve is open daily, March to October from 9am to 5pm, and November to February from 10am to 4pm; admission charge for non-RSPB members. There are seven birdwatching hides at Ynys-hir including those on the estuary, at Marion Mawr, the saltings and the heronry, and one in the woodland at Penrhyn Mawr. One mile (1.6 km) of the nature trail is wheelchair accessible.

Glandyfi lay-by: For those not venturing into the reserve, views of the estuary can be obtained from the small lay-by at Glandyfi along the A487 towards Machynlleth, road improvements have put in a lay-by on the East bound side only at SN691967. This gives an excellent view up the upper Dyfi, from the edge of Ynys-hir reserve towards the Dyfi Osprey Project and Cors Dyfi (Site 98).

53 RHEIDOL VALLEY

OS Landranger Map 135
OS Explorer Map 213
Grid Ref: SN697796
Website: http://www.welshwildlife.org/coed-simdde-lwyd-rheidol-valley-ceredigion/

Habitat

The Rheidol river, just 19 miles long – 'short but tempestuous' is how one author describes it – meanders through quite a wide valley, from the hills down to Aberystwyth. Its source is on the western slopes of Plynlimon, at 2,467 feet (752 metres) the highest peak in the Cambrian Mountains, from which both the Rivers Severn and Wye also flow. The main feature of nature conservation interest here are the large area of Sessile Oak woodlands; these still cover many of the steep hillsides, despite clearances for agriculture, overgrazing and the planting of conifers.

Halfway down the valley is a small reservoir, part of Powergen's hydroelectric scheme. Water is brought from the Nant-y-moch and Dinas reservoirs in the hills to the north, and travels by pressure tunnel to drive the generators of the Cwm Rheidol power station before being released into the river. Mining for lead and zinc took place from at least the late 16th century, finally ceasing in 1933. A special reminder of those far-off days being one of the 'Great Little Trains of Wales', which runs the 10 miles between Aberystwyth and Devil's Bridge climbing some 650 feet (198 m) and affording spectacular views of the valley.

Species

Small numbers of diving ducks use the Cwm Rheidol Reservoir in winter, while Common Sandpiper, Kingfisher, Sand Martin, Grey Wagtail and Dipper can be seen along the river in summer. Look out also for Goosander, which have come to breed on the Rheidol. The woodlands contain all the traditional Welsh woodland species, including summer visitors such as Tree Pipit, Redstart,

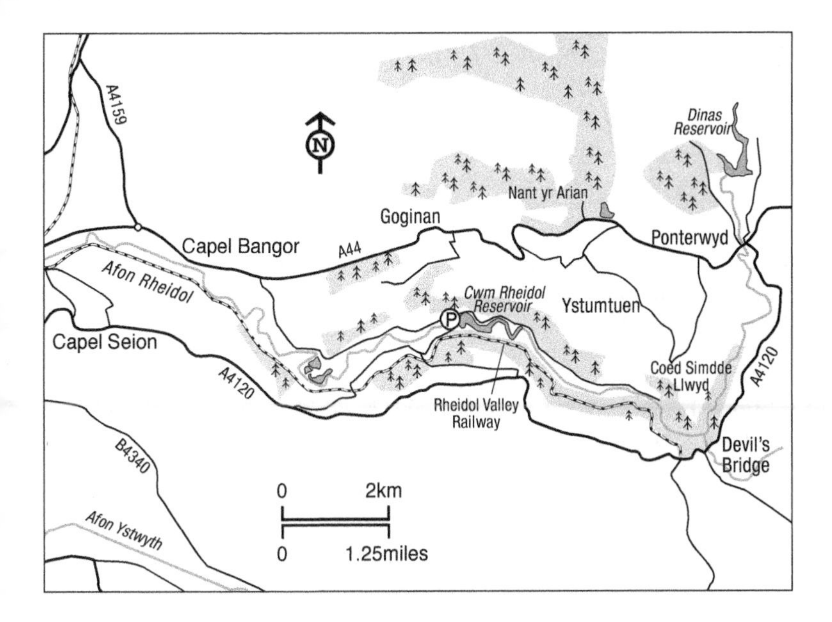

Wood Warbler and Pied Flycatcher. Ravens are usually to be seen, while Sparrowhawk, Red Kite and Buzzards are the most likely birds of prey.

Timing

Any time of day.

Access

A minor road, extremely narrow in places from the A44 in Capel Bangor traverses the length of the valley. On the high ground to the south, the A4120 has several vantage points for viewing the valley and looking down onto birds of prey and Ravens. The Coed Simdde Lwyd nature reserve of the Wildlife Trust for South & West Wales, the name meaning 'Wood of the Grey Chimney', provides access into a section of the oak woodlands, but be prepared for steep climbs. When opening the reserve in 1985, David Bellamy hugged a moss bank in exhilaration at experiencing such a brilliant woodland habit.

CALENDAR

Resident: Grey Heron, Goosander, Red Kite, Sparrowhawk, Goshawk, Buzzard, Kestrel, Coot, Great Spotted Woodpecker, Grey and Pied Wagtails, Dipper, Wren, Dunnock, Robin, Blackbird, Song Thrush, Mistle Thrush, Goldcrest, tit family, Nuthatch, Treecreeper, Jay, Raven.

December–February: Woodcock, Fieldfare, Redwing, Brambling.

March–May: Common Sandpiper, Cuckoo, Sand Martin, Tree Pipit, Redstart, Whitethroat, Garden Warbler, Blackcap, Wood Warbler, Chiffchaff, Willow Warbler, Pied Flycatcher.

June–July: Kingfisher likely on river. All summer migrants and residents still present but become more difficult to locate as breeding season draws to a close.

August–November: Woodcock, Fieldfare, Redwing and Brambling during late October.

54 BWLCH NANT-YR-ARIAN

OS Landranger 135
OS Explorer 213
Grid Reference SN717813
Website: bnya@naturalresourceswales.gov.uk

Habitat

Bwlch Nant-yr-Arian – the Pass of the Silver Spring – at 1,000 feet (305 m); eastwards a magnificent sweep of upland scenery, the highest peak being Plynlimon at 2,467 feet (752 m), the very heart of the Cambrian Mountains; to the west, some 10 miles beyond the Melindwr Valley, the open waters of Cardigan Bay. Much of the ridge extending north for nearly 3 miles, part of the National Forest for Wales, is clothed with coniferous plantations, the result of the massive planting of Welsh uplands after the Second World War. Two small lakes close to the visitor centre attract winter wildfowl, while the steep escarpment fringing the narrow valley of the Nant-yr-arian river far below is ideal for watching soaring birds of prey, especially if the wind is from the west creating the updraught that Buzzards and Red Kites so enjoy, and which they frequently have to share with boisterous Ravens.

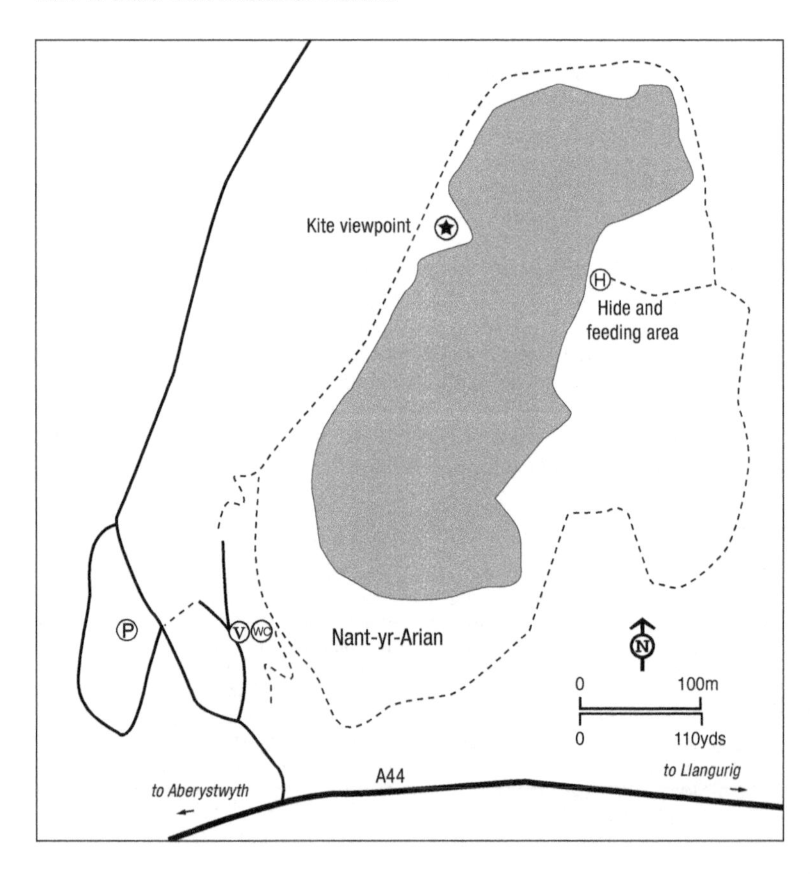

Species

Red Kites are the star attraction with a feeding station beside the lake, just a short distance from the visitor centre, a must on any visit, with sometimes up to 150 individuals coming in to feed each afternoon – and to think that half a century ago barely 16 pairs nested in the whole of Ceredigion. Buzzards and Sparrowhawks are present, the former soaring high over the plantations, the latter perhaps no more than a brief sighting as one flashes along the edge of a plantation on a hunting foray. The plantations usually have Crossbills, listen for their deep purring notes and do not be too surprised to find a small party at the lakeside, Crossbills being partial to fresh water. Great Spotted Woodpeckers, Goldcrests and Siskins are resident, while summer visitors include Redstarts and Pied Flycatchers, and the open uplands attract Stonechats and Tree Pipits. Tufted Ducks and Little Grebes on the lakes are often joined in winter by Teal, Goldeneye and Goosander.

Timing

All year, with Red Kite feeding daily at 2pm.

Access

Open at all times with a choice of waymarked trails for walkers, mountain bikers and runners.

The Bwlch Nant-yr-Arian Visitor Centre, Ponterwyd, Aberystwyth SY23 3AB (Tel: 0300 065 5470) is open throughout the year save for Christmas, Boxing Day and New Year's Day.

CALENDAR

Resident: Mallard, Tufted Duck, Red Kite, Buzzard, Goshawk, Wood Pigeon, Tawny Owl, Meadow Pipit, Grey Wagtail, Stonechat, Goldcrest, Wren, Raven, Lesser Redpoll, Siskin, Crossbill, Chaffinch.
December–February: Goldeneye, Goosander, Little Grebe.
March–May: Common Sandpiper, Cuckoo, Tree Pipit, Chiffchaff, Willow Warbler, Blackcap, Redstart, Wheatear and Whinchat from mid-April
June–July: Common Sandpiper and other summer visitors.
August–December: Summer visitors depart in September, Redwings and Fieldfares from October onwards.

55 CORS CARON

OS Landranger Map 135 and 146
OS Explorer Map 199 and 187
Grid Ref: SN687636
Website: https://naturalresources.wales/days-out/places-to-visit/mid-wales/cors-caron-national-nature-reserve/?lang=en

Habitat

Travel back in time from the last ice age and look north from the outskirts of present-day Tregaron. A huge but shallow lake, the result of a glacial moraine damming the valley, would be seen; perhaps it teemed with waterfowl, the envy of the hunter people who huddled on the shore. Gradually, the lake infilled, first with the plants of open water, then tall reeds and other emergent vegetation, and finally trees such as Willow, Alder and Birch. Later, during a wet period the trees died and Sphagnum mosses, master builders of peat bogs, began to dominate. Several thousand years elapsed and now like a huge upturned saucer, the bog's centre rises about 30 feet (9 m) above the original lake bed.

Cors Caron, one of the most important examples of a raised bog in Great Britain, is designated as a wetland of international importance under the Ramsar Convention of 1971. Most is protected as a National Nature Reserve managed by Natural Resources Wales; this extends over 2,083 acres (843 ha) and is a classic site for studies into the development sequence of raised mires. Fortunately, much west of the Teifi has not been exploited for peat. Cross-leaved Heath, Bog Asphodel, Bog Rosemary, Cranberry and Round-leaved Sundew are just a few of the special plants that occur. Large Heath Butterfly is at the southern extremity of its range, while the reserve supports a range of animals including Water Shrews, Harvest Mice, Water Voles, Polecats and Otters.

Species

Cors Caron has been a National Nature Reserve since 1955. Some 170 species have been seen here, of which 40 breed including Teal, Mallard, Water Rail, Snipe, Curlew, Redshank, Black-headed Gull, Tree Pipit, Redstart, Grasshopper Warbler and Lesser Redpoll. Regular visitors include Grey Heron – most likely birds from the heronry above Maesllyn just beyond the eastern boundary of the reserve – Little Egrets, Whooper Swan, Wigeon, Gadwall, Greenshank and Green Sandpiper.

Raptors are a particular feature of the reserve, with Red Kite and Buzzard being especially numerous. During the winter, Hen Harrier, Merlin and Short-eared Owl occur, with up to five of the former roosting on the bog. In summer months Hobby are often hunting dragonflies over the peat cuttings and probably breed locally. Both Willow and Marsh Tits are found along the disused railway line nature trail. Cuckoos are common here in spring, as are Whinchat, Whitethroat, Grasshopper and Sedge Warblers.

Occasional visitors have included Green-winged Teal, Bittern, Great White Egret, Night Heron, the first record for Ceredigion being here in June 1986, while the first Purple Heron was here for several days in May 1970, White-tailed Eagle which according to one excited report 'darkened the sky', Ring-necked Duck, Osprey, Lesser Yellowlegs, Bluethroat, Great Grey Shrike and Rustic Bunting, the latter the day before the White-tailed Eagle, all grace the reserve list, as does the Spotted Crake – one in 1970 was detected while the warden was constructing a hide, the bird calling each time a nail was hammered.

Timing

Visit in the mornings, especially when using the observation hide situated along the boardwalk at in the southern end, for then the light is behind you as you look across the bog.

Access

General views are obtainable from the A485 which runs northwest from Tregaron. The reserve car park is just south of Maesllyn lake (SN692625) and from here there is a circular boardwalk and paths to an observation hide. There

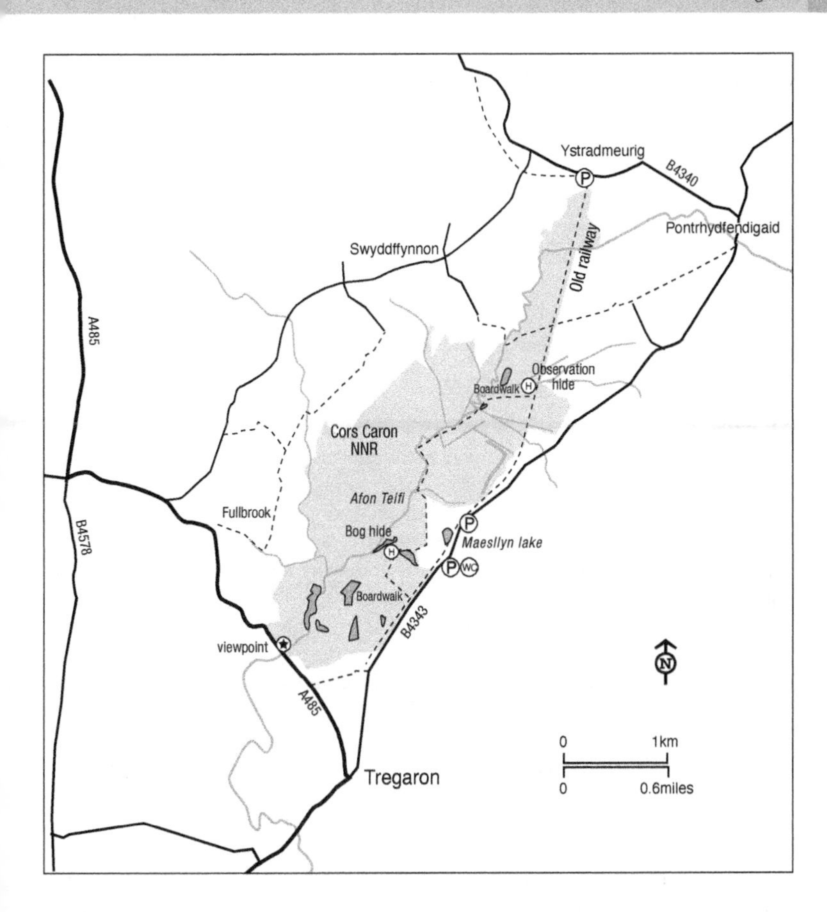

is limited access to the rest of the reserve, taking the path along the river, which extends for over 4 miles (6.4 km).

CALENDAR

Resident: Grey Heron, Teal, Mallard, Red Kite, Buzzard, Kestrel, Coot, Moorhen, Water Rail, Snipe, Skylark, Meadow Pipit, Carrion Crow, Raven, Willow Tit, Lesser Redpoll, Reed Bunting.
December–February: Whooper Swan, Wigeon, Shoveler, Red Kite, Hen Harrier, Merlin, Fieldfare, Redwing, Siskin.
March–May: Winter visitors depart, but sometimes Hen Harriers and even Whooper Swans remain until late April. Breeding Lapwing, Curlew and Redshank return, while summer migrants include Hobby, Cuckoo, Tree Pipit, Redstart, Whinchat, and Grasshopper, Sedge and Willow Warblers, while Swifts, Swallows, House and Sand Martins often feed over the reserve.
June–July: All breeding species present. Passage waders from mid-July, usually Whimbrel and Green Sandpiper but Greenshank and Wood Sandpiper have been recorded.
August–November: Summer visitors depart, wildfowl numbers increase during late September and early October, and raptors begin to feed more frequently over the bog. First Whooper Swans usually appear in early November.

56 LLANRHYSTUD–LLANON COAST

OS Landranger Map 135
OS Explorer Map 199 and 213
Grid Ref: SN521685

Habitat

Situated about halfway between Aberystwyth and Aberaeron are the small villages of Llanon and Llanrhystud. This flat agricultural land, close to the sea between these two villages is unique in the county, as rocky cliffs are the norm, and regular observation has shown that this area attracts a good number of species.

Species

Worth a look at any time of the year. Although few species breed here, this area continues to surprise. In winter, the arable fields may hold small numbers of wintering pipits and finches and occasionally both Snow and Lapland Buntings, both rare in the county. Large numbers of Grey and Golden Plover winter along this section of the coast, often right under the low sea cliffs, which makes observation difficult. Small numbers of Black-headed and Mediterranean Gulls winter all along the coast and these may attract the very occasional Iceland, Little and Glaucous Gulls.

Spring passage is restricted to Whimbrel, Pied and White, occasionally Yellow, Wagtails, Meadow and Rock Pipits and Wheatears, although a few Grasshopper Warblers, Whinchats and Redstarts may also be seen. Offshore there are daily movements of Gannets and Manx Shearwaters, best viewed in the evening, from their breeding colonies in Pembrokeshire.

Numbers of Mediterranean Gulls start to increase from early July onwards as they make their way across from their continental breeding grounds, and over the last few years have reached up to 200 birds during late July – more than any other gull present. Autumn wader passage may attract unusual species and in the past Pectoral Sandpiper, Stone-curlew and American Golden Plover have been seen. Other rarities recorded here include Great White and Cattle Egrets, Bonaparte's Gull and Caspian Gull.

Timing

Any time of day, although morning is usually most productive.

Access

Llanrhystud beach: Turn opposite the petrol station, towards the sea and follow the narrow lane down to the coast. Often there is a small floodwater pool here which is good in passage periods for waders, pipits, wagtails and Wheatears. From here follow the coastal path south towards Llanon.

Llanon: Turn off the main coast road A487 at Llanon through its narrow roads towards the church at **Llansantffraid** (SN512674), here there is a small car park, or carry on down the lane and park where you can. Then follow the coastal path north towards Llanrhystud, approximately 2 miles (3 km).

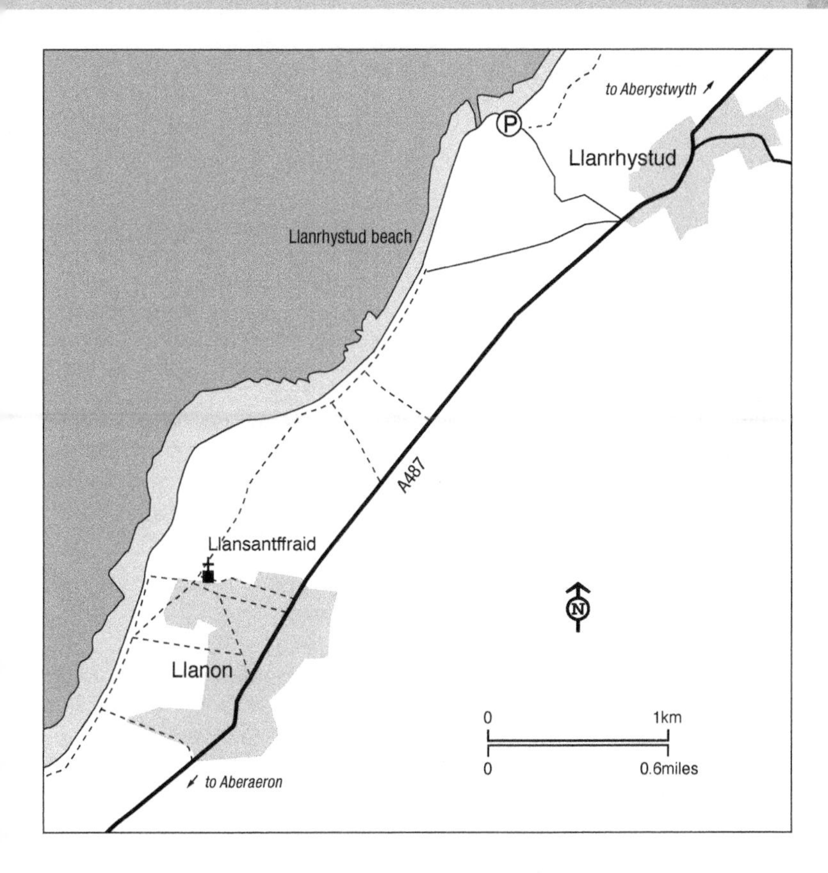

CALENDAR

Resident: Rock Pipit, Stonechat, Choughs which breed along the cliffs to the north are always likely on feeding forays.

December–February: Wintering waders include both Grey and Golden Plovers, Lapwing, Curlew and Turnstone. Small groups of finches including Linnet and Goldfinch among the arable fields.

March–May: Wheatears usually the first migrant to arrive, while from late April into May those of the Greenland race occur on passage. Sand Martins and Swallows on passage from late March, followed by House Martins, White and Yellow Wagtails and Whinchat.

June–July: Passages of Manx Shearwaters and Gannets offshore with large numbers gathering if fish are shoaling. Mediterranean Gulls begin to arrive.

August–November: Waders on passage when Ruff, Green and Wood Sandpipers are possible. Large movements south of hirundines in September, followed from early October by finches, thrushes and Starlings.

57 SOUTH CEREDIGION COAST

OS Landranger Map 145
OS Explorer Map 198

Habitat

The 20 miles (32 km) or so of coastline between Aberaeron and the mouth of the Teifi estuary contains a number of sites of interest to birdwatchers. Aberaeron has a picturesque harbour surrounded by colourful terraced houses. The harbour itself doesn't attract many species due to the levels of disturbance, but the rocky shore is favourable to waders and gulls. There are various bays and coves, some hidden from view and rarely visited, others the haunt of holiday-makers. The peninsula of Ynys Lochtyn, just north of Llangrannog, and the aptly named Bird Rock west of New Quay are the most interesting places on the seaward coast for birdwatchers. One of only two large groups of Bottlenose Dolphins in Great Britain frequents this part of Cardigan Bay, some animals often occurring close inshore, thus adding a further dimension to one's experience. Because of their importance, a significant section of the bay has been notified as a Special Area of Conservation (SAC).

The first nature reserve of the then Pembrokeshire Bird Protection Society, later the West Wales Field Society and now the Wildlife Trust of South & West Wales, was Cardigan Island, leased in 1944 and purchased in 1963. In the first operation of its kind in Great Britain, the Brown Rats, which had come ashore from a shipwreck in 1934, were eradicated. Unfortunately, despite much effort, including transplanting fledging Manx Shearwaters from Skomer and placing model Puffins along the cliff slopes, neither species has established a colony. With the Pembrokeshire colonies of Manx Shearwaters continuing to increase, is it too much to hope Cardigan Island will be colonised?

Species

Seabirds nest on this section of coast but not in the numbers that can be found in Pembrokeshire. Cardigan Island has around 500 pairs of both Lesser Black-backed and Herring Gulls, a handful of Great Black-backed Gulls and small numbers of Fulmar, Shag, Guillemot and Razorbill. Quite a large colony of Canada and up to 30 pairs of Barnacle Geese also breed on the island, the latter an offshoot from the breeding flock on the Dyfi, which itself originated from a feral population in the Lake District.

Further along the coast, Guillemots and Razorbills can be seen throughout the year and breed at a scattering of sites, mostly in small, generally inaccessible colonies of fewer than 30 pairs, with larger numbers at New Quay Head and Ynys Lochtyn. The New Quay Head colony is hard to watch from land, with better views coming from taking one of the many boat trips available from the harbour. Fulmar can be found in small numbers all along the coast, while the only Kittiwake colony in Ceredigion is at New Quay Head. Chough, Peregrine and Stonechat occur all along the coast, particularly where there are larger cliffs, such as at New Quay, Lochtyn and Mwnt.

During the winter months, there are large flocks of gulls all along the coast, mainly of Herring, Lesser Black-backed and Black-headed but they may attract Mediterranean, Glaucous and Iceland. The various bays usually hold a handful of Red-throated Divers, Great Crested Grebes and Common Scoter.

Occasionally, Great Northern and Black-throated Divers, Slavonian Grebe may be present. Wintering Black Redstarts can be found all along the coast with luck, particularly at Mwnt, Aberporth, Tresaith and New Quay.

Thousands of Manx Shearwaters occur offshore on feeding movements from the Pembrokeshire colonies, while autumn watches from New Quay Head, Mwnt and Gwbert have revealed passage of Balearic Shearwater, Leach's Petrel, Common Scoter, Sabine's Gull and skuas – although not in the numbers that have been recorded off Strumble Head, Pembrokeshire.

Scarce birds along the coast have included Ring Ouzel, Richard's Pipit, Snow and Lapland Buntings. A dead Common Nighthawk was found on the tideline at Mwnt in October 1999, its skin is preserved in the National Museum of Wales, Cardiff.

Timing

Morning is best on this coast to ensure the light is behind you.

Access

The Ceredigion Coastal Path provides access to virtually the whole coast save for a section overlooking Cardigan Island.

Gwbert (SN161499): coastal path goes inland, to avoid Cardigan Island Park (a private country park, giving views across to Cardigan Island), then onto the coast towards Mwnt. There is also a short public path by the Cliff Hotel and its golf course, on which Choughs forage, giving views of Cardigan Island and frequent sightings of Bottlenose Dolphins close inshore.

Mwnt (SN194518): National Trust car park – charges apply.

Aberporth (SN259514) – **Tresaith** (SN278515): park in either village and take the coastal path to the next village and back.

Llangrannog: couple of footpaths out of the village to Ynys Lochtyn (SN314553), owned by the National Trust.

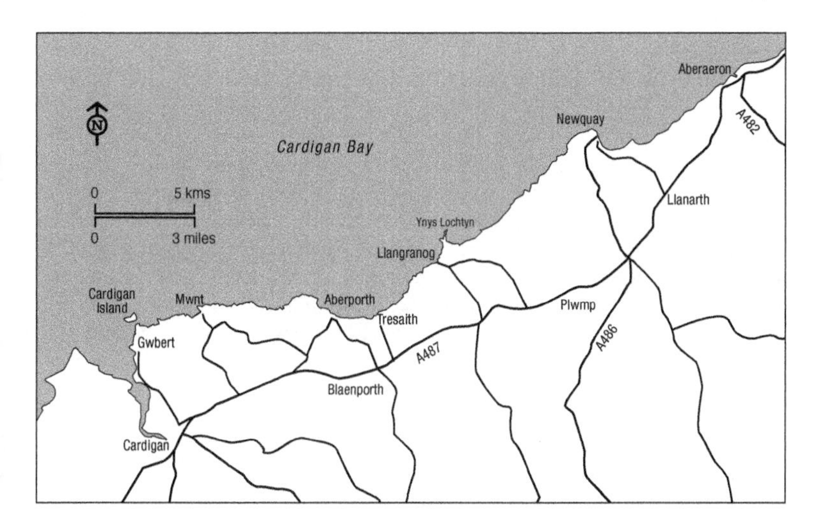

New Quay Harbour & Head (SN390600): parking is available at a couple of places in the village, charges apply.

Aberaeron (SN457628): access to coastal footpath, either side of the harbour, charges apply.

CALENDAR

Resident: Cormorant, Shag, Buzzard, Kestrel, Peregrine, Oystercatcher, Herring and Great Black-backed Gulls, Rock Pipit, Stonechat, Chough, Jackdaw, Raven.

December–February: Red-throated Diver, Fulmars return to cliffs in early January, Common Scoter, Ringed Plover, Lapwing, Dunlin, Curlew, Redshank, Black-headed and Lesser Black-backed Gulls, Kittiwakes return in February, Guillemots and Razorbills offshore throughout whole period with occasional brief visits to the cliffs.

March–May: Guillemot and Razorbill commence egg-laying about the end of April. Passage migrants from late March, Wheatear and Ring Ouzel being among the first to arrive, later arrivals include Grasshopper Warbler and Whitethroat.

June–July: Manx Shearwaters and Gannets frequent offshore, first terns may be seen on return passage, wader numbers increase in the Teifi estuary from late July.

August–November: Fulmars leave in September, and although Kittiwakes will have left in mid-August some remain offshore throughout period. Red-throated Divers may arrive as early as September, though usually not until well into October. Manx Shearwater passage dies away as September proceeds. Arctic and Great Skuas may be seen offshore, also terns during August and September. Always a chance late in the autumn of a Snow Bunting or two on the coast or a Black Redstart in the towns and villages.

58 TEIFI ESTUARY

OS Landranger Map 145
OS Explorer Map 198 and OL35
Grid ref: SN165480

Habitat

The Afon Teifi, one of the longest rivers wholly within Wales, 50 miles (80 km) in length, rises in the high wilderness above Strata Florida, flows through the great peat bog of Cors Caron, then meanders south and west to mingle with salt water at Llechryd or thereabouts before passing through the Teifi gorge to become truly estuarine. Part of the estuary is in Ceredigion, part in Pembrokeshire. The estuary may be conveniently divided into two distinct sections separated by the historic town of Cardigan. The lower estuary, St Dogmaels to Poppit on the Pembrokeshire side, and Cardigan to Patch and then Gwbert on the Ceredigion side, with exposed tidal mudflats and a small area of saltmarsh at the Webley where most of the waders, wildfowl and gulls are to be seen, with good access on both sides of the river.

Species

Small numbers of Little Grebes winter on the lower estuary and occasionally Great Crested Grebes, occur in the estuary. Grey Herons are always to be seen, they breed near Llechryd, as are Little Egrets which probably breed nearby.

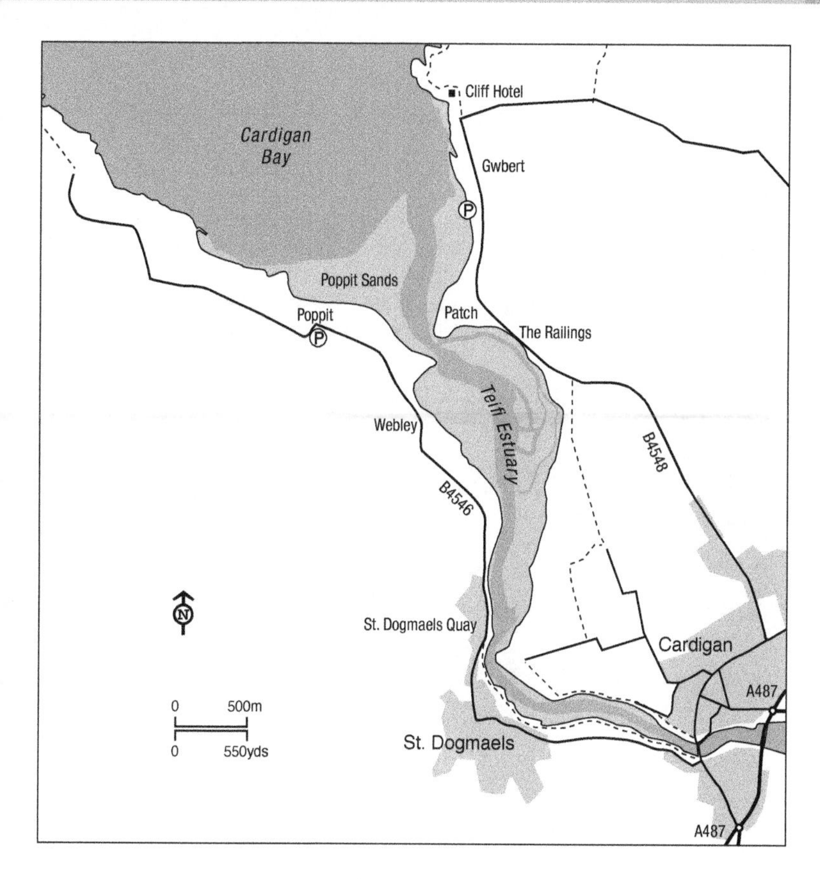

Small numbers of waders and wildfowl winter on the estuary, with Wigeon, Teal and Curlew numbering around 100 each, along with up to 40 Oystercatcher, Dunlin and Redshank respectively, and a dozen or so Ringed Plover. Occasionally a couple of Black-tailed and Bar-tailed Godwits, Greenshank and Grey Plover winter. The surrounding fields in winter host up to 2,000 Lapwing and Golden Plover, both of which sometimes make use of the estuary. The Canada Goose flock has steadily increased over the last 50 years to around 2,000, and there is a small flock of Barnacle Geese, numbering around 120, that breed on Cardigan Island that also frequent the estuary throughout the year. The Mute Swan herd at St Dogmaels can rise on occasions to 40 or so strong, while Whooper Swans are occasional visitors, usually passing through to their wintering grounds in the Tywi Valley, Carmarthenshire, in late October–November. Spring and autumn passage increase the number and variety of waders, with near-annual records of Curlew Sandpiper and Little Stint. There is always a large gull congregation in the estuary, numbering anything up to 2,000, mainly of Herring and Black-headed, with smaller numbers of Lesser and Great Black-backed. Among these Mediterranean Gulls are regular and both Iceland and Glaucous Gulls are seen most years.

Timing

As even the lower Teifi is not that wide, timing as regards low-water and waders and wildfowl being at a distance is less critical here than on some larger estuaries.

Access

Southern/western, Pembrokeshire side, of the estuary can be watched from St Dogmaels Quay (SN163467), Webley (SN158480) and Poppit (SN152484) where parking charges apply.

Northern/eastern, Ceredigion side, of the estuary can be watched from road ('The Railings' at SN166483) and Patch (SN162485), where there is limited on-road parking.

CALENDAR

Resident: Cormorant, Grey Heron, Little Egret, Mute Swan, Canada Goose, Shelduck, Mallard, Red Kite, Sparrowhawk, Buzzard, Kestrel, Peregrine, Oystercatcher, large gulls.
December–February: Little Grebe, Wigeon, Teal, Goosander, Golden Plover, Grey Plover, Ringed Plover, Dunlin, Snipe, Curlew, Redshank, Greenshank, Turnstone, Black-headed, Mediterranean and Common Gulls, Kingfisher.

March–May: Waders and winter wildfowl depart. Passage waders move through, Whimbrel, Bar-tailed and Black-tailed Godwits, Common Sandpipers, Sanderling.
June–July: First passage waders put in appearance.
August–December: Wader passage in full swing in first part of period, chance of terns on the lower estuary. Wildfowl and wader numbers increase, while Shelduck return in October from their moult migration.

DENBIGHSHIRE (DINBYCH)

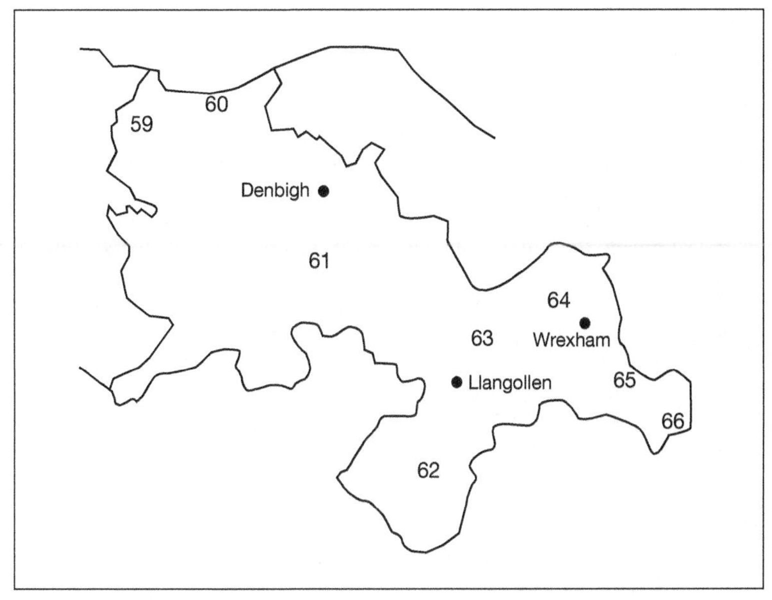

59 RSPB Conwy
60 Rhos-on-Sea to Kinmel Bay
61 Mynydd Hiraethog and Llyn Brenig
62 Berwyn Range
63 Horseshoe Pass and World's End
64 Wrexham area
65 Hanmer Mere
66 Fenn's, Whixall and Bettisfield Mosses

The county of Denbigh is upland and wild, with only a small access to the sea on its north coast. The two upland areas are the Berwyn range and Mynydd Hiraethog, both offering the observer the usual upland species, Red Grouse, Buzzard, Peregrine, Ring Ouzel and Raven. Choughs can still be found if you are lucky, as can Black Grouse. The deep wooded valleys also provide summer homes for Redstart, Wood Warbler and Pied Flycatchers, while the numerous mountain streams are ideal for Dipper and Grey Wagtail.

Off the north coast, an area bounded by Rhos-on-Sea in the west (near to the Little Orme) and Kinmel Bay, off Towyn, in the east has large numbers of wintering Common Scoter and Great Crested Grebes.

The Afon Conwy forms the western boundary between Denbigh and Caernarfonshire. At its mouth is RSPB Conwy, the reserve having turned up many notable additions to the county list and now one of the top sites in Wales, while not too far away Ospreys have nested at Llyn Brenig since 2018.

59 RSPB CONWY

OS Landranger Map 115
OS Explorer Map OL17
Grid Ref: SH800771
Website:
https://www.rspb.org.uk/reserves-and-events/reserves-a-z/conwy/

Habitat

Across the estuary from the walled town of Conwy with its famous castle, and lying almost entirely in Denbighshire, is a splendid RSPB reserve, opened in 1996, following the construction of the Conwy tunnel carrying the A55 beneath the river. Some 80 acres (32.4 ha) of saltmarsh were unfortunately destroyed during the tunnel construction; the reserve is welcome recompense for such a loss. Two large lagoons with islands, together with several smaller pools, beyond which lies the Conwy estuary, have proved once again what can be achieved on a relatively small area, the reserve being just 125 acres (50 ha) in extent. Reedbeds were created, while more open areas, especially of grassland, provide resting and feeding places for many waterfowl. Some 370 different plants, 280 moths, 28 species of butterfly and 16 dragon and damselflies have found the new habitats to their liking, while the 12 mammals recorded include Otter, Water Vole and Stoat, the latter being regularly seen.

Species

Breeding species include Mute Swans, Canada Geese, Gadwall, Great Crested and Little Grebes, Coot, Oystercatcher, Common Sandpiper, Cetti's, Sedge and Reed Warblers, Lesser Whitethroat, Whitethroat and Reed Bunting. Grey Herons and Little Egrets nest in Coed Benarth on the opposite bank of the estuary and feed regularly on the reserve; in late summer, more than 100 Little Egrets occur, as do a few Great White Egrets.

The reserve really comes into its own as both passage birds and winter visitors ebb and flow in numbers. Oystercatchers peak between April and June when up to 700 occur on the estuary, some frequenting the reserve. Curlew numbers reach almost 500 between late June and September, while the highest Lapwing counts, up to 500, and Redshank, usually up to 800 or so, occur in midwinter. All other waders occurring regularly in western Britain pass through annually on passage, including Little Stints and Curlew Sandpipers, while some such as Spotted Redshank may overwinter.

Large numbers of Canada Geese visit Conwy from across northwest Wales to moult, while Whooper Swans visit annually, usually for brief periods in early winter. Around 150 Shelduck, which breed farther up the valley, feed in the estuary in early spring, prior to nesting, and broods can be seen on the estuary through the summer. Up to a hundred Wigeon overwinter together with as many as 350 Teal and 150 Mallard. Gadwalls have become more common in recent years, with up to 80 in winter, and a few pairs breed. Pintail, Shoveler, Pochard, Tufted Duck, Goldeneye and Red-breasted Merganser are regular in small numbers, one or two Garganey occur in most years, and Black Duck and Smew are among the rarer wildfowl on the reserve list. Kingfishers are seen regularly and Firecrests have been reported.

Regular passage visitors include White Wagtail, Osprey, Mediterranean Gull,

Sandwich, Common and Arctic Terns. In the breeding season it is one of the most reliable places locally to see Water Pipit and Firecrest which are present through most winters. Choughs are recorded regularly over the reserve outside the breeding season, after all they breed not too far away. The reedbeds attract a Starling roost most winters, usually in January and February, and this has numbered up to 250,000 birds, which in turn lures raptors such as Sparrowhawk and Peregrine. Water Rails overwinter and can be easily seen close to the reserve café.

Notable species that have graced the reserve include American Wigeon, Slavonian Grebe, Bittern, Spoonbill, a Little Crake in September 2001 was the seventh for Wales, Marsh and Hen Harriers, Goshawk, Baird's, Pectoral, Broad-billed, Stilt and Terek Sandpipers, Yellow-legged and Iceland Gulls, Alpine Swift, Wryneck, Shorelark, Black-headed and Citrine Wagtails, Bluethroat, Twite, Common Rosefinch, Black-headed, Snow and Ortolan Buntings.

Timing

The reserve is open daily throughout the year from 9:30am to 5pm or dusk if earlier. If you have a choice, visit when high tide is in the morning as more wildfowl and waders are likely on the lagoons, with the sun behind you from the main viewpoints.

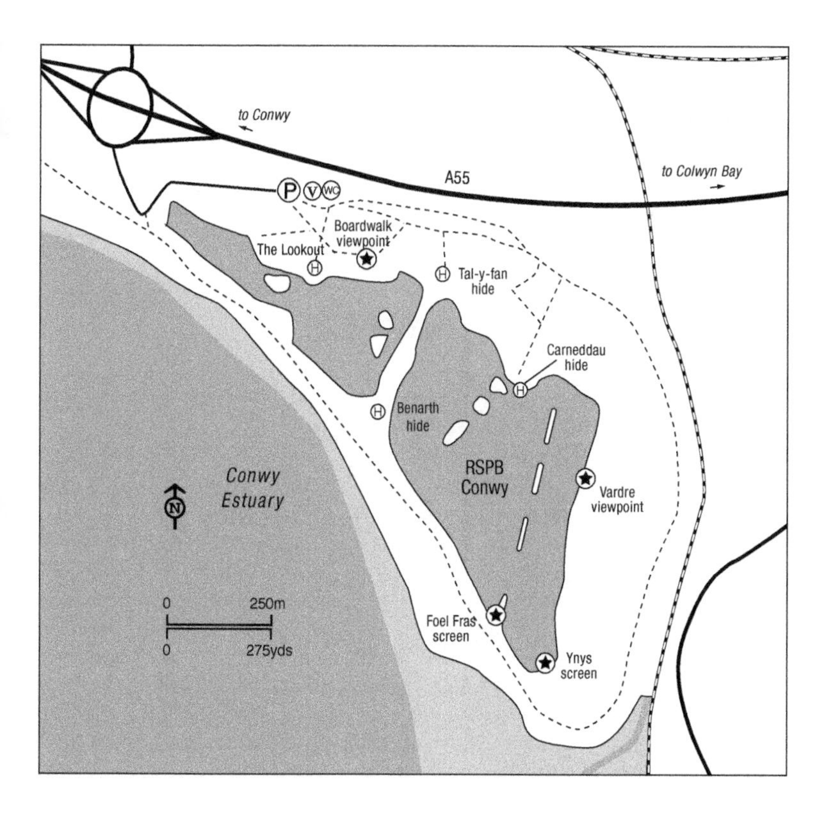

Access

Members of the RSPB and Wildlife Explorers are admitted free, non-members pay an admission fee. Leave the A55 at the exit signed Conwy and Deganwy, the reserve entrance is on the south side of the roundabout above the expressway and quickly leads to a car park. Close by are the visitor centre and café, which includes toilets, from where a series of footpaths, the shortest half-a-mile (0.8 km), the longest 2 miles (3.2 km), provide access to the reserve and its three observation hides, each being wheelchair accessible. There are regular guided walks and events, and the reserve has a strong programme of activities for schools and family visitors.

CALENDAR

Resident: Little Grebe, Cormorant, Grey Heron, Mute Swan, Canada Goose, Mallard, Moorhen, Coot, Oystercatcher, Black-headed and Herring Gulls, Reed Bunting.

December–February: Wigeon, Gadwall, Teal, Pintail, Pochard, Tufted Duck, Goldeneye, Red-breasted Merganser, Peregrine, Dunlin, Lapwing, Curlew, Snipe, Common and Great Black-backed Gulls. A striking feature of this period is a huge Starling roost, often attracting predators such as Sparrowhawks and Peregrines. Choughs overhead in early morning and mid-afternoon.

March–May: Winter visitors depart, Wheatears and Sand Martins arrive, large numbers of White Wagtails and smaller numbers of Black-tailed Godwit, Ringed Plover and Whimbrel pass through. Shelduck numbers build prior to nesting farther up the valley. Breeding waterfowl under way.

June–July: Herring Gull and Lesser Black-backed Gulls forage and loaf on the reserve. Breeding season in full swing with plenty of well-developed young songbirds and waterbirds to excite the visitor. First returning waders such as Common and Green Sandpiper put in an appearance and Curlews reach peak numbers.

August–November: Wader passage reaches its peak. September is the time when large groups of hirundines catch insects over the pools and roost in the reedbeds. Visible migration of Skylarks, Meadow Pipits and finches from late October. Winter wildfowl numbers begin to increase.

60 RHOS-ON-SEA TO KINMEL BAY

OS Landranger Map 116
OS Explorer Map OL17 and 264

Habitat

Virtually the whole 11-mile (17.6-km) coastline of Denbighshire is one long sandy shore backed by holiday resorts, as at Kimnel Bay, Towyn, Abergele and Colwyn Bay. The shingle ridge at Llanddulas is the fourth-largest such coastal feature in Wales, though alas Oystercatcher and Ringed Plover no longer nest here; however, it does offer a good vantage point for offshore watching. In the extreme west, the coast at Rhos on-Sea provides the first sheltered area from westerly and southwesterly winds after the Dee Estuary 20 miles (32 km) eastwards. The shore habitat here is hotels and houses, so quickly turn your back on these and search out to sea or along the rocky beaches of Penrhyn Bay right up to the Caernarvonshire border. If the shoreline proves unproductive then head for Pentremawr Lake in Abergele with its small number of wintering wildfowl and gulls, the flock of which occasionally attracts rarer visitors, or search the Clwyd estuary which the county shares with Flint to the east.

Species

Coastal birds such as Oystercatcher and Little Tern have long since ceased to breed. The only cliff-nesting seabirds have colonised the limestone quarries nearly half-a-mile (0.8 km) inland at Llanddulas, where 10–15 pairs of Fulmars nest. It is always a magic moment as one drives along the A55 at this point to see a Fulmar pass over the road as it returns to, or sets off, from its breeding cliff to the south of the road. Herring Gulls nest in the same area, also on rooftops at a number of localities along the coast, a habitat not high on a birdwatcher's priorities but one that needs considered attention so that we have a measure of such changes in distribution and numbers. There are few birds in summer save for ubiquitous gulls, Cormorants fishing offshore and Shelduck, which breed on the Clwyd Estuary where up to 70 or so young can be seen, while the number of adults can rise to well over 200.

From early August through to March it is well worth spending time at Rhos-on-Sea and indeed other spots eastwards along the shore. All the divers are regularly seen, though Red-throated Diver, as elsewhere, occurs in the greatest numbers. Great Crested Grebes winter offshore in small numbers, and there are occasional records of Slavonian and Black-necked Grebes. Manx Shearwaters occur, most especially in late summer when occasional Storm Petrels pass, while small numbers of Leach's Petrels occur in most years during October gales along with Great and Arctic Skuas. Gannets appear in late summer, and Fulmars and Cormorants, no doubt from the colony on the Little Orme, just over the county boundary to the west, are seen almost throughout the year.

At the opposite end of the county, the Clwyd estuary has a roost where up to 100 Cormorants may be counted, highest numbers occurring in the autumn. Common Scoters are resident offshore at Rhos-on-Sea or Pensarn, where even in June the flock may total several hundred. Numbers in winter may rise to 20,000 part of the immense flock which frequents Liverpool Bay both sides of the English Welsh border and they are occasionally joined by Velvet Scoter and other winter visitors, including Scaup, Eider, Long-tailed Duck and even Surf Scoter. Up to six males and a female have been counted from here in the past.

The Clwyd estuary, mostly in Flint, is a good spot for gull watching, and usually turns up scarce species such as Iceland and Glaucous Gulls each year. Guillemots and Razorbills are occasionally seen in winter close inshore, while one is of course not far from the colonies at the Little and Great Ormes. There have been several records of Black Guillemots, and occasionally Little Auks are observed.

Despite the limited shoreline and feeding areas, a range of waders can usually be seen at Rhos-on-Sea. Up to 1,000 Oystercatchers winter here, together with about 300 Knots, 400 Curlews, and most dramatic of all in terms of numbers, Turnstones, which find the rocky shores to the west of Rhos-on-Sea much to their liking and where up to 200 have been counted. This is one of the key wintering areas in Wales for this bird and also Purple Sandpiper whose numbers can rise to about 100 in midwinter – this is an exceptional number, but birds can so easily be overlooked. Sanderlings find the sandy beaches much to their liking and several hundred can overwinter; other waders occurring in smaller numbers include Grey Plover, Bar-tailed Godwit, Redshank and

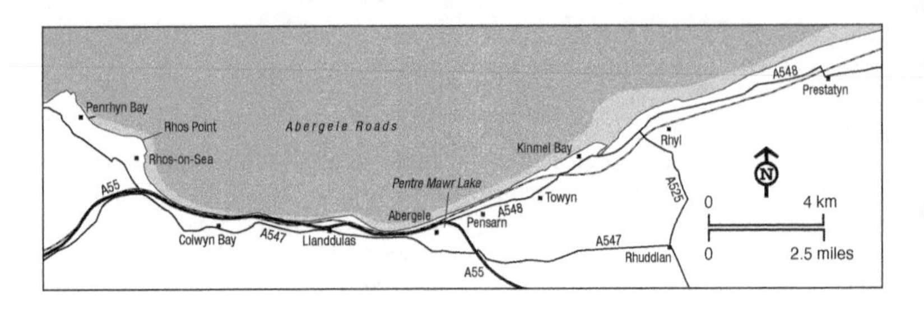

Greenshank, while Spotted Redshank and Curlew Sandpiper have occasionally been recorded.

Timing

Early morning is best, whatever the state of the tide, in order to ensure the least disturbance from other beach users.

Access

A road follows the whole of the seafront, and during the winter months, observations from this using the car are usually possible, especially at Rhos Point for seawatching and its Purple Sandpipers, which are also sometimes at the Llanddulas and Pensarn beaches.

CALENDAR

Resident: Cormorant, Shelduck, Common Scoter, Little Egret, Oystercatcher, Ringed Plover, Redshank, Turnstone, Black-headed, Lesser Black-backed, Herring and Great Black-backed Gulls, Rock Pipit, Pied Wagtail.

December–February: Red-throated Diver, Great Crested Grebe, Fulmar, Wigeon, Scaup, Eider, Long-tailed Duck, Velvet Scoter, Red-breasted Merganser, Grey Plover, Knot, Purple Sandpiper, Dunlin, Curlew, Bar-tailed Godwit, Common Gull, Guillemot, Razorbill.

March–May: Winter visitors depart, spring passage brings new arrivals including Sanderling, Whimbrel, Common Sandpiper, Kittiwake, Sandwich and Common Terns.

June–July: Manx Shearwater, Gannet and Shag appear offshore, while terns become more evident late in period. Shelduck numbers in the Clwyd estuary reach a peak in June before rapidly falling away through July, but will return in the autumn.

August–November: Seabird passage, with last terns usually in late September, when winter visitors such as divers, grebes and sea-ducks reappear as do the Purple Sandpipers with Snow Buntings possible at the Rhos-on-Sea golf course.

61 MYNYDD HIRAETHOG AND LLYN BRENIG

OS Landranger Map 116
OS Explorer Map 264
Grid Ref: SH973555
Website: http://www.llyn-brenig.co.uk/

Habitat

Travel south-west from the town of Denbigh by way of the A543 and after about 5 miles (8 km) the road climbs on to Mynydd Hiraethog, a large moorland area with several lakes – Llyn Aled, Llyn Alwen and Llyn Bran – and numerous boggy areas and Sphagnum moss flushes. Much of the land is a rolling upland of about 1,400 feet (427 m) altitude, with the highest point south of the road where Mwdwl-eithin ridge rises from 1,570 (479 m) to 1,745 feet (532 m).

To the south-east, in the headwaters of the Afon Alwen, a tributary of the Dee, are the massive Alwen and Llyn Brenig reservoirs. Both are largely surrounded by coniferous plantations, the western extremity of Clocaenog Forest, one of the largest areas of coniferous woodland in Wales. Llyn Brenig has been developed for water sports, fishing and walking, along with a large visitor centre.

Species

Ospreys have summered on Llyn Alwen and Llyn Brenig since the mid-2010s and have nested at Llyn Brenig since 2018, rearing single chicks that year and again in 2019 and 2020. Unfortunately, the nest platform was cut down in an act of vandalism in 2021, the nest and eggs destroyed, but the pair remained for the summer. Hopefully a different platform will be occupied and offspring from this pair will colonise other suitable lakes in the future. An osprey observation hide has been erected, with CCTV and telescopes in an RSPB and North Wales Wildlife Trust partnership project giving excellent views of the nest platform.

The moorland pools and reservoirs attract small numbers of wildfowl. Great Crested Grebes remained a scarce breeder until the early 1970s, when up to five pairs nested in the county, Llyn Brenig being one of their regular sites, as are smaller waters to the west such as Llyn Alwen and Llyn Aled. Little Grebes also nest on several of these upland waters. The Grey Heron population in Denbighshire is small, probably fewer than 30 pairs, but some come regularly to the reservoirs, and to the hilltop bogs during the summer months. Common Sandpipers nest on suitable waterside stretches such as the banks of the Alwen reservoirs and the rivers. Teal have nested in the past, but this is a notoriously difficult species to prove breeding and without a doubt some are overlooked; the broods hide away rather than coming to open water, so that injury-feigning and distraction displays by the adults are usually the first and only sign that they have nested. Mallard breed widely, and both this and the previous species, together with Wigeon, Pochard, Tufted Duck, Goldeneye and Goosander, frequent Llyn Brenig during the winter. Small numbers of Whooper Swans also come here and to some of the larger upland pools.

The decline of moorland nesting waders has been well publicised throughout Britain. By 2019 there were only around 25–35 pairs of Golden Plover, four or five pairs of Dunlin, a dozen Snipe and fewer than 30 Curlew breeding on the moorland of Mynydd Hiraethog. Breeding raptors include Hen Harrier, Peregrine and Merlin with the occasional pair of Short-eared Owls. Meadow Pipits are plentiful, and where there is upland grassland, Skylarks and Wheatears. A few pairs of Ring Ouzels are usually present, while other species of the moorland edge include Tree Pipit, Whinchat, Mistle Thrush and Reed Bunting. There are Red Grouse in the heather. Black Grouse, however, seem to prefer areas where conifers front on to the moorland, the population on Mynydd Hiraethog being small.

The developing plantations support a changing bird population. When the trees are small, look for Short-eared Owl, Whinchat and Grasshopper Warbler. Later, the main species in the mature plantations will be Woodpigeon, Goldcrest, Coal Tit and Chaffinch, Buzzard, Tawny Owl, Sparrowhawk and possibly Goshawk. Look out also for Siskins and Crossbills, both of which nest here. It was near Llanfihangel Glyn Myfrfyr at the southern edge of the forest in March 1991 that a male Two-barred Crossbill was located, only the second for Wales, the first being found dead at Llandrindod Wells in 1912. Hawfinch is a scarce bird in Wales, though undoubtedly overlooked, but has been described as widespread but local in Denbighshire where one of the places to search is the quiet valley of the Alwen, below Pentre-llyn-cymmer (SH979525).

Timing

Visit in early morning, indeed dawn if you wish to watch and listen to Golden Plover and Dunlin on the hill. An early morning visit will also be necessary if you intend to search for and hopefully see a Goshawk. The time of one's visit is otherwise not critical.

Access

The A453 traverses the area, with an unclassified road going north from this at SH929567, near Pont y Clogwyn to the high point of Foel Lwyd, passing as it does so Llyn Aled and Aled Isaf Reservoir, from which the Afon Aled flows north eventually to reach the sea at Rhyl.

For the forests, Alwen Reservoir and Llyn Brenig, turn south at SH959592 on the B4501 for Cerrigydrudion. You are immediately in the forest, and there are numerous picnic sites and parking places. There is a nature trail at SH962572, a visitor centre at SH967547 on the shore of Llyn Brenig, and another at

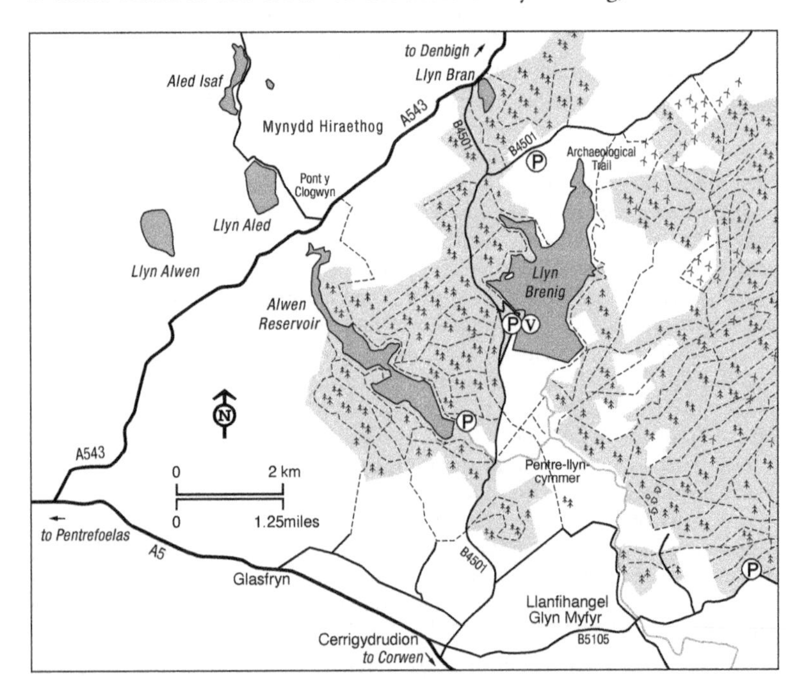

SH951537 on the shore of Alwen Reservoir. Do not neglect to visit the Archaeological Trail at SH984573, on the shore of Llyn Brenig, where there is also a hide. The valley north of Llanfihangel Glyn Myfyr is worth exploring for its woodland birds which can include Hawfinch.

CALENDAR

Resident: Great Crested Grebe, Cormorant a regular visitor to the larger waters even in winter, Mute Swan, Teal, Mallard, Hen Harrier, Goshawk, Sparrowhawk, Goshawk, Buzzard, Kestrel, Merlin, Red and Black Grouse, Pheasant, Moorhen, Coot, Black-headed Gull, Woodpigeon, Tawny and Short-eared Owls, Great Spotted Woodpecker, Skylark, Meadow Pipit, Grey and Pied Wagtails, Dipper, Dunnock, Robin, Blackbird, Song and Mistle Thrushes, Goldcrest, Coal Tit, Carrion Crow, Raven, Chaffinch, Goldfinch, Siskin, Linnet, Redpoll, Reed Bunting.
December–February: Whooper Swan, Wigeon, Pochard, Tufted Duck, Goldeneye, Peregrine, Jack Snipe, Woodcock, Common Gull, Fieldfare, Redwing, Snow Bunting.
March–May: Winter visitors depart, Golden Plover, Lapwing, Dunlin, Snipe and Curlew return to the moorland, soon followed by Wheatear and Ring Ouzel, while on the lower ground Cuckoo, Tree Pipit, Whinchat and Redstart take up their territories. Ospreys return at the end of March–early April.
June–July: All breeding species present, sometimes large feeding movements of Swifts over the hills in hot weather while Sand Martins are over the reservoir. Raven and Mistle Thrush, having completed breeding, in family parties or even small flocks.
August–November: Summer visitors largely slip away unnoticed, the Ospreys in early September, occasional passage waders such as Green and Common Sandpipers. Winter visitors from early October, though by then the hills seem almost deserted by birds.

62 BERWYN RANGE

OS Landranger Map 125
OS Explorer Map 255
Grid Ref: SJ076335

Habitat

The Berwyn Range in the south of the county runs southwest as a more or less rectangular block of high ground, bounded to the north and west by the valleys of the Afon Dyfrdwy, which includes Llyn Tegid (Bala Lake, Site 97), to the south by Lake Vyrnwy (Site 100), while to the east the ground falls away more gently, intersected by river valleys such as those of the Ceiriog and Tanat. Nearly half of the range is in Montgomeryshire. The Berwyns contain one of the largest areas of heather-dominated heath and blanket mire remaining in Wales, though even here much has been lost to both agriculture and forestry. The ground is higher than at Mynydd Hiraethog to the north, on the summit ridge reaching peaks of 2,590 feet (798 m) at Cadair Bronwen and 2,730 feet (832 m) at Moel Sych, where there is a line of west-facing cliffs.

It is on these summits, and nowhere else in Wales, that Cloudberry is located – this its most southerly outpost in Great Britain, the nearest site being in the Pennines. Other upland plants include the graceful white-plumed, wind-blown Cotton-grass, Bog Asphodel, Marsh St John's-wort, Round-leaved Sundew and if you are prepared to search, Marsh Clubmoss. With luck you may also discover the tiny Bog Orchid, though it will be purely by chance, for this plant rarely exceeds 2 inches (5 cm) in height, has rather insignificant yellow-green flowers, to suit its retiring nature, and flowers in late summer.

Species

Red Grouse are still found where the heather moorland remains intact, for without Heather *Calluna vulgaris*, their chief food source, these birds quickly move on. A visitor to the Berwyns in 1797 noted that 'kites, moor-buzzards and other birds here make their nests in security; and the long heath shelters the grouse, a race that would have been extinct here but for the wide range of these wild mountains'. Undoubtedly, the deterioration of heather moorland throughout much of central Wales has reduced Red Grouse numbers, though the Berwyn range remains a stronghold for this species. H.E. Forest writing in 1907 said that 'there are few finer grouse moors in the kingdom than those which stretch across the broad back of the Berwyn range'.

Look out also for Black Grouse, there are few finer upland birds. Conservation measures since the late 1990s have seen a welcome increase at least in its northern strongholds. Other scarce breeding birds of these uplands include Hen Harrier, Merlin, for which the moors have always been noted, Peregrine, Golden Plover, Dunlin and the occasional pair of Short-eared Owls. Tree Pipit, Whinchat, Wheatear, Raven and Ring Ouzel all occur on the higher ground, the latter more likely found where there are quarries and ravines. Among waders, both Dotterel and Whimbrel may occur on spring passage.

Do not neglect the woodland and valley areas as you make your way towards the high ground. Common Sandpiper, Grey Wagtail and Dipper on the rivers. Sparrowhawk and all three woodpeckers, Redstart, Wood Warbler, Goldcrest, Pied Flycatcher, the tit family, Nuthatch, Treecreeper, Siskin, Redpoll and, on occasion, even Crossbills in the woodlands. Goshawks are reported from the larger blocks of conifers.

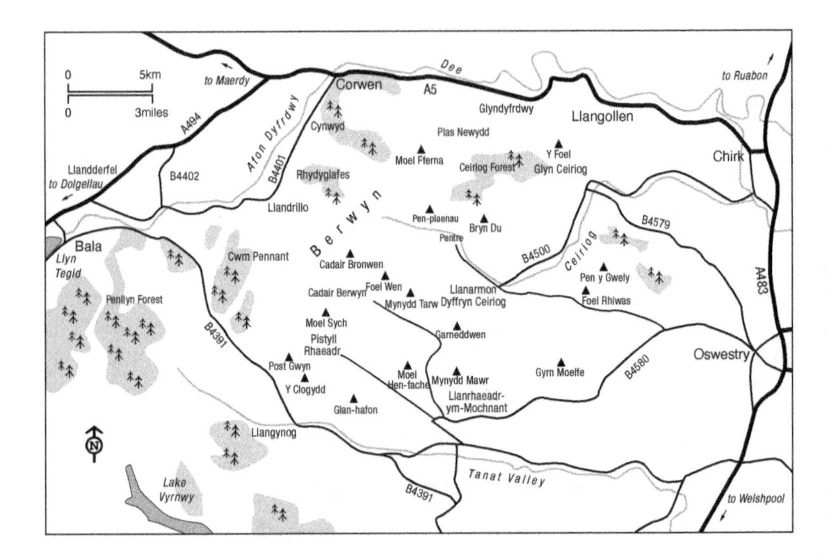

Timing

Early morning especially if hoping to encounter Black Grouse, or late evening is by far the best time.

Access

Not easy, and for the most part an area only for those able and well prepared for much walking. Close scrutiny of the map will show various routes onto or leading to the high ground, among them the two unclassified roads running towards Moel Sych from the village of Llanrhaeadr-ym-Mochnant; the southern one offers the best opportunity even for those who do not wish to leave the car, for there is a car park and public toilets at Pistyll Rhaeadr (SJ074295). Other points of access on to the Berwyns are at the picnic site (SJ166384) in the Ceiriog Forest from where a track leads through the woods and eventually to the open moor, and at Pentre (SJ136348) in the east. The minor road from the B4401 at SJ035370 in Llandrillo takes one south through Cwm Pennant and eventually to the B4391 Llandderfel to Llangynog road. A little north of Llandrillo at Rhydyglafes (SJ049397), a minor road leads onto the moors with parking at SJ075375, beyond which one can continue on foot if one desires right over the high ground to Pentre and the valley of the Ceiriog. At Cynwyd, on the B4401, a minor road leads southeast to a reservoir from where tracks continue through the plantations to the open moor. A northern approach is possible from the A5 in Glyndyfrdwy (SJ148426) to Plasnewydd where there is parking and beyond which a footpath to Moel Fferna.

CALENDAR

Resident: Hen Harrier, Buzzard, Kestrel, Merlin, Peregrine, Red and Black Grouse, Short-eared Owl, Skylark, Meadow Pipit, Grey and Pied Wagtails, Dipper, Mistle Thrush, Carrion Crow, Raven, Linnet.

December–February: Jack Snipe, Fieldfare, Redwing, Brambling in the valleys, chance of Snow Bunting on the tops, though probably few birdwatchrs go to look.

March–May: Black-headed Gull, Lapwing, Golden Plover, Dunlin, Snipe, Common Sandpiper, Curlew, Cuckoo, Tree Pipit, Wheatear, Redstart, Whinchat, Ring Ouzel all arrive to breed, Dotterel and Whimbrel occasionally on passage.

June–July: All breeding species present.

August–November: Summer visitors depart, the uplands seem rapidly to become deserted; even the resident birds are hard to find, especially at the onset of colder weather.

63 HORSESHOE PASS AND WORLD'S END

OS Landranger Map 116 & 117
OS Explorer Map 256
Grid Refs: SJ184469 and SJ227479

Habitat

The Horseshoe Pass on the A542 from Llangollen to Ruthin passes through the heart of Maesyrychen Mountain, with Llantysilio Mountain to the west and Eglwyseg Mountain skirted by Offa's Dyke Path to the east. Offa was King of

Mercia between 757 and 796, during which he defined the western boundary of his kingdom with the Dyke which stretches 177 miles from the Mersey to the Severn Estuary. The minor road north from Llangollen passes through one of the most remarkable valleys in North Wales where the Eglwyseg River, a tributary of the Dee plunges on its southerly course. Here is World's End, beyond which another area of moorland extends north to Esclusham Mountain before descending to lower ground near the village of Minera.

Species

Both sites are good for seeing the commoner upland species of Wheatear, Meadow Pipit and Ravens, but lucky observers may stumble across Peregrine, Red Grouse and Ring Ouzel. Chough may also be seen in the Horseshoe Pass area. The larch trees around the car park at World's End can produce many interesting species such as Whitethroat, Redpoll and Crossbill, but most bird-watchers come here to see the Black Grouse lekking in early spring. As many as 30 males can be seen on the hillsides either side of the car parks. Please don't approach the leks as this will disturb the birds, and you are advised to observe from your vehicle. The wonderful woodland walks around the area can produce good numbers of Cuckoo, Wheatears, Redstarts and Hen Harriers on the moorland. Despite the long-term decline of the Ring Ouzel in Wales, there is always a chance of encountering birds on passage on these high moors, which perhaps are not high enough to encourage this summer visitor to breed.

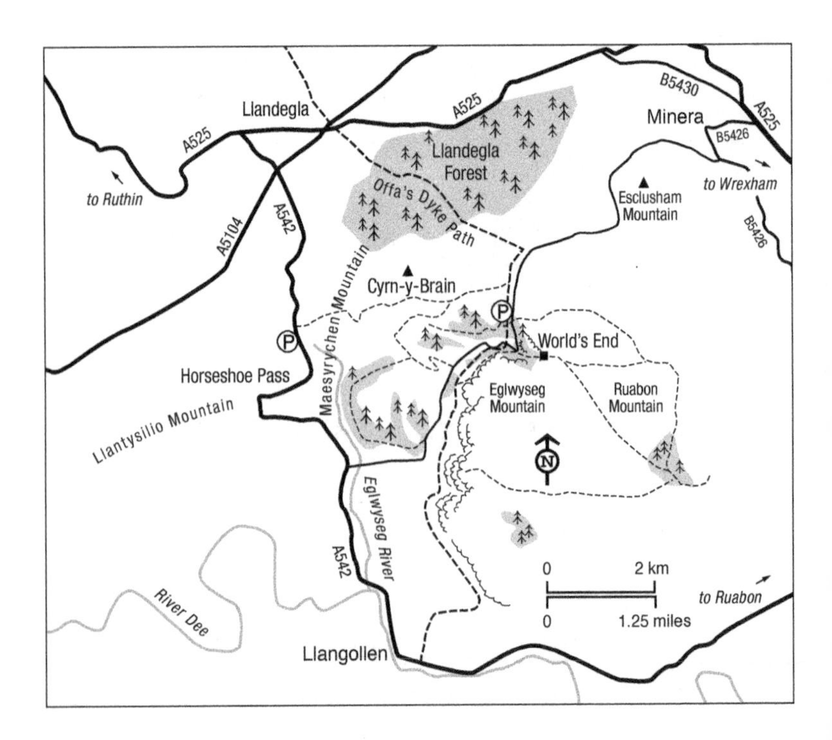

Timing

Spring and summer. It is advisable to avoid the Horseshoe Pass area in peak holiday periods or visit early in the morning before the day-trippers arrive. For World's End, the earlier in the morning the better.

Access

There are many footpaths leading from the car parks and lay-bys on the Horseshoe Pass, from which birdwatchers can explore the uplands and quarries. For World's End, the easiest access is from the A525 Ruthin road turning to World's End in the village of Minera.

CALENDAR

Resident: Red and Black Grouse, Buzzard, Sparrowhawk, Great Spotted Woodpecker, Meadow Pipit, Treecreeper, Raven, Lesser Redpoll, Chaffinch, Siskin and Crossbill.

November–February: Hen Harrier, Merlin and Peregrine.

March–June: Wheatears arrive in late March early April, other migrants include Cuckoo, Ring Ouzel, Redstart, Whitethroat and Willow Warbler.

July–October: A few Wheatears remain until August, by which time the woods are alive with roving tit flocks.

64 WREXHAM AREA

OS Landranger Map 117
OS Explorer Map 256 and 241
Grid Ref: SJ335505

Habitat

There are no fewer than 10 birdwatching sites in or in close proximity to Wrexham, the most important being Gresford Flash (SJ346536) – a 10-acre natural lake on the edge of the Wrexham Industrial Estate. Many of these sites feature or are close to kettle holes of which there are 41 around Wrexham, each formed when giant blocks of ice separated from the main glacier during the last ice age.

Species

At Gresford Flash, Mute Swan, Canada Goose, Mallard, Teal, Shoveler, Tufted Duck, Pochard, Goldeneye, Great Crested and Little Grebes, Moorhen and Coot. The gull flock at Gresford Flash always merits attention as Mediterranean, Yellow-legged, Glaucous, Iceland and Caspian Gulls have all been recorded. In the wooded locations Green, Great Spotted and Lesser Spotted Woodpeckers, Chiffchaff, Marsh Tit, Lesser Redpoll, Yellowhammer. Lapwings and Sand Martins breed at Borras.

Timing

All year.

Access

Alyn Waters Country Park SJ332550: The largest of the Wrexham country parks includes a visitor centre including the all-important café and toilets, cycle way woodland and riverside walks.

Acton Park SJ345520: Originally part of the grounds of Acton House, includes a lake, wetland and wildflower meadow and easily reached from the roads on the east and south sides.

Borras SJ354530: Former sand and gravel workings, a private airfield and the nearby pools can be reached.

Cox Lane Wood SJ370555: Described as a hotspot for Lesser Spotted Woodpecker, part of a small population in this far north-east corner of Wales is reached by way of the minor road from Marford.

Erddig Country Park SJ327482: Just a few wing-beats from the centre of Wrexham, has several access points and car parks for the National Trust property of Erddig House and close by to Erddig Flash.

Fagl Lane Quarry SJ301588: Extraction ceased 2004, the site being unique in North Wales for using a floating dredger and barges to carry the product, the result a 35-acre lake now set in woodlands and wetlands by the gorgeous River Alyn. There are exciting plans for a visitor centre a replica Iron Age meeting hall, farmstead and village and a recreated Roman fort.

Gresford Flash SJ346536: Leave the A483 Wrexham bypass at the Nantwich turnoff onto the Gresford roundabout, then take a left onto the A5156 into the industrial estate, down a narrow road; the flash is on the right after about half a mile, with easy viewing.

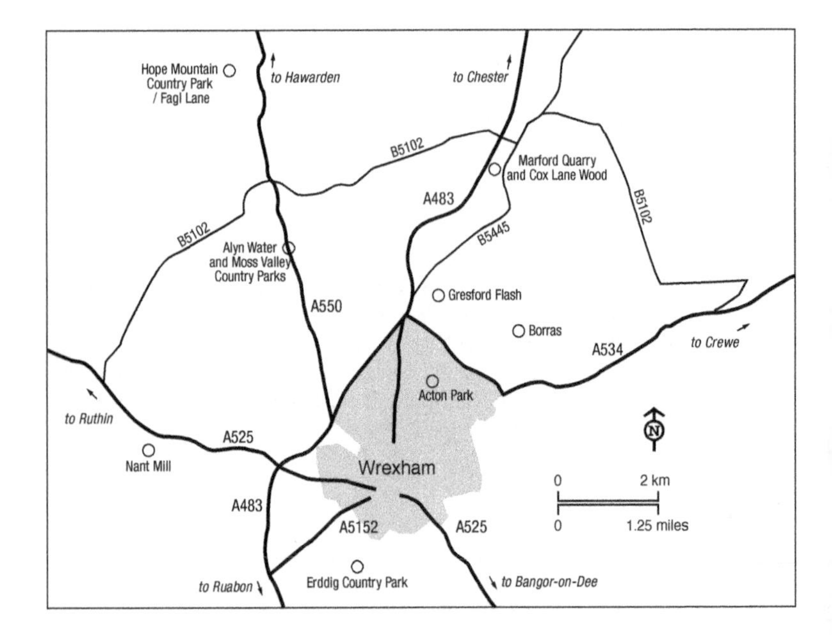

Hope Mountain Country Park SJ285577: A hilltop site with steep paths reached from the A550 by the minor road signposted Horeb. Despite the name, no refreshments or toilets are available.

Marford Quarry SJ357560: Some 2.5 miles north ofWrexham and adjacent to the A483, this North Wales Wildlife Trust nature reserve is particularly renowned for its ants, bees and wasps. There are spaces for six cars just beyond the railway bridge (SJ357563), then walk under the bridge and look for a public footpath sign.

Moss Valley Country Park SJ311523: Some 3 miles (4.8 km) north of Wrexham and another example of reclamation from mining tips, the result woodlands, two small lakes and reedbeds with a series of footpaths.

Nant Mill SJ288501: To the west of the town and close to Esclusham Mountain, the visitor centre and mill provide a good starting point for the valley or, if feeling energetic, the nearby moorlands.

CALENDAR

Resident: Mute Swan, Canada Goose, Greylag Goose, Mallard, Grey Heron, Coot, Moorhen, Buzzard, Lapwing, Kingfisher, Great Spotted and Lesser Spotted Woodpeckers, Grey Wagtail, Dipper, Stonechat, Treecreeper, Nuthatch, Great, Blue and Coal Tits, Nuthatch, Bullfinch, Lesser Redpoll, Yellowhammer, Reed Bunting.

November–February: Whooper Swans occasionally, Teal, Shoveler, Tufted Duck, Pochard, Goldeneye, Goosander, Great Crested and Little Grebes, Lapwing, Herring and Lesser Black-backed Gulls.

March–June: Ospreys on passage, Little Ringed Plover, Tree Pipit, Sand Martin, Wheatear, Sedge Warbler, Chiffchaff, Willow and Wood Warblers.

July–October: Passage waders may be attracted including Curlew, Common and Green Sandpipers.

65 HANMER MERE

OS Landranger 126
OS Explorer 241
Grid ref: SJ452391

Habitat

Hanmer Mere lies just inside the Welsh border with England, an outlier of the Shropshire meres 4 miles to the south-west, which were formed during the last ice age. Hanmer being described as a 'spacious mere deriving a great degree of picturesque beauty from the rich woodlands in its immediate vicinity', at one time its dubious claim to ornithological fame being that it was one of the best places in Wales to find Ruddy Ducks. This North American species which had escaped from waterfowl collections quickly became established in Great Britain, the first summering pair in Wales being in 1974 in Montgomeryshire, and soon began to occur widely until an eradication programme began in September 2005, successfully eradicating the species by 2020.

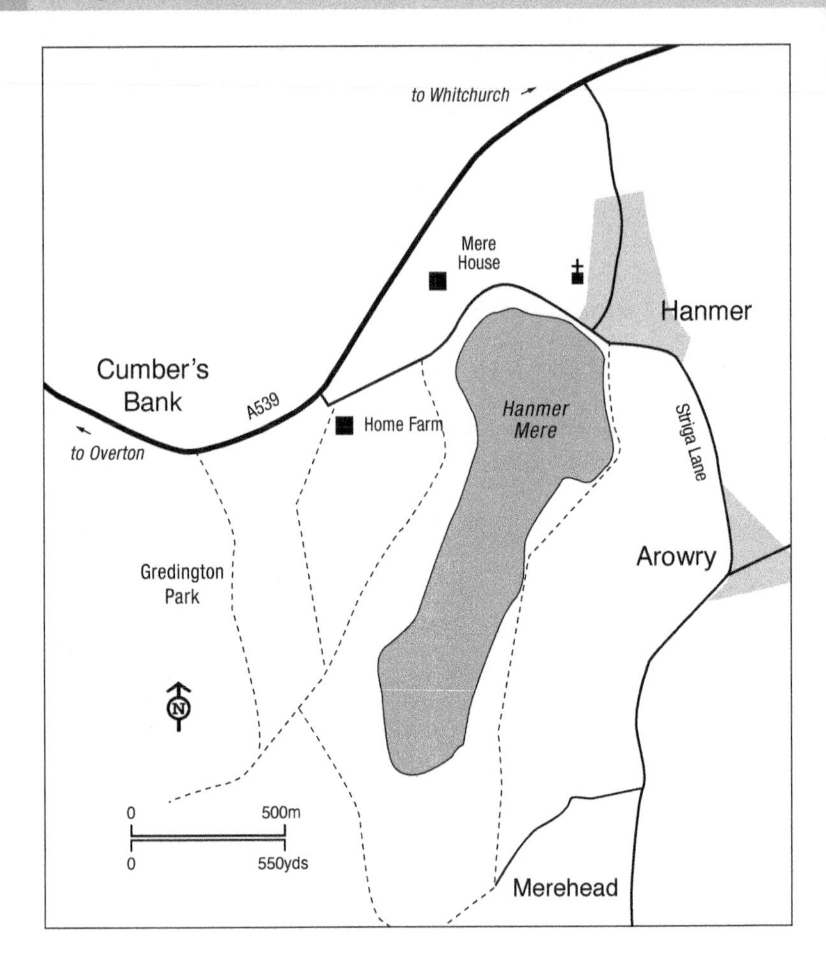

Species

In winter, Canada Geese and Mallard often number hundreds along with Wigeon, Teal, Shoveler, Tufted Duck, Goldeneye, Goosander, Great Crested, Little Grebes and Coot. In summer, Reed and Sedge Warblers, while the late summer gatherings of hirundines can prove spectacular. Scarce visitors have included Slavonian Grebe. The woods around the Mere contain expected species including all three woodpeckers, Treecreeper and Nuthatch.

Timing

All year.

Access

A footpath from Hanmer runs beside then close to the eastern shore of the Mere.

CALENDAR

Resident: Mute Swan, Canada Goose, Greylag Goose, Mallard, Coot, Moorhen, Buzzard, Kingfisher, Green, Great Spotted and Lesser Spotted Woodpeckers, Grey Wagtail, Treecreeper, Nuthatch, Great, Blue and Coal Tits, Nuthatch, Bullfinch, Reed Bunting.
November–February: Teal, Shoveler, Tufted Duck, Pochard, Goldeneye, Goosander, Great Crested and Little Grebes.
March–June: Ospreys on passage, Sedge Warbler, Chiffchaff, Willow and Wood Warblers.
July–October: Swifts depart early August, Swallows, House and Sand Martins in September, with winter visiting Fieldfares and Redwings in October.

66 FENN'S, WHIXALL AND BETTISFIELD MOSSES

OS Landranger Map 126
OS Explorer Map 241
Grid Ref: SJ494369
Website: http://publications.naturalengland.org.uk/publication/30029

Habitat

Straddling the Wales/England Border, just to the south-west of Whitchurch is an area of lowland raised bog. At 948 hectares, it is the third-largest in the UK and as such it is of international importance for wildlife – the majority of similar habitat has been drained for peat cutting or for conversion into farmland or forestry. Despite near destruction by commercial peat cutting, Fenn's, Whixall and Bettisfield Mosses have been acquired as a National Nature Reserve by Natural Resources Wales/Natural England. Large-scale peat extraction has been stopped, and the conservation bodies are now in the process of mending the Mosses by clearing the smothering trees and bushes and damming the peat cuttings to restore water levels. Now bog-mosses and cotton-grasses cover the restored habitat again, over 1,900 species of invertebrates thrive and dragonflies are plentiful.

Species

The mosses are an important site for breeding ducks and waders, with small numbers of nesting Teal, Curlew and Snipe, all of which have declined dramatically in Wales over the last 50 years. Meadow Pipits, Skylarks and Yellowhammers are plentiful in the area, while in summer there are Nightjar and Hobby. Crane, Osprey, Montagu's and Marsh Harrier have also occurred. The Canal Floods, near Moss Farm, on the English side, attract migrant waders and gulls. This site has also attracted a Green-winged Teal so that as elsewhere, anything is possible.

Timing

Spring and summer.

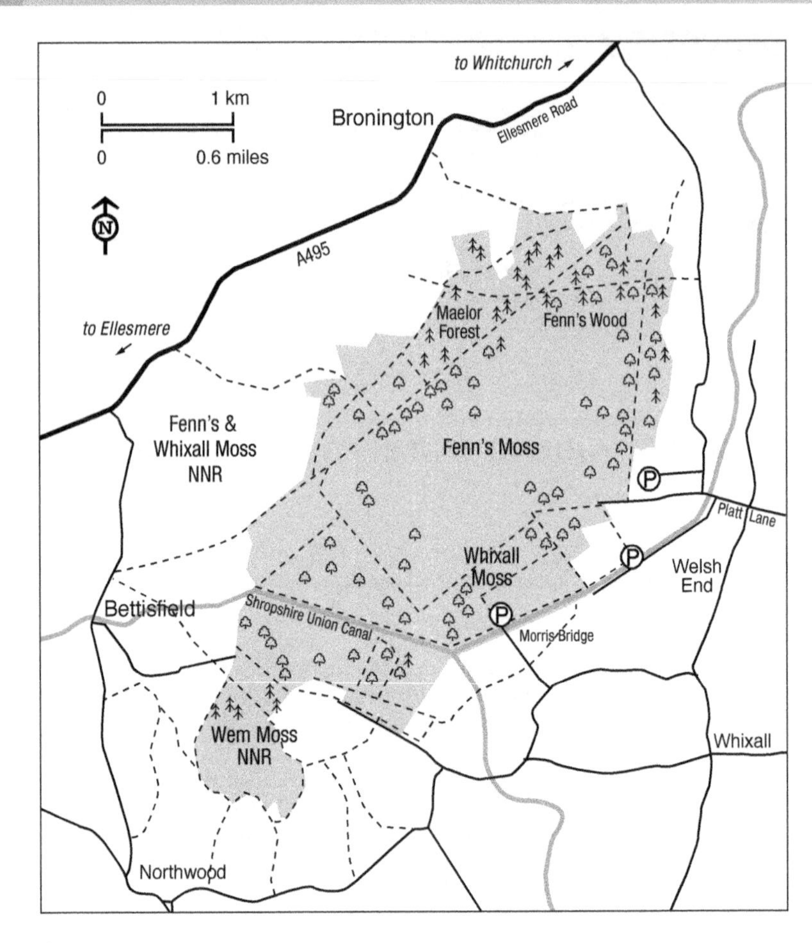

Access

From the A495 at Welshampton, turn onto the B5063, then left towards the village of Welsh End. The National Nature Reserves are signed, on the left, with car parks at Morris Bridge (SJ492354) or by Moss Cottage (SJ504359) and (SJ503364). There are many trails, fanning from the car parks across the moss. Please keep to the waymarked routes – the mosses are riddled with deep, flooded and partially vegetated ditches.

CALENDAR

Resident: Mallard, Teal, Grey Heron, Kestrel, Water Rail, Snipe, Curlew, Stonechat, Meadow Pipit, Yellowhammer and Reed Bunting.
November–February: Wigeon, Peregrine, Merlin, Hen Harrier, Black-headed Gull and Jack Snipe.
March–June: summer visitors arrive from the end of April, including Hobby, Nightjar, Cuckoo, Swallow, Swift, Whinchat, Grasshopper, Sedge and Willow Warblers.
July–October: Summer visitors depart. Green Sandpipers maybe pass through. Other waders and gulls occur in small numbers, including Little Ringed Plover, Ruff, Wood Sandpiper, Whimbrel, Little and Mediterranean Gulls.

EAST GLAMORGAN (DWYRAIN MORGANNWG)

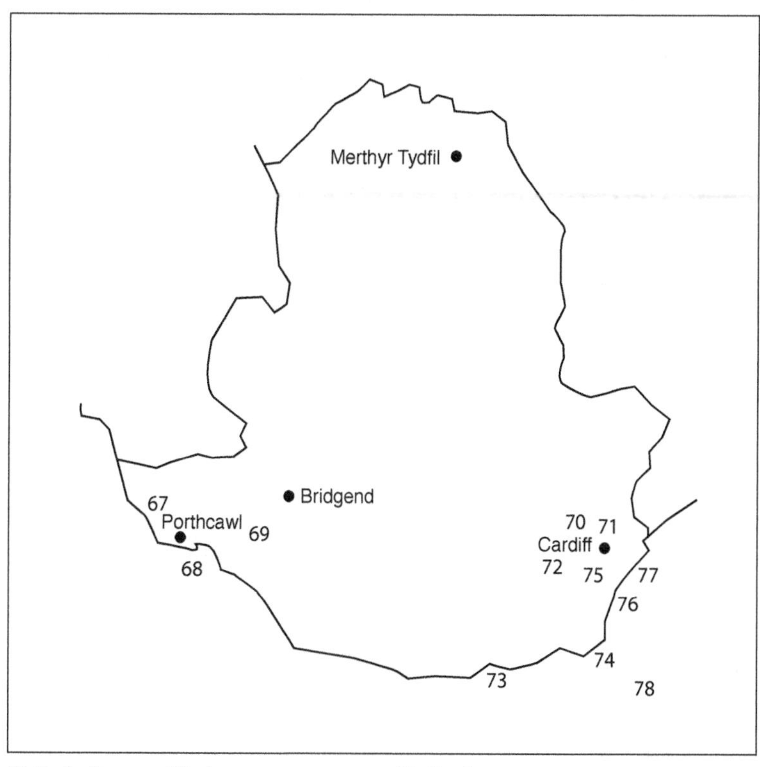

67 Kenfig Dunes and Pool
68 Ogmore Estuary and Southerndown
69 Parc Slip
70 Roath Park Lake
71 Lisvane and Llanishen Reservoir
72 Forest Farm
73 Aberthaw
74 Lavernock Point
75 Cosmeston Lakes
76 Cardiff Bay
77 Rhymney Estuary and Great Wharf
78 Flat Holm

East Glamorgan by far is the most industrialized and urbanized county in Wales, but due to the larger numbers of birdwatchers in the county, there are many better-known sites, mainly around Bridgend and Cardiff. The coastline is highly contrasting, moving from the sand dunes of Kenfig, along the small cliffs of Southerndown and finally to Cardiff and the start of the Severn Estuary.

Kenfig National Nature Reserve is one of the best birdwatching sites in Wales; its dune system, scrub and large pool are a magnet for birds and bird-watchers alike. Detailed, almost daily observations from here have turned up numerous rarities, including Ring-necked Duck and Pied-billed Grebe from

North America. Kenfig is also one of the best places to see Bittern in Wales. Several winter here and with effort and time spent in the hide, you stand a very good chance of seeing one.

The dockland in Cardiff has been transformed into plush apartments, offices and restaurants. It is here, in the 'Bay', that the Welsh Assembly sit, in the Senedd. Unfortunately, all these developments have resulted in the loss of nationally important mudflats, a small compensation being the Cardiff Bay Wetland Reserve, which attracts wintering wildfowl and gulls. All but one of the sites are on or close to the coast, the exception being Parc Slip near Bridgend, headquarters of the Wildlife Trust of South & West Wales.

67 KENFIG DUNES AND POOL

OS Landranger Map 170
OS Explorer Map 151
Grid Ref: SS797814
Websites:
https://www.first-nature.com/waleswildlife/e-nnr-kenfig.php
https://www.kenfigcorporationtrust.co.uk
https://naturalresources.wales

Habitat

Most of the east side of Swansea Bay has been lost to industrial development, but one sizeable remnant remains to the south of where the Kenfig river enters the sea. Humans have lived and worked here since Mesolithic times, but most traces of early occupants are hidden under the calcareous sand dunes which have accumulated, swallowing up many centuries ago the prosperous town of Kenfig which nestled hard by what was a navigable river in pre-Conquest times. Now all that remains above ground is the top of the keep of the Norman castle, but what a treasure house for future archaeologists lies hidden away beneath the sand! Several more recent dwellings are of interest, such as Mawdlam Church, Kenfig Farm and Sker House, the latter the grange of the Cistercian monks of Neath Abbey. The novel *The Maid of Sker*, by R.D. Blackmore of *Lorna Doone* fame, is based on the house and the surrounding area.

Kenfig, a local nature reserve of 1,557 acres (630 ha) since 1978, now a National Nature Reserve, is administered jointly by the Kenfig Corporation Trust and Natural Resources Wales, though its importance for birds and plants had been recognised since at least the beginning of the 20th century. Here the father of Welsh ornithology, always known as Morrey Salmon, came as a young man in the years before the First World War to photograph the nesting Merlins and other dune birds.

Immediately after the First World War, the coast was still relatively inaccessible to the majority of people, except at the terminal points of the railways; but it soon changed, and wherever a roadway led down to a beach or cliff, however remote, it was open to a motor car. The quiet solitudes of the great sand-dune areas and adjacent beaches, with their abundant bird life, were the first to suffer: Little Terns last nested at Kenfig in 1936 when a single chick was reared, while

Merlins nested until the Second World War as did several pairs of Black-headed Gulls – at that time the only colony in Glamorgan.

At Kenfig you can see a superb succession from the embryonic and mobile dunes fronting the shoreline through to the older dunes and slacks. Here some of the botanical treasures can be found, as well as a wealth of common species such as Bird's-foot-trefoil, Evening Primrose, Carline Thistle, Viper's Bugloss, and in the slacks Marsh-orchids, Marsh Helleborines and Round-leaved Wintergreen. In the drier areas, look for Yellow Rattle, Green-winged Orchid and Bee Orchid. More than 90 per cent of the Fen Orchids which occur in Great Britain are now to be found at Kenfig. Further inland, there are areas of dune grassland with patches of scrub here and there, and close to the river an Alder carr. Of special importance is the 70 acres (28 ha) of Kenfig Pool, one of the largest natural freshwater sites in South Wales. It is not just birds at the pool: 20 species of dragonfly and damselfly have been recorded, nearly half the total on the British list.

Species

Breeding species at the pool are small in number, but have included Great Crested Grebe, Mute Swan, Mallard, Tufted Duck, Water Rail, Moorhen and Coot. Garganeys have nested occasionally at this, their westernmost site; this is a scarce species in Wales, most frequently seen as an early spring passage migrant, with Kenfig one of the regular haunts. Lapwing, Snipe and Redshank all nested in the past but as in so many other localities, the tale is of declines and eventual loss as breeding birds. Will we ever be able again to watch the marvellous display flights of male Lapwings, to hear Common Snipe drumming and the calls of Redshank over these dunes and wetlands?

The pool comes into its own in winter, when up to 150 Tufted Duck and 50 Pochard take up residence, while the Coot population swells to as many as 400, largely with immigrants from Eastern Europe. Wigeon, Teal and Mallard also number several hundred as they dabble in the shallows or graze on nearby fields. The flock of Gadwall, which on one occasion reached 80, was the largest in Wales, though numbers have not been so high in recent years. Visitors in smaller numbers include Scaup, Long-tailed Duck, Goldeneye and the occasional Smew and Whooper Swan. The reedbed that surrounds the pool plays host to Bitterns during the winter months. Up to five have been recorded here, with the Pool Hide being one of the best places in Wales to see this most reclusive of birds.

It is worth seawatching in late summer from Sker Point. Small numbers of Manx Shearwaters are usually to be seen, increasing from late July onwards. Other species recorded have included Storm and Leach's Petrels, especially after stormy weather, Gannet, Pomarine Skua, Long-tailed Skua, Sabine's Gull and Roseate Tern.

Rare visitors to Kenfig from North America include Pied-billed Grebe, American Wigeon, Ring-necked Duck, Surf Scoter, Pectoral, Buff-breasted, White-rumped and Baird's Sandpipers, Lesser Yellowlegs, Bonaparte's and Laughing Gulls. The river mouth at Sker Point was also the site of the first British record of Little Whimbrel, which was discovered on the afternoon of 30 August 1982 near Sker Farm and enthralled many observers until it departed on 6 September.

Rare or scarce visitors from closer to home have included Little Bittern,

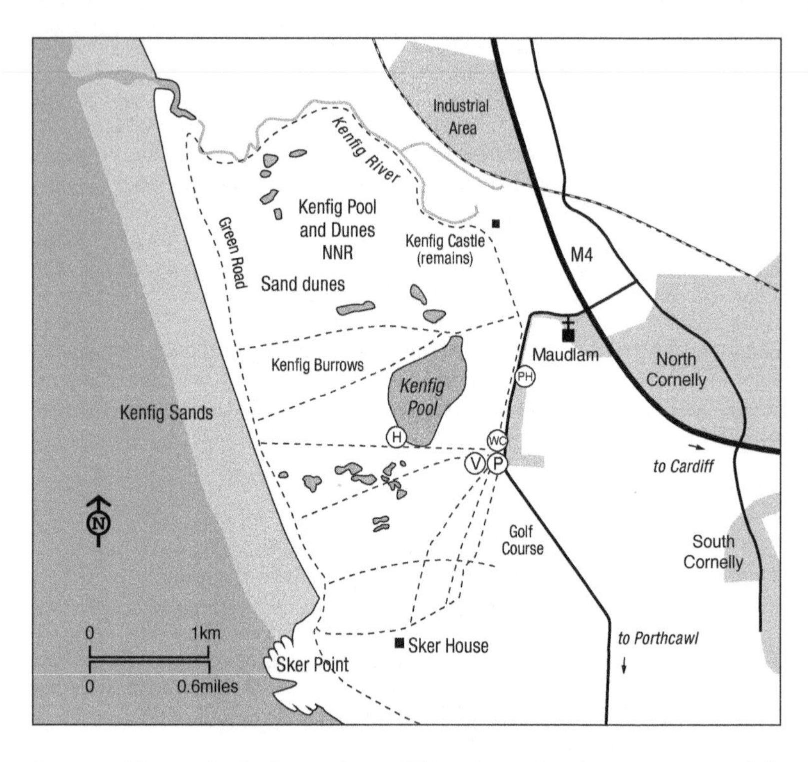

Squacco Heron, Cattle Egret, Great White Egret, Purple Heron, Spoonbill, White Stork, Ferruginous Duck, Black Kite, Honey Buzzard, Spotted Crake, Crane, Stone-curlew, Kentish Plover, White-winged Black, Whiskered, Caspian and Royal Terns, Alpine Swift, Hoopoe, Short-toed Lark, Dartford, Aquatic, Barred and Yellow-browed Warblers, Red-breasted Flycatcher, Bearded Tit, Penduline Tit, Golden Oriole, Woodchat Shrike and Lapland Bunting.

CALENDAR

Resident: Great Crested Grebe, Cormorant, Grey Heron, Mute Swan, Canada Goose, Teal, Mallard, Tufted Duck, Sparrowhawk, Kestrel, Moorhen, Coot, Oystercatcher, Ringed Plover, Lapwing, Herring Gull, Great Black-backed Gull, Meadow Pipit, Skylark.
December–February: Bittern, Shelduck, Wigeon, Gadwall, Pintail, Shoveler, Pochard, Goldeneye, Hen Harrier, Merlin, Peregrine, Water Rail, Dunlin, Jack Snipe, Curlew, Common Gull, Short-eared Owl.
March–May: Winter visitors depart. Summer migrants arrive, passage Sandwich Terns, Swallows and Sand Martins among the first, a chance of Wheatears among the dunes, while Garganeys are seen in late March and early April in some years. Whimbrel, Cuckoo and Swift in late April.
June–July: All resident species present, some seabird movement offshore, chiefly Manx Shearwaters and occasional Gannets. Large gatherings of Swifts and hirundines come to feed over the Pool. First returning waders, even a chance of Sanderling on the beach by late July.
August–November: Passage movements well underway until mid-September, when winter visitors begin to arrive at the Pool, chance of the occasional diver and grebe offshore.

Timing
Visit at any time of day, but early mornings are strongly recommended.

Access
Leave the M4 at junction 37 and head for North or South Cornelly, from where unclassified roads are signed to Kenfig. A car park is at SS802809, and adjacent to this is the reserve centre. There are numerous footpaths which head towards Sker Point and other places on the shore, and to Kenfig Pool, where there are observation hides at the south-west corner and at the north inlet (reached by a boardwalk through the reeds).

68 OGMORE ESTUARY AND SOUTHERNDOWN

OS Landranger Map 171
OS Explorer Map 151
Grid Ref: SS860757

Habitat
The Ogmore Estuary is a relatively small estuary near to Bridgend and Ogmore-by-Sea. Due to its size, it is easily covered and this means that many birdwatchers visit the site. The vast sand-dune system of Merthyr Mawr Warren National Nature Reserve, aptly described as a plant paradise, dominates the northern side of the estuary, which has two small islands and near its head Ogmore Castle where a deep ditch around the inner ward was designed to fill with sea water at high tide. Close by, a series of ancient stepping stones provides a crossing to the opposite bank.

The coast to the east of the river to Southerndown is part of the Glamorgan Heritage Coast which stretches 14 miles (22.5 km) from Aberthwaw west to Porthcawl and provides coastal heath habitat with its characteristic species and is part of Wales's Jurassic coast, with excellent fossil beds and even the occasional dinosaur specimen. The wooded valleys extending inland from Dunraven Bay offer an excellent contrast to the seacoast and estuary.

Species
The estuary is best for gulls and waders, with hundreds of Black-headed and Herring Gulls in winter, among which can be found a few Mediterranean Gulls, and more rarely nowadays Ring-billed Gull. Little Gulls may be seen on passage, while Glaucous, Iceland and Sabine's Gulls have all enchanted observers. Winter visitors offshore include Common Scoter, Red-throated Diver and Shag. Estuary wading birds include Grey Heron and Little Egret, Little Ringed Plover, Oystercatcher, Dunlin, Redshank, Greenshank, Bar-tailed and Black-tailed Godwits, Little Stint – the estuary being described as 'probably the best spot in the county for this wader' – Curlew Sandpiper, Green Sandpiper. Roosting Purple Sandpipers can share the rocks at the southern edge of the car park with Turnstones, Rock Pipits and the occasional Water Pipit. In winter, particularly from January to March, a flock of Goldeneye are often found downstream of the island, and small numbers of Goosander are to be expected – they prefer the island itself or upstream at Ploran Mawr. Seawatching from

the car park overlooking the mouth of the estuary can be rewarding, with Fulmars, Manx Shearwaters, Storm Petrels and skuas when the weather brings such species close inshore. Scarce species seen on the estuary in recent years include Whooper Swan, Great White Egret, White Stork, Spoonbill, Osprey, Avocet, Ruff, Spotted Redshank, Grey Phalarope, Laughing Gull, while Semi-palmated Sandpipers have been seen at the Island in 1990 and 2001.

Ravens and Buzzards make use of the air currents over the estuary-side slopes, while the heathland close to the estuary car park supports breeding Stonechat, Whitethroat and Goldfinch, as does the coast east to Southerndown. Dunraven has been described as one of the best locations for wintering Black Redstart in Glamorgan; check the vicinity of the visitor centre and the walled garden. Choughs, which ceased to breed about Dunraven in the 1860s, have happily now returned and are likely to be encountered anywhere along the Heritage Coast. A few pairs of House Martins nest on the cliffs, after all this was their traditional home, while occasional visitors include Wryneck, Ring Ouzel, Firecrest, Red-backed Shrike, Great Grey Shrike, Lapland and Snow Buntings. The valley woodlands are worth exploring in winter for Woodcock and in summer to enjoy a wealth of breeding species.

Timing

Winter and passage periods.

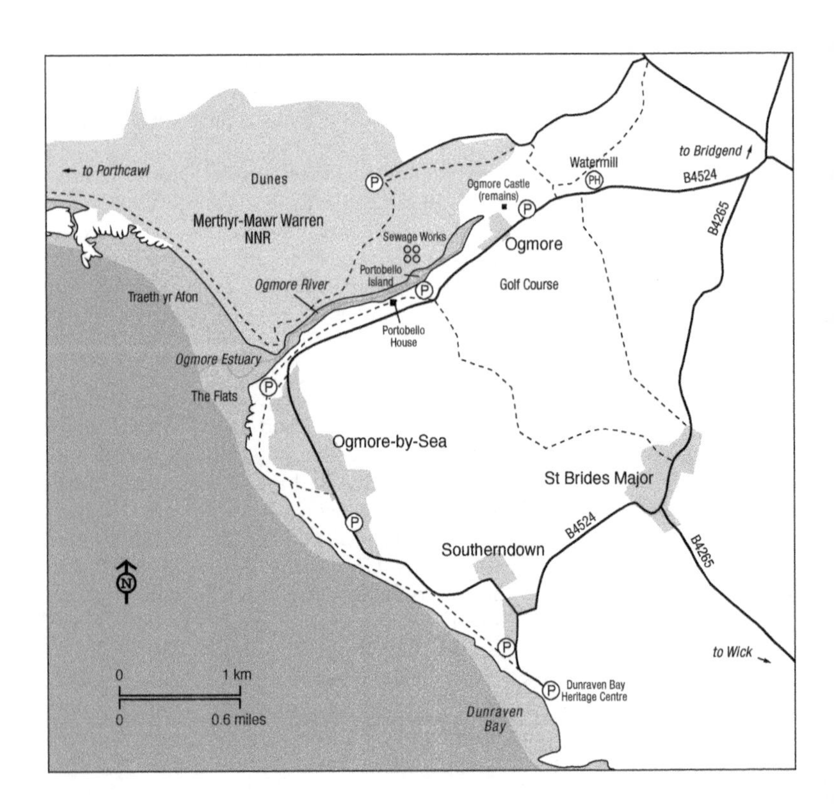

Access

Ogmore Estuary can be accessed from the east along the coast road from Ogmore-by-Sea, or from the east by following the B4265 out of Bridgend (signposted Ewenny and Ogmore-by-Sea) and stopping off at several view-points. The best place to start is at Portobello House (SS874762), which is an area of free parking and which allows you to view birds on Portobello Island, or upstream towards the sewage works, or downstream towards the Ogmore river mouth – both ways offer good walking. Further upstream there is a small, flooded field near the Watermill Pub (SS890771). There are often wildfowl here during the autumn and winter and sometimes waders on passage.

There are two car parks at Southerndown, one at the top of the hill (SS882734) and another close to the shore by the Dunraven Bay Heritage Centre (SS885731), fees payable. From here there is access by a short climb to the clifftop and onwards using the coastal path, which is well worth the 4 miles (6.4 km) and back to reach Nash Point (SS916683), a walk of seascapes, spectacular cliffs and sheltered valleys and another superb location for autumn seawatching.

CALENDAR

Resident: Fulmar, Cormorant, Sparrowhawk, Buzzard, Kestrel, Herring Gulls, Meadow and Rock Pipits, Mistle Thrush, Stonechat, Chough, Raven, Goldfinch, Linnet.
November–February: Black Redstart, Shelduck, Ringed Plover, Dunlin, Redshank, Curlew, Mediterranean and Common Gulls, Water Pipit.
March–June: Shelduck, Lesser Black-backed Gull, House Martin, Grasshopper Warbler, Whitethroat and passage migrants such as Ring Ouzel and Wheatear.
July–October: Shelduck, return by the end of October when Red-throated Divers and Great Crested Grebes appear offshore, passage waders throughout, while during rough weather a chance of Manx Shearwaters and skuas at sea, Mediterranean and Lesser Black-backed Gulls.

69 PARC SLIP

OS Landranger Map 170
OS Explorer Map 151
Grid Ref: SS877838
Website: https://www.welshwildlife.org/visitor-centres/parc-slip-visitor-centre/

Habitat

Parc Slip is a 305-acre (123-ha) nature reserve, owned by the Wildlife Trust of South & West Wales, which has its Headquarters next door. Formerly an open cast mine, Parc Slip is a 305-acre (123 ha) nature reserve, owned by the Wildlife Trust of South & West Wales. The reserve has a memorial to the 112 lives that were lost in the mining disaster in 1892. Varied habitats including ponds, rushy meadows, wetlands and woodlands have since been created and the park supports an increasing diversity of wildlife, abundant flowers among which seven species of orchid, in the pools both Smooth and Great Crested Newts. Parc Slip is also renowned for its dragonflies with 20 species having been recorded.

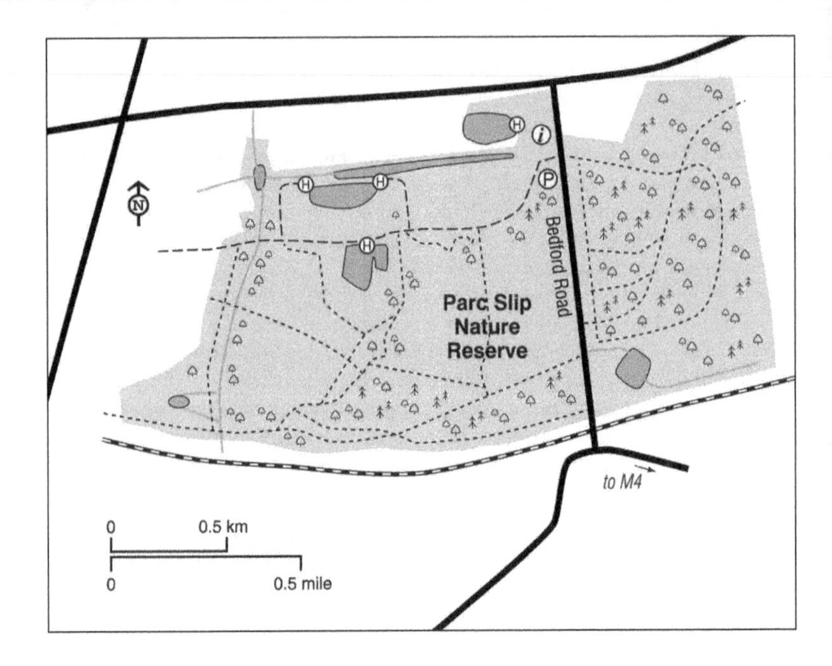

There is a visitor centre and information points along the trails and paths providing further details about the surrounding wildlife. There are several picnic sites and outdoor activity areas for children, and cycling is encouraged as part of the National Cycle Network passes through the park.

Species

There are two wetland areas, both overlooked by hides, and the site has developed a good bird life, including breeding Tufted Ducks, Little Grebes and Lapwings, wintering flocks of Golden Plover, and scarcer species such as Kingfisher along the waterway, and Lesser Spotted Woodpecker in the poplars.

Alongside Parc Slip Nature Park is Park Pond Nature Reserve, which has an old millpond at its heart, surrounded by a small area of mature woodland. Here a different variety of birdlife can be seen, such as Chiffchaff, Garden Warbler, Blackcap and Redpoll, thanks to the existence of a set of habitats associated with this woodland's greater age.

Timing

All year.

Access

The Nature Park and reserve are open all year, but the visitor centre only opens from April to October. There are three trails and two hides, with many paths suitable for wheelchairs.

CALENDAR

Resident: Mute Swan, Canada Goose, Mallard, Tufted Duck, Little Grebe, Water Rail, Buzzard, Sparrowhawk, Lapwing, Green and Great Spotted Woodpeckers, Reed Bunting.
November–February: Teal, Gadwall, Goosander, Golden Plover, Snipe, Kingfisher, Mistle Thrush, Siskin.
March–June: Reed and Sedge Warbler, Chiffchaff, Blackcap.
July–October: Passage Green and Common Sandpipers, while winter visiting Fieldfare and Redwings arrive.

70 ROATH PARK LAKE

OS Landranger Map 171
OS Explorer Map 151
Grid Ref: ST185797

Habitat

In the centre of Cardiff (ST185801) is Roath Park, with its fine trees, flower-beds, greenhouses, play area and most importantly a large artificial lake, the land having been donated by the Third Marquess of Bute to the City of Cardiff in 1887. A dam was then constructed across the marshy valley through which the Nant Fawr stream meandered, resulting in a lake of 30 acres (12 ha) with several islets. Due to its ease of access, circular footpath and central location, Roath Park is very popular with the public. It was here that pioneer ornithologist R.M. Lockley aged just four years was taken by his father afloat for the first time. A fine memorial lighthouse to Captain Scott of Antarctic fame stands at the southern end, which with its rowing boats and pedaloes is always busy, while the northern end is less disturbed for the wildfowl.

Species

There are the resident feral Canada Geese which first nested in 1982 and Greylag Geese, Mute Swans, Mallards, Moorhens and Coot. The number of Mute Swans generally peaks in July, at around 120, with about 80 present most of the year, while Coot may number in their hundreds from September to March, peaking in October at around 350, with around 80 present at other times. Tufted Duck and Pochard are also attracted to the lake, with up to 120 of the former but far fewer, around 10 of the latter, during the winter months. Great Crested and Little Grebes can be seen throughout the year, although only a couple of pairs of each remain to breed. Other wildfowl, including Teal, Shoveler and Gadwall are regular visitors, while Goldeneye occasionally makes an appearance.

Gulls, as always, dominate the lake, mainly Black-headed, Herring and Lesser-Black-backed Gulls among with which Mediterranean and Yellow-legged are often seen, while Iceland and Little are more unusual visitors. At the northern end, an island provides a roosting site for Cormorants and Grey Herons, while within the wild area, including damp woodland, careful observation should locate residents like Great Spotted Woodpecker, Mistle Thrush,

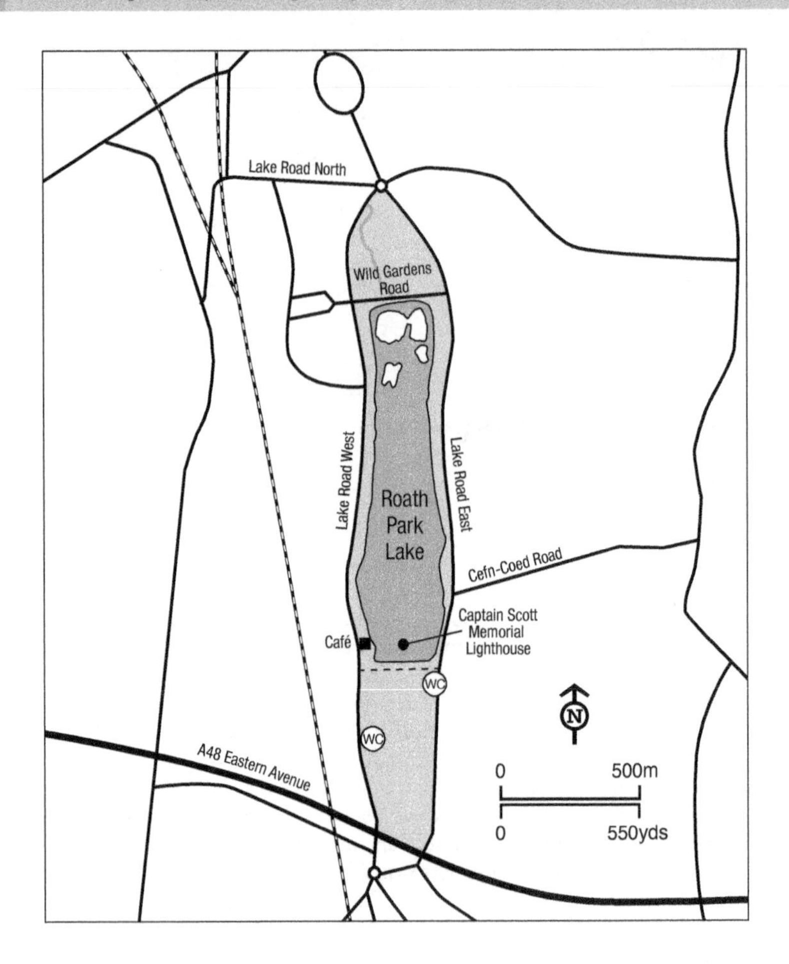

Nuthatch, Goldcrest, Treecreeper and Siskin, joined in summer by Blackcaps and Chiffchaffs.

Timing

Any time but as elsewhere, the earlier in the day, the better.

Access

There is plenty of on-street parking each side of the lake, which is situated between The Heath and Cathays.

CALENDAR

Resident: Canada Goose, Greylag Goose, Mute Swan, Great Crested and Little Grebes, Mallard, Grey Heron, Cormorant, Black-headed and Herring Gulls, Coot, Moorhen, Kingfisher, Great Spotted Woodpecker, Nuthatch, Treecreeper, Goldcrest and Grey Wagtail.

December–February: Tufted Duck, Pochard, Shoveler, Gadwall, possibility of other grebes, ducks and gulls.

March–May: Swallows and Sand Martins arrive to feed above the lake, Chiffchaff, Willow Warbler and Blackcap in the wooded northern section. Possibility of passage wildfowl and Common Sandpiper.
June–July: Build-up of numbers of Mute Swans and Mallard.

August–November: Numbers of Coot peak. Possibility of passage wildfowl. Do not overlook making a careful search of the gull flock. Winter thrushes feed on the berry-laden bushes at the southern end of the lake.

71 LISVANE AND LLANISHEN RESERVOIRS

OS Landranger Map 171
OS Explorer Map 151
Grid Ref: ST187818

Habitat

Separated by a narrow embankment and extending for some 110 acres (45 ha), these two stone-banked reservoirs were completed in 1886 to supply water to Cardiff – a role they played until the mid-1970s, when after several ownership changes, they have since 2016 been leased to Welsh Water who in their own words 'became guardians of its amazing ecology and heritage'. Lisvane Reservoir is designated a Site of Special Scientific Interest (SSSI) for overwintering birds, as are the embankments, while much of the reservoir grassland and scrub woodland outside the SSSI is designated a Site of Importance for Nature Conservation (SINC). Welsh Water, while understanding the ecological importance of the 110 acres, recognises the potential to create a hub for recreation, health and wellbeing and is aiming to maximise public access. Plans include a two-storey building affording spectacular views across both reservoirs with changing rooms, showers and toilets to enable water-sports use, paths around the reservoirs, bird hides and a nature trail. Birdwatchers may rightly feel apprehensive about this, but restoration of the former water levels and access after many years is much welcomed.

Species

Tufted Ducks are present throughout the year with over 200 present in some midwinters and over 100 during the late summer moult; numbers during the rest of the year are far lower than in the past. Pochard, a declining winter visitor in Wales, is usually present from October to March with a recent peak of 40 at the reservoirs, where 10 times that number once wintered. Teal, Shoveler, Goldeneye, Great Crested and Little Grebes, Great Northern Diver and Slavonian Grebe occur as winter visitors, while Black, Common and Arctic Terns are occasionally reported on late summer passage. Lesser Black-backed, Herring and Black-headed Gulls are a regular feature, joined on occasions during the winter by Common and Mediterranean Gulls. Rare visitors have included Lesser Scaup, Ring-necked Duck, Ferruginous Duck, White Stork, Black-throated Diver, Leach's Petrel, Spotted Sandpiper, Wood Sandpiper, Alpine Swift, Firecrest, Yellow-browed Warbler and Ortolan Bunting. Hopefully, with all that is planned, this site will retain the capacity to surprise.

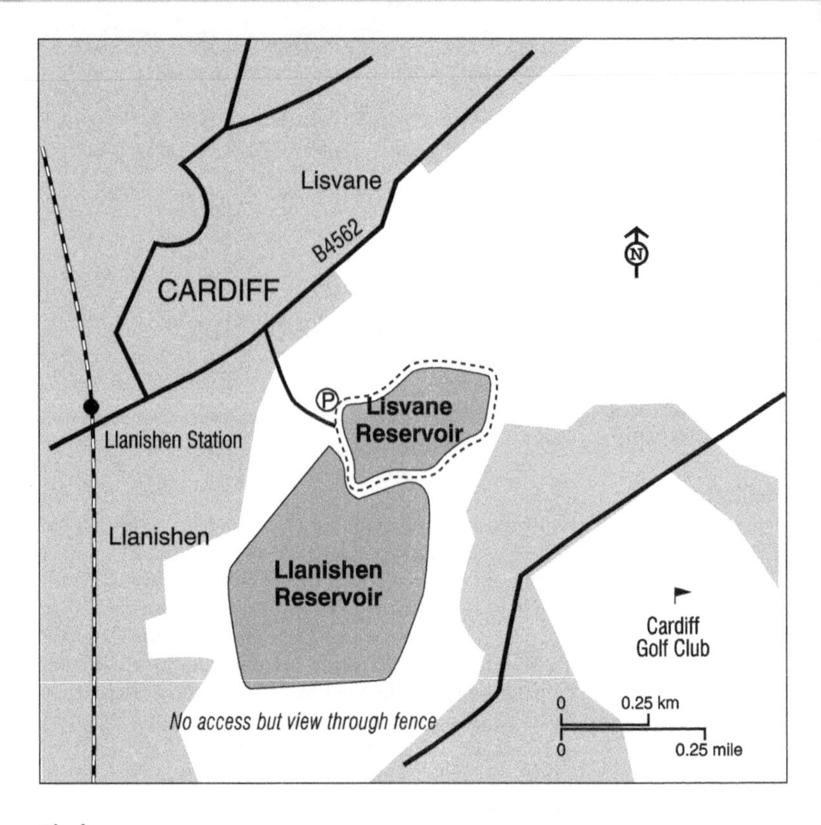

Timing

Winter and passage periods.

Access

Presently uncertain until the site reopens.

CALENDAR

Resident: Mute Swan, Canada Goose, Mallard, Grey Heron, Coot, Moorhen, Lesser Black-backed, Herring and Black-headed Gulls.

November–February: Teal, Shoveler, Tufted Duck, Pochard, Goldeneye, Great Crested and Little Grebes, Black-headed, Common and Herring Gulls.

March–June: Spring migrants including Swift, Swallow, Sand and House Martins, Chiffchaff, Willow Warbler.

July–October: Passage wildfowl and waders including Wigeon, Garganey, Little Ringed Plover, Golden Plover, Green and Common Sandpipers, Whimbrel.

72 FOREST FARM

OS Landranger 171
OS Explorer 151
Grid Reference ST138805
Website: https://forestfarm.org.uk/

Habitat

Described as 'a little gem of a reserve' and a 'haven of peace in a heavily built-up area', the woodlands and marsh beside the last remaining stretch of the Glamorganshire Canal, which originally ran from Merthyr Tydfil to Cardiff, is where R.M. Lockley spent many hours studying the wildlife as a schoolboy. He built a simple hut on his first island, which he called 'Moorhen Island', where he knew 'every flower and bird and tree'. The inspiration from those early days led him in 1927 to live on Skokholm off the coast of Pembrokeshire, see pages 270–274. A Local Nature Reserve since 1967 and a Country Park since 1992 with habitats, in addition to the canal, ranging from meadows to Alder carr and adjacent woodlands all bordered on the west by the River Taff along which Otters regularly travel. Alas, vandalism is a major hazard here – indeed, even Lockley's hut was wrecked, with him later writing that 'I could willingly have murdered the hooligans as I stared at the havoc'. More recently, in 2019 a hide was burnt down; now, as these words are written in the summer of 2021, there is news that the stable block and Warden's Hide have been torched.

Species

Local people regularly put food out at several points along the canal-side footpath for small songbirds like Robin, Blue and Great Tits, Nuthatch and Chaffinch, though Moorhens often take advantage as well. Water Rails will be more often heard than seen while Grey Herons, Little Grebes and Kingfishers visit the small lake. Reed Buntings are resident at Forest Farm, being joined in summer by Reed and Sedge Warblers, Blackcaps, Garden Warblers, Chiffchaffs and Willow Warblers. All three woodpeckers have been recorded, though only the Great Spotted can be considered resident. Winter visitors include Fieldfares and Redwings, while you should never pass by a group of Alder or Birch trees without pausing to add Lesser Redpoll and Siskin to your list. Kingfishers, Grey Wagtails and Dippers are likely on the River Taff, over which in summer Swallows and martins hawk for insects. Ravens, which have nested on the clocktower of City Hall, Cardiff, frequently fly over, their 'kronking' calls stirring memories of hearing the same on the magical uplands and coasts of Wales.

Timing

No particular time, but the earlier the better is recommended, especially during good weather; the site is easily well covered in two hours.

Access

One way to reach Forest Farm is by train to Radyr Station, from where a footbridge over the River Taff provides immediate access. By car, leave the M4 at Coryton Junction 32 taking the third exit south of the motorway and head past the ASDA store to the end of Longwood Drive parking at ST135808. Another option is to take the A4054 at the second exit south off the M4, near the

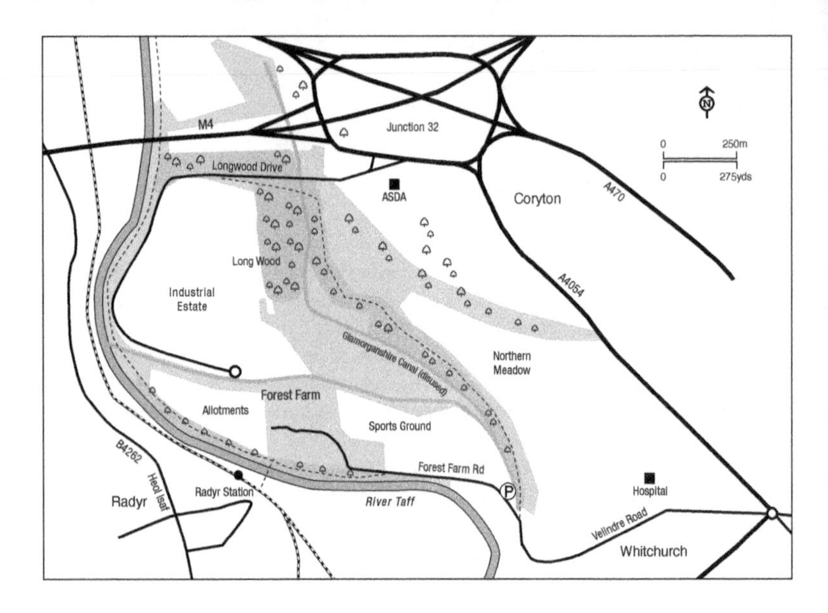

hospital bear right into Velindre Road followed by Forest Farm Road to park at ST137806 at the foot of the hill. The Taff Trail on the eastern bank of the river makes a splendid walk, while another route follows the canal and into the main woodlands all with good, mostly firm surfaces.

CALENDAR

Resident: Little Grebe, Cormorant, Grey Heron, Water Rail, Moorhen, Green and Great Spotted Woodpeckers, Grey Wagtail, Pied Wagtail, Dipper, Wren, Dunnock, Robin, Blackbird, Song Thrush, Mistle Thrush, Long-tailed Tit, Marsh, Coal, Blue and Great Tits, Nuthatch, Jay, Chaffinch.
December–February: Snipe, Kingfisher, Fieldfare, Redwing.
March–May: Summer migrants begin to arrive, has the first singing Chiffchaff reached Forest Farm from winter quarters in Portugal, Spain or Morocco or quietly overwintered hereabouts? Swallows and House Martins on passage throughout April with Sand Martins using an artificial bank in which to nest. Other visitors include Sedge and Reed Warblers, Willow Warblers and Whitethroats.
June–July: The breeding season slowly comes to an end and the birdsong, so much a part of the woodlands since April, is now silent though the birds are still present, if you are patient.
August–November: Summer visitors depart while by November winter visiting Fieldfares, Redwings, Lesser Redpolls and Siskins have arrived.

73 ABERTHAW

OS Landranger Map 171
OS Explorer Map 151
Grid Ref: ST041659
Website: https://post60travelogue.com/walks-3/aberthaw-nature-reserve/

Habitat

Breaksea Point, Aberthaw well described as 'the rockiest beach along this coast' and marking the eastern end of the Glamorgan Heritage Coast, is the most southerly point in Wales and 12 miles from Minehead on the Somerset shore. A small but important shipbuilding port flourished until the mid-19th century where the River Thaw enters the sea. The Blue Anchor Inn on the hill above, where many a thirst was quenched, dates from 1380 and, not surprisingly, there are many tales of the smuggling which took place hereabouts. The East Aberthaw Nature Reserve embraces the lagoon (ST037660) and remnants of a once-extensive area of fresh water and saltmarsh at the mouth of the Thaw, though much has been lost to tipping of pulverised fuel ash from the coal-fired power station that dominates the skyline at the site. Do not despair: the fragment of open water and surrounding wetlands are easily viewed and always merit attention, as does the shoreline of Limpert Bay, so Aberthaw remains a popular birdwatching site and is well worth a visit, especially during the spring and autumn migration periods.

Species

The scrub, reed fringes and extensive areas of Hawthorn surrounding the lagoon should be checked for migrants and in late summer and early autumn regularly host Whinchat, Reed and Sedge Warbler, Garden Warbler, Blackcap and Spotted Flycatcher. In common with other sites along the Glamorgan coastline, visible migration in late autumn is often quite spectacular, with large movements of Wood Pigeon, Skylark, Meadow Pipit, Redwing, Fieldfare, hirundines and finches noted in favourable conditions in the early morning. Scarcities have occasionally occurred and in recent years have included Avocet, Red-backed Shrike, Woodchat Shrike, Hoopoe, Golden Oriole, Wryneck and Yellow-browed Warbler. The lagoon attracts a few wildfowl, notably Wigeon, plus Little Grebe and Water Rail. The shoreline can be watched from the seawall, where Ringed Plover and Rock Pipit are found, while after prolonged

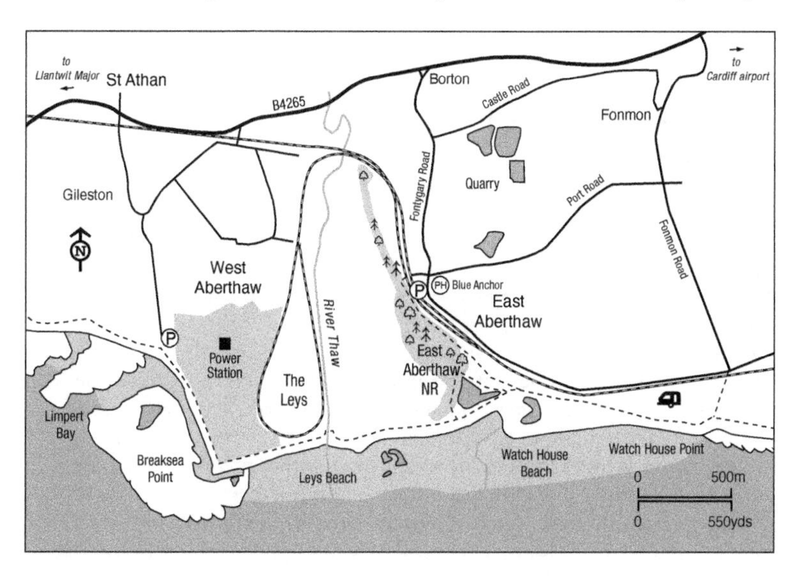

westerly winds it is worth looking out to sea for the seabirds, such as divers, Gannet, Manx Shearwater, skuas, Kittiwake and Guillemot. By continuing west along the seawall, the footpath crosses the River Thaw and skirts the perimeter fence of the power station, where Black Redstart regularly occur in winter. The warm-water outfall from the power station occasionally attracts Little Gull, so search the gull flock carefully. Aberthaw's claim to fame was the first and only Welsh record of Least Sandpiper in September 1972.

Timing

All year but particularly during passage periods.

Access

Car parking is to be found either in the public car park opposite the historic Blue Anchor Pub in East Aberthaw or in a lay-by approximately a quarter of a mile (400 m) south-east of the village, from where footpaths lead through the steep woodlands directly to the lagoon and shore. For Limpert Bay, pass through West Aberthaw and thence to the shore.

CALENDAR

Resident: Cormorant, Grey Heron, Little Egret, Oystercatcher, Rock Pipit, Linnet, Reed Bunting.
November–February: Red-throated and Great Northern Divers, Little Grebe, Wigeon, Merlin, Peregrine, Water Rail, Ringed Plover, Dunlin, Bar-tailed Godwit, Redshank, Curlew and Water Pipit.
March–June: Passage waders including Sanderling, Whimbrel and Greenshank. Summer visitors in the shape of Wheatear, Whinchat, Reed, Sedge, Garden and Willow Warblers, Blackcap, Whitethroat.
July–October: Ringed Plover, Dunlin, Greenshank, Redshank, Whimbrel, with Kittiwake, Gannet and Manx Shearwater offshore especially during stormy weather. Late in October many Woodpigeons and passerines like Skylark, Meadow Pipit, Redwing, Fieldfare, Starling and Chaffinch may move through in large numbers.

74 LAVERNOCK POINT

OS Landranger Map 171
OS Explorer Map 151
Grid Ref: ST187680

Habitat

Just over 1 mile (1.6 km) beyond the southern outskirts of Penarth, the coast, which has been following a north-east to south-west line, takes a right-angle towards the west at Lavernock Point, well described as a prime seawatching site. Here the Liassic limestone cliffs are about 50 feet (15 m) high, and some 14.5 acres (5.8 ha) of clifftop grassland and dense Hawthorn scrub are a reserve of the Wildlife Trust of South & West Wales. This consists of two fields, that on the clifftop being unimproved limestone grassland with a rich flora and at least 25 species of butterfly. Towards the western edge of the reserve is Lavernock Battery. Originally built in the 1860s, this military installation was one of the Palmerston Forts, a series of defences designed to protect access to the Bristol

Channel. It also served as a gun emplacement during the Second World War and is now a scheduled ancient monument. A major scientific achievement took place here in 1897 when Guglielmo Marconi successfully transmitted the first ever radio broadcast across open sea between Lavernock Point and Flat Holm Island 3 miles (4.8 km) distant.

Species

Breeding birds include most of the common passerines such as Wren, Dunnock, Blackbird, Whitethroat and Linnet. It is, however, as an observation point for watching migrants and for seawatching that Lavernock Point is most renowned. Being in the extreme east of the Bristol Channel one is much dependent on strong southwesterly winds bringing birds into the mouth of the Severn

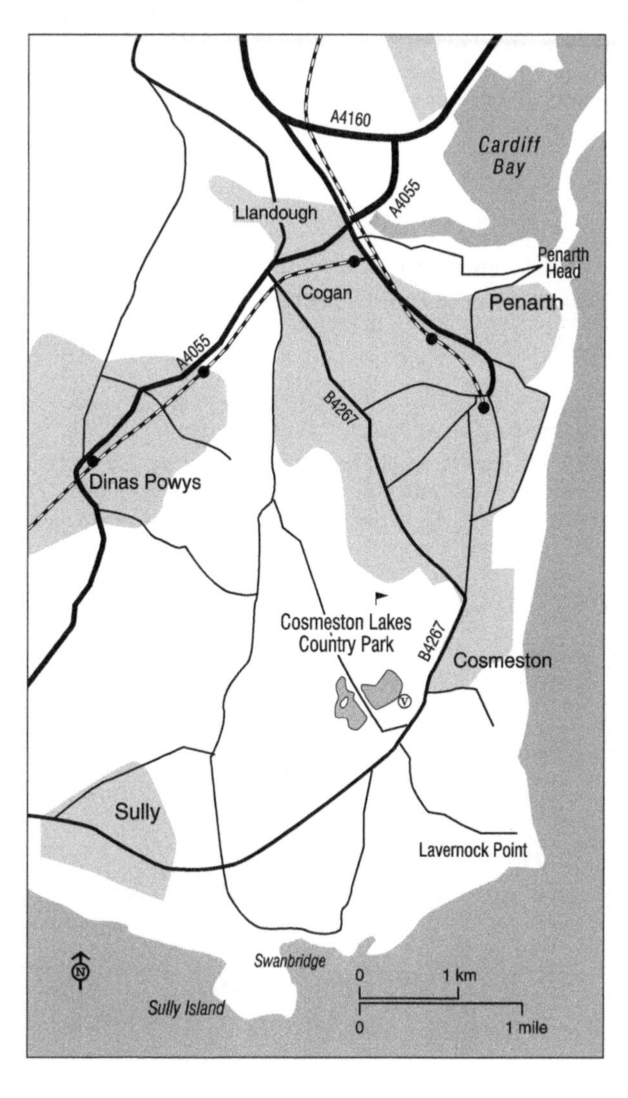

Estuary, then, as the wind drops or swings to the north-west, so they push back out to sea, many coming close inshore to delight the watchers. The most regular species are Manx Shearwater, Fulmar, Gannets and Kittiwakes, together with occasional Guillemots and Razorbills, the latter two species mostly in midwinter. Terns pass in small numbers from late July to late September. Other seabirds occasionally recorded here include Red-throated, Black-throated and Great Northern Divers, Fulmar, Balearic and Sooty Shearwaters, Storm and Leach's Petrels, Common and Velvet Scoters, Pomarine, Arctic and Great Skuas, Mediterranean and Sabine's Gulls.

Lavernock Point is also an excellent place for observing visible migration throughout the autumn, when many thousands of Woodpigeons and smaller numbers of Skylarks, Meadow Pipits, Fieldfares, Redwings, Starlings and Chaffinches can be seen following the coastline. Yellow Wagtails and Pied/White Wagtails, Tree Pipit, Redstart, Whinchat, Lesser Whitethroat and Pied Flycatcher are typical early September migrants, while a little later Firecrest, Brambling and Lesser Redpoll occur. Scarce species at the Point have included Spoonbill, Honey Buzzard, Long-tailed Skua, Woodlark, Richard's Pipit, Icterine Warbler, Red-breasted Flycatcher, Red-backed Shrike, Great Grey Shrike, Ring Ouzel, Lapland, Snow and Ortolan Buntings.

Timing

Being close to the towns of Penarth and Barry, and with Cardiff only a few miles away, the Point with its camping site is a popular spot. Birdwatchers are well advised to make their visit in the early morning, and this will be essential at migration time.

Access

Leave the B4267 Penarth to Barry road at ST177688 and take the unclassified road which leads directly to the Point, with car parking at the end of Fort Road. There is also a café and toilets at Cosmeston Lakes Country Park, about a mile away (see site 76). There are cliff-top footpaths, part of the Welsh Coast Path.

CALENDAR

Resident: Woodpigeon, Skylark, Meadow Pipit, Rock Pipit, Wren, Dunnock, Robin, Blackbird, Song Thrush, Jackdaw, House Sparrow, Greenfinch, Linnet, Yellowhammer.

December–February: Possible divers, sea-ducks and Guillemots and Razorbills offshore.

March–May: Summer passage migrants much in evidence from late March onwards, Swallow, Sand Martin and Chiffchaff among the first to arrive, Turtle Dove and Swift the last.

June–July: Small numbers of Manx Shearwaters offshore, especially from July onwards, when a few Gannets may also be seen.

August–December: August and September are the best months for observing seabird passage but much depends on weather conditions, those willing to be patient being rewarded by a good range of species. Autumn passage of passerines dominated during September by large numbers of Swallows and Sand Martins, while birds such as Yellow Wagtail, Pied/White Wagtail, Whinchat, Redstart, Lesser Whitethroat, Pied Flycatcher, Ring Ouzel, Tree Pipit, may be located; main movements from late October with Wood Pigeon, Skylark, Meadow Pipit, Starling and Chaffinch by far the most numerous. Among these late migrants look out for Woodlark and Lapland Bunting.

75 COSMESTON LAKES

OS Landranger Map 171
OS Explorer Map 151
Grid Ref: ST175692
Website: cosmestonlakes@valeofglamorgan.gov.uk

Habitat

Cosmeston Lakes is a country park, some 250 acres (100 ha) in extent, at the south-west edge of Penarth centred on two lakes created from quarry workings which commenced about 1890 and finished up 80 feet (24 m) deep before closure. The resultant lakes and surrounding area were opened in 1978 as Cosmeston Country Park which includes a reconstructed medieval village and visitor centre and café. The park has its own website, with details of access. A circuit of the site is about one and a quarter miles (2 km).

Species

Water sports are prohibited on the western lake, which is the better of the two for birds and can be overlooked from several observation platforms and a hide. Up to 300 Tufted Duck winter on the western lake joined ten years or so ago by a similar number of Pochard, alas the latter is now much reduced in number, no more than 65 in 2019. Scaup, Goosander, Gadwall are regular visitors. Winter is also the time to look out for a Kingfisher while Bittern is a speciality of the park that lurks in the reedbed on the west lake. Though seldom seen during daytime, Bittern is often more active at dusk so if you want to improve your chances of glimpsing this elusive bird, a visit late afternoon on a cold, still day is best. Listen for Water Rails wintering in the ditches and reedbeds. Although heavily disturbed by human activity, the east lake is popular with gulls that gather close to the visitor centre for scraps of food thrown to the swans. Breeding birds include Great Crested and Little Grebes, Tawny Owl, Marsh Tit and Reed Bunting. Among the regular Black-headed, Herring and Lesser Black-backed Gulls, it is worth checking for the occasional Yellow-legged, Little and Mediterranean Gulls and rarities such as Ring-billed, Iceland and Caspian Gulls. Both Sandwich and Common Terns have been reported.

Being located only a mile away from Lavernock Point on the coast, Cosmeston Lakes is also a good place for migrants, with heavy passage of Sand Martins and Swallows in spring and autumn, together with Willow Warbler, Whitethroat, Lesser Whitethroat, Sedge and Reed Warbler, and Redstart to name but a few. Both Blackcap and Chiffchaff have wintered here. Recent rarities have included Night Heron, Lesser Scaup, Ring-necked Duck, Ferruginous Duck, Isabelline and Woodchat Shrikes, Bearded Tit and Wryneck.

Timing

All year.

Access

The site has a dedicated car park and visitor centre that is signposted off the B4267 between Penarth and Sully. The park caters for various recreational activities, including sailing, which means that it gets very busy especially at weekends, making early mornings the best time for birdwatching. There is no charge for parking at present, but there is if you enter the medieval village.

CALENDAR

Resident: Mute Swan, Canada Goose, Mallard, Grey Heron, Great Crested and Little Grebes, Coot, Moorhen, Black-headed and Herring Gulls, Tawny Owl, Reed Bunting.
November–February: Tufted Duck, Pochard, Goosander, Goldeneye, Gadwall, Great Crested Grebe, Bittern, Mediterranean Gull, Common Gull, Fieldfare and Redwing.
March–June: Lesser Black-backed Gull, Sand Martin, Swallow, Whitethroat, Lesser Whitethroat, Reed, Sedge and Willow Warblers.
July–October: Mediterranean and Lesser Black-backed Gulls, passage migrants. Swifts depart in early August, Swallows and Martins in September, Fieldfares and Redwings arrive mid-October.

76 CARDIFF BAY

OS Landranger Map 171
OS Explorer Map 151
Grid Ref: ST188740

Habitat

Cardiff was the departure point for countless ships including Captain Scott's *Terra Nova* in 1910 when waterside areas like Bute Town, usually known as Tiger Bay, were notorious for their bars and other entertainment. Although first mooted in 1920, it was not until 1987 that plans were proposed to regenerate over 2,700 acres (1,100 ha) of Cardiff's derelict docklands. The plans included a barrage at the mouth of the bay where the Ely and Taf estuaries merge. This proceeded, not without controversy and strong opposition, before being completed in 1999. Cardiff Bay and its birds had been changed forever, a 1,235-acre (500-ha) freshwater lake in place of tidal mudflats. In addition, now there was a sweep of apartments close to the waterfront including the Senedd – the Welsh National Assembly – and the Wales Millennium Centre. As part of the compensation measures, the Cardiff Bay Wetland Reserve was created, some 35 acres (14 ha) of marshland and lagoons together with a substantial boardwalk.

Species

The wetland reserve has been laid out in the north-east corner of the Bay in front of the imposing St David's Hotel, with boardwalks and observation platforms. Tufted Duck are present throughout the year reaching peak numbers of up to 500 in midwinter. What a contrast to the Pochard, which in Wales has declined in numbers by 85% and now only occurs irregularly on the Reserve. Better news concerning Great Crested Grebes, which nest here and are seen throughout the year, while Little Grebes winter. The gull flock is impressive and not surprisingly vigilant observers have been rewarded with sightings of Bonaparte's, Caspian, Little, Glaucous, Mediterranean, Ring-billed and Yellow-legged Gulls. Bearded Tits are seen from time to time, they nest not many miles away in the Newport Wetlands Reserve, while Cetti's Warblers are frequently heard. Grey Herons and Little Egrets are regular visitors, while Great White Egret and Cattle Egrets have been recorded on several occasions. Seabirds blown far up the Bristol Channel during stormy autumn weather are always possible offshore. Rarities have included Lesser Scaup, Ring-necked Duck, American Wigeon and Penduline Tit.

At the mouth of the bay lies the Barrage, which provides an excellent vantage point to admire the ever-changing skyline of Cardiff, but also now regularly hosts Black Redstart in winter among the huge boulders each side of the sluice gates where Cormorants may be seen fishing. Rock Pipits along the shoreline and the occasional Snow Bunting may occur in late autumn.

Timing

All year but winter and passage periods the most productive.

Access

Several car parks within the Bay, fee payable. The complete circuit is about 8 miles (13 km); bicycles are available to hire. The wide pathways within the bay are appropriate for wheelchairs. There is a regular bus service seven days a week from central Cardiff to Cardiff Bay.

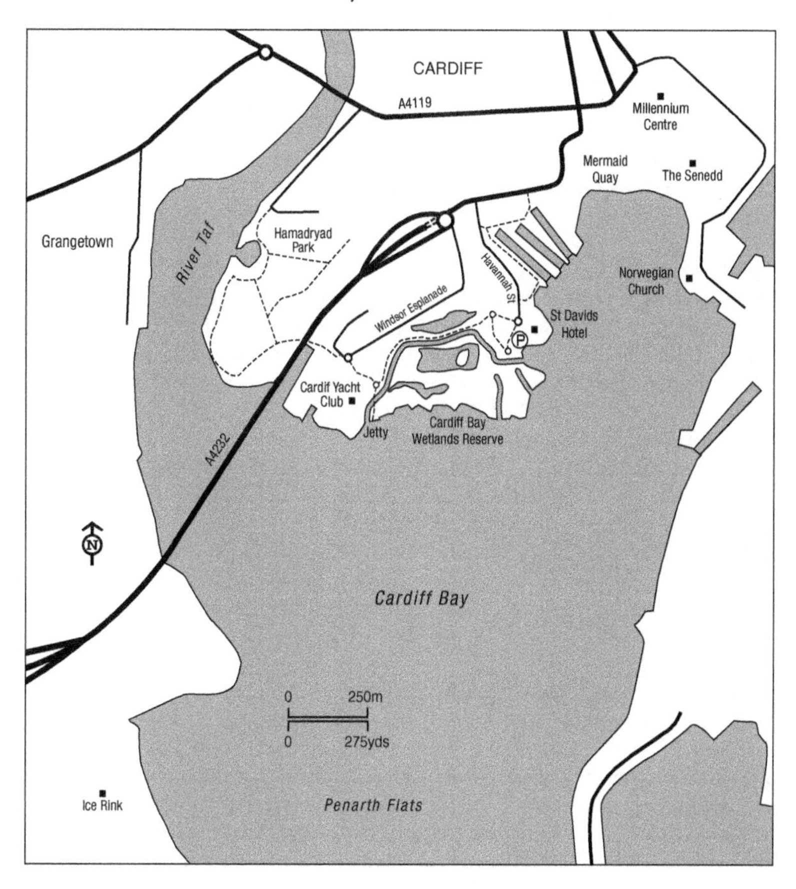

CALENDAR

Resident: Mute Swan, Grey Heron, Mallard, Tufted Duck, Gadwall, Great Crested Grebe, Coot, Moorhen, Black-headed, Herring, Great and Lesser Black-backed Gulls.
November–March: Large gull roost with Common and Mediterranean Gulls also present, Black Redstarts, Rock Pipits.
May–July: Passage waders may include Common Sandpiper and Whimbrel, also Swift, Swallow, Sand and House Martins.
July–October: Common Sandpiper, Swallow and passing waders with seabirds offshore.

77 RHYMNEY ESTUARY AND GREAT WHARF

OS Landranger Map 171
OS Explorer Map 151
Grid Ref: ST222773

Habitat

What a superb birdwatching area this must have been a century ago when Cardiff Moors stretched virtually unbroken between the Taf and Ely Estuaries and the Rhymney Estuary, the latter now the sole area of extensive intertidal mudflats that remains in Cardiff. Its importance as a refuge for wintering and passage waders, wildfowl and gulls cannot be underestimated. A waste tip towards the head of the estuary has been transformed into Parc Tredelerch, the centerpiece of which is Lamby Lake, some 10 acres in extent and easily accessed. Rhymney Great Wharf also easily accessed provides excellent opportunities for viewing the Severn Estuary.

Species

Winter is the best time to visit, when up to 600 Shelduck, 200 Pintail, 900 Redshank, 2,000 Dunlin, along with Teal, Oystercatcher, Lapwing, and gulls, mostly Herring and Lesser Black Backed, are encountered. Occasionally, Common Scoter and Scaup are found offshore, though these will invariably be distant. The estuary is flanked by the disgustingly smelly municipal landfill site at Lamby Way, which attracts thousands of Herring and Lesser Black-backed Gulls. Not surprisingly, Iceland and Glaucous Gulls are occasionally detected, though they can be frustratingly difficult to observe from the perimeter of the site as there is no public access to the landfill.

Spring often produces a sprinkle of early migrants, such as Wheatear, along the shoreline, and under favourable conditions Arctic Terns and the occasional Arctic and Pomarine Skua are seen migrating offshore. Breeding birds at Lamby Lake include both Reed and Sedge Warblers, while Cetti's Warblers are occasionally heard.

Autumn wader passage varies from year to year, though in a good one it is quite possible to find Little Stint, Curlew Sandpiper, Spotted Redshank and Bar-tailed Godwit among the more frequently occurring Ringed Plovers, Dunlin, Knot and Turnstone, while Green Sandpipers make use of the narrow reens (ditches). Both Ruff and Spotted Redshank have occasionally been recorded. Rarities have included Great White Egret and Squacco Heron, but the star must be a Long-billed Dowitcher which not only remained from 10

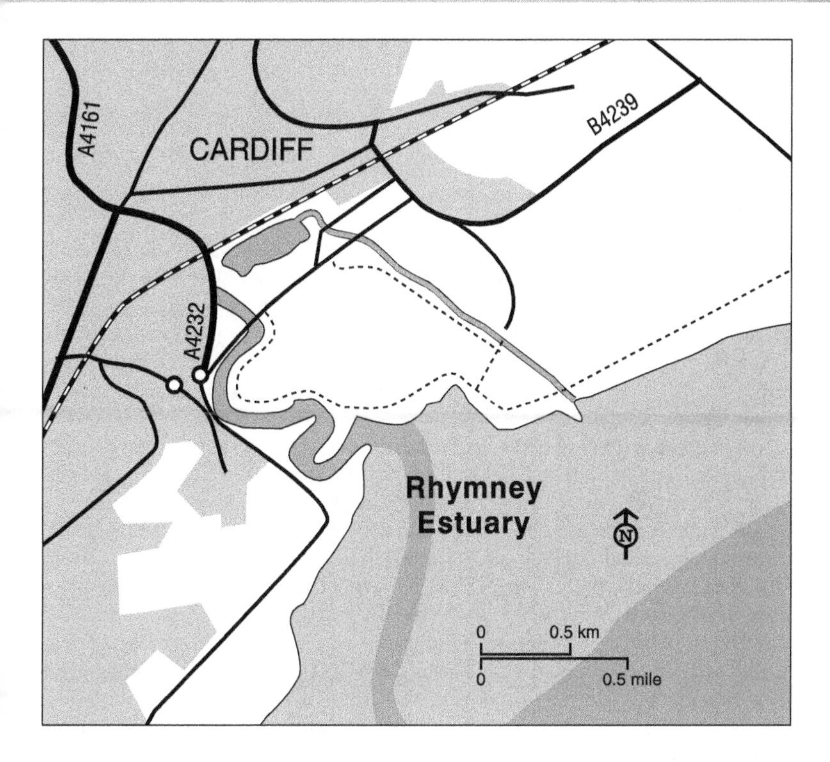

March to 18 April 1989 but from time to time most conveniently moved across the border into Gwent.

Timing

Winter and passage periods.

Access

Owing to the considerable tidal range of the upper Severn, a visit to Rhymney Estuary and the nearby open coast is recommended in the period of three hours prior to high tide. Views over the estuary can be obtained from the corner of Rover Way on the western side and from the public footpath that runs along the top of seawall on the eastern side. Lamby Lake is best visited early morning before too many folk are about.

CALENDAR

Resident: Cormorant, Grey Heron, Little Egret, Mallard, Oystercatcher, Curlew, large gulls.
November–February: Shelduck, Pintail, Teal, Lapwing, Redshank, Knot, Dunlin, Ringed Plover, Turnstone, Black-headed, Common and Mediterranean Gulls.
March–June: Shelduck and passage birds such as Whimbrel, Bar-tailed Godwit, Common Sandpiper and Wheatear.

July–October: Wader passage may include Ringed Plover, Dunlin, Redshank, Whimbrel, Bar-tailed Godwit, Black-headed, Common and Mediterranean Gulls. Stormy weather may result in Manx Shearwaters and other oceanic seabirds appearing offshore. Shelduck return from moult migration while Black Redstarts and Snow Bunting are possible at the end of October.

78 FLAT HOLM

OS Landranger Map 171
OS Explorer Map 151
Grid Ref: ST220648

Habitat

Flat Holm, the only true island in the whole of Glamorgan, lies 3 miles (4.8 km) out to sea from Lavernock Point, between Penarth and Barry. To the south, some 2.5 miles (4 km) distant, is Steep Holm, which belongs to Somerset. Barely 56 acres (23 ha) in extent, and less than half a mile (0.8 km) in length, Flat Holm, as its name suggests, is a low-lying island – the only cliffs rise to little more than 105 feet (32 m) on its south-eastern side, on the top of which stands a lighthouse.

Farming activity is recorded as far back as the 15th century, when William Philip paid an annual rent of 10 shillings for the farming and warren rights. As on the Pembrokeshire islands, Rabbits would have been a valuable source of income for many centuries as large numbers were sold in mainland markets. Farming continued until the island was evacuated in 1940, the last people being the Harris family, with Fred Harris living on Flat Holm for over 40 years. Particularly noteworthy is the work here by Marconi in 1897, when the first wireless message to be received over water was sent from Lavernock Point. Other activities included the establishment of an isolation hospital, and a great deal of military activity, first during the reign of Queen Victoria and then during the early years of the Second World War as part of the defences for the Bristol Channel ports. All these things have shaped the island's habitats.

A lighthouse was built in 1737, this being automated in 1988, while since 1975 the island has been leased by Trinity House to the South Glamorgan County Council, who designated it a Local Nature Reserve in 1977 and set up the imaginative Flat Holm Project, which involved renovation of the buildings and the establishment of facilities for visiting naturalists.

The vegetation of Flat Holm has been modified by human habitation, by countless generations of Rabbits, and in more recent years by the changing gull populations. There are large areas of low, windswept scrub, rough pasture and maritime swards. The chief item of botanical interest is Wild Leek, a plant of the Mediterranean region, found in Wales only on Flat Holm. Elsewhere in Great Britain it occurs on neighbouring Steep Holm, a few locations in Cornwall and the Channel Islands.

Species

The most striking feature is the large gull colony, where the numbers of Herring and Lesser Black-backed Gulls have risen and fallen dramatically across the past 70 or so years. In 1954 there were five pairs of Lesser Black-backed Gulls and a similar number of Herring Gulls; 10 years later there were 616 and 380 pairs respectively, and by 1974 a teeming colony of just over 6,000 pairs, divided equally between the two species. Subsequently there was a decline, and by 1980 the population numbered 2,380 pairs of Lesser Black-backed and 1,300 pairs of Herring Gulls, and in 1989, 1,200 and 70 pairs respectively. Another 10 years later, there had been a recovery to 3,700 pairs of Lesser Black-backed Gulls and 500 pairs of Herring Gulls, and in 2019 there were 2,000 or so pairs of Lesser Black-backed Gulls and 500 pairs of Herring Gulls. Great Black-backed Gulls were first found nesting here in 1962, though there have never been more than five pairs.

Flat Holm is an important site for nesting Shelduck in East Glamorgan, along with several pairs of Oystercatchers, Meadow and Rock Pipits, Wren, Robin, Blackbird and Song Thrush. In autumn, large numbers of migrating Woodpigeon, Skylarks, Meadow Pipits, thrushes and finches pass over as they move south-west along the coast of Wales or cross to England. Merlins are often present at such times. Consistent observations in due season will without question reveal Flat Holm as an excellent place for finding some of our scarce migrants, as Black Redstart, Firecrest and, in August 1995, an Icterine Warbler. Furthermore, in 2005 a Frigate bird was seen flying over the island. Unfortunately, it could not be specifically identified, yet only a day later a Magnificent Frigatebird, the first record for Britain, was found dying in a field in Shropshire. Intriguingly both records related to a bird lacking a tail feather and therefore one can only surmise on the likelihood that they were the same bird.

Timing

The island is open from April to early October, so visit from April to late July if you wish to experience the hectic clamour of gull colonies, from early August onwards to observe migrants.

Access

Because of the tidal range in the upper reaches of the Bristol Channel and the limited periods when landings can be accomplished, there are only certain days

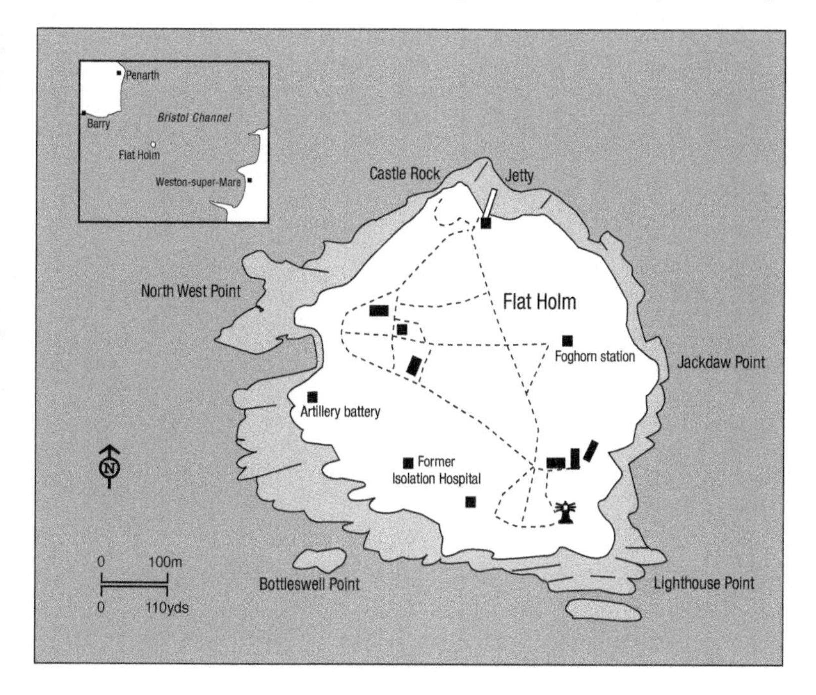

each month when the boat crosses from Barry with day visitors, who have about three hours ashore. There are two companies at present taking day visitors across to Flat Holm: Cardiff Cruises and Bay Island Voyages. For more information, please contact them:

https://www.cardiffcruises.com/discover-flatholm-island/

https://www.bayislandvoyages.co.uk/booking-flat-holm-visit/

CALENDAR

Resident: Cormorant, although not nesting, always present offshore; Shelduck, Oystercatcher, Herring and Great Black-backed Gulls, Woodpigeon, Skylark, Meadow and Rock Pipits, Wren, Dunnock, Robin, Blackbird, Song Thrush.

December–February: Occasional divers and grebes offshore, Common Scoter.

March–May: Lesser Black-backed Gulls return to the colonies early in period. Summer passage migrants from late March, with Wheatear and Ring Ouzel the earliest, quickly followed by Swallow and Sand Martin and the occasional Chiffchaff. Small numbers of most other migrants throughout April and early May, with Swift the last to arrive.

June–July: All resident breeding birds present, Shelduck ducklings leaving the island to complete fledging elsewhere in the Severn Estuary.

August–November: Seabird movement offshore, particularly when stormy weather drives birds such as skuas and terns well up the Bristol Channel, while some undoubtedly cross England and move down the Severn Estuary. Warblers appear in August, large passages of Swallows and Sand Martins in early September, and winter visitors in October, with Fieldfares and Redwings passing through, by which time the Shelduck have returned.

FLINTSHIRE (FFLINT)

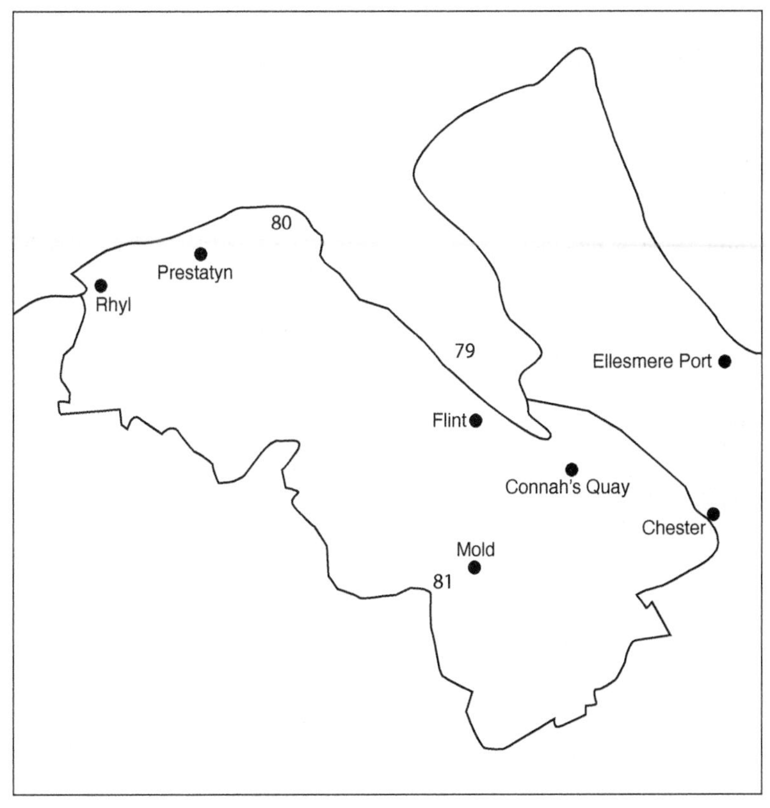

79 Dee Estuary
80 Point of Ayr and Gronant
81 Nercwys Mountain

The Dee estuary is the prominent feature in this county, which lacks the upland sites of neighbouring Denbigh. The Dee is internationally important for wintering wildfowl and waders, one of the top five estuaries in the UK in this respect. Birds obviously use both sides, depending on the tide, and move up and down the estuary depending on food availability. On the English side, there is Heswall, Neston, RSPB Burton Mere Wetlands (of which one pool is on the Welsh side of the border) and at the Dee's mouth the Hilbre Islands. On the Welsh side, key sites are Connah's Quay, Bagillt Bank, Flint and at the mouth of the Dee opposite Hilbre is RSPB Point of Ayr. All these may hold large high-tide wader roosts. Just round the Point of Ayr, among the sand dunes and shingle is Gronant, the last Little Tern colony in Wales.

Inland there are few birdwatching sites, the only one of note is Nercwys

Mountain – an area of conifer plantations a few miles south-west from Mold (yr Wyddgrug). Here, careful studies of Nightjar breeding behaviour have been carried out in the past.

79 DEE ESTUARY

OS Landranger Maps 116/117
OS Explorer Map 266, 265

Habitat

Originally gouged by a glacier, this funnel-shaped estuary is about 5 miles (8 km) wide at its mouth, from where it extends south-southeast some 15 miles (24 km) to Connah's Quay at its head. In the Middle Ages, the Dee was navigable right up to Chester, 22 miles (35 km) from the open sea, and was a major port until at least the 14th century. The good citizens of the city were, however, fighting a losing battle: vast quantities of silt were accumulating in the estuary, making the journey hazardous and eventually impossible. Now the city stands 7 miles (11 km) from the head of the estuary.

Most of the saltmarshes (those on the Dee represent about 5% of all British saltmarshes) are situated on the Cheshire bank, but there are fragments on the Flintshire shore. Everywhere there is Common Cordgrass *Spartina* first introduced in 1928 and now covering vast areas, colonising the open flats in a thick sward, reducing feeding areas and excluding some of the less vigorous estuarine plants such as Glasswort, Sea Aster, Scurvygrass and Sea Purslane. Move to the upper marshes about the high-water mark and Sea-milkwort, Sea Arrowgrass and Common Reed will be encountered. Although 150 miles (240 km) from their nearest large breeding colonies on the coast of Pembrokeshire, the haul-outs of Grey Seals on the West Hoyle Bank at the mouth of the Dee can rise to as high as 200 animals in summer.

The estuary was designated a Ramsar site in 1985 as the wetlands are deemed to be of international importance in terms of ecology, botany, zoology, limnology and hydrology. Further highlighting the supreme importance of the Dee was its designation in the same year as a Special Protection Area (SPA) under the European Community Directive on the Conservation of Wild Birds.

Starting at the bottom left right-hand corner of the estuary and moving clockwise on the Welsh bank to the mouth, key nature reserves are Burton Mere Wetlands (RSPB), Oakenholt Marsh (RSPB and local volunteers), Flint Marsh, Bagillt Bank, Greenfield, Ffynnongroyw and finally Point of Ayr (RSPB). Right at the base of the Dee Estuary, to the south of the A548 are a series of pools and marshes which often attract birds away from the estuary. These include Shotton Lagoons, within the perimeter of TATA Steel's Shotton steelworks, which has a sizable Common Tern colony, Shotwick Boating Lake and Shotwick Fields, on which large numbers of grey geese can often be found.

Species

The attention of many birdwatchers was first drawn to the Dee, and to its immense wader flocks, by the superb photographs of densely packed Oystercatchers, Knots, Dunlins, Curlews, Redshanks and others, taken by Eric

Hosking on the Hilbre Islands – Hilbre, Little Hilbre and Hilbre Eye – off West Kirby on the north-west corner of the Wirral. These rocky islands, in total 15 acres (6 ha) in extent and now a nature reserve, can be approached on foot by watchers who wish to see the wader flocks as the tide rises or to wait for migrating landbirds. We digress, however, for this is not Wales but England – though the birds do not observe human boundaries, suffice to say they roam at will across the sands of the Dee from the English to the Welsh banks.

Since the 1960s, detailed counts have been made of waders on estuaries throughout Britain as part of the Wetland Bird Survey organised by the British Trust for Ornithology. So, it is now possible to assess the importance of the Dee in a national, indeed international, context. The information collected clearly shows this to be one of the most important estuaries for wintering birds in north-west Europe. Oystercatchers are the most numerous, and although only about 16 pairs nest around the estuary, the huge flocks that gather here are one of the most impressive sights. Numbers generally peak in August–September, to around 8,500 off Point of Ayr, 3,000 off Mostyn and 2,500 off Oakenholt at high-water. Many disperse over nearby low-lying fields to feed on earthworms, as well as seeking cockles on the estuary. Lapwings breed at some five sites in estuary-side fields, the most important being at RSPB Burton Mere, where in 2018, 71 pairs were present. Midwinter numbers on the estuary usually reach about 4,000 birds.

The numbers of waders using the estuary has declined dramatically in the last 50 or so years, probably linked to a change in migration due to global warming. Knot may number up to 2,500, Dunlin up to 7,000 and Bar-tailed Godwits as few as a hundred. Curlew numbers do not seem to have changed much over the last 30 years, with around 1,500 being present, particularly at Point of Ayr in the autumn. On the plus side, there has been a steady increase in the number of Black-tailed Godwits, now up to around 5,000. Although the number of breeding Redshank on the Dee have declined from 55 pairs in 1991 to 25 in 2018 (all at RSPB Burton Mere) there has been a rise in the number using the estuary in autumn, with 6,000 counted in 2018, declining to around 1,500 in the winter. Much smaller numbers of several other species occur, particularly on passage; these include Ringed Plover, Little Ringed Plover, Little Stint, Curlew Sandpiper, Ruff, Whimbrel, Spotted Redshank, Greenshank and Turnstone, the latter largely at Mostyn Dock. Scarce waders noted have included Kentish Plover, Temminck's Stint and Pectoral and Buff-breasted Sandpipers.

Not a wader, but a wading bird, Little Egrets have increased, as elsewhere in the UK, and now up to 120 may be present on the estuary in autumn. Great White Egret numbers have also increased in the last 20 years and have even bred at Burton Mere, although on the English side of the border. Spoonbill and Marsh Harriers are regular at this site, with individuals also being seen on the Welsh side of the estuary, such as at Connah's Quay.

Small numbers of Mute Swans breed around the estuary, while a flock at Burton Mere Wetlands can rise to about 100 by late winter. Small numbers of Whooper Swans, less than 40, overwinter on the same marshes. Up to 10,000 Pink-footed Geese use the estuary in winter, grazing the farmland mainly on the English side and around Shotwick Fields. As elsewhere, the number of Canada Geese has increased dramatically, with around 7,000 frequenting the estuary. Shelducks breed at quite a few points in the vicinity of the estuary, with autumn numbers reaching 4,000 in October. Wigeon, Teal and Mallard are recorded in some hundreds each through the winter, with on occasions several thousand being present. Almost half the Pintail wintering in Great Britain are on the Dee, with nearly 8,000 birds in 2018; many concentrate at RSPB Burton Mere Wetlands at the head of the estuary, while large numbers also occur at Oakenholt on the opposite bank, with smaller numbers and wandering parties elsewhere. Fewer than 30 pairs of Pintail nest in Britain, and most of those that winter here come from Iceland, Scandinavia and western Russia. In contrast to Pintail, Shoveler barely reach 50 in winter, with a few remaining to breed.

The open estuary, around Point of Ayr, attracts small numbers of Scaup, Common Scoters (as many as 2,000 in October in recent years), Goldeneyes and Red-breasted Mergansers, these being joined in hard weather, when the Cheshire Meres and other inland waters are frozen, by Pochard, Tufted Ducks and Coots. Small numbers of Great Crested Grebes are present throughout the winter, and occasional Red-throated Divers move into the estuary, though most prefer to remain at sea.

Breeding birds on the saltings and marshes, in addition to those already mentioned, include Mallard, Cuckoo, Skylark, Meadow Pipit, Stonechat, Grasshopper, Sedge and Reed Warblers, Linnet, Yellowhammer and Reed Bunting. In winter these saltings attract flocks of Twite (as many as 150 at times), usually around Connah's Quay and Flint Castle, and even occasional flocks of Lapland and Snow Buntings and even Shorelark. Scarce visitors have not been lacking and have included Lesser Scaup, American Wigeon, Pacific Golden Plover, Pallid Harrier and Bonaparte's Gull.

The provision of nesting rafts on the pools at Shotton allowed the rapid colonisation of Common Terns. Previously, the birds had struggled to breed on the Dee marshes in the face of increasing human disturbance, predation by foxes and flooding by high tides. In the safety of the rafts, the colony grew rapidly from 13 pairs in 1970 to 762 pairs in 2007, with more recently an average of 373 pairs each year. Other breeding species within the reserve include Little Grebe, Shoveler, Kestrel, Grey Partridge, Black-headed Gull, Cuckoo, Whinchat, Grasshopper Warbler and Reed Warbler. Huge numbers of Swallows and House Martins (with some Sand Martins) roost in the reedbeds at Shotton in the late summer before leaving on autumn migration. Other passerine migrants noted include many Skylarks, Meadow Pipit, Pied and White Wagtails, and Wheatears. There have been few recent scarce species reported from here, but such birds have included Black-necked Grebe, Bittern, Garganey, Long-tailed Duck and Hen Harrier.

Timing

With such a vast area for birds to disperse over at low tide, it is imperative that a visit is planned so as to take advantage of the high-tide concentrations of the flocks.

Access

RSPB Burton Mere Wetlands (SJ313736): Straddling the border between Wales and England, this RSPB reserve complete with visitor centre is just 10 minutes from the M56 and accessed via the A540 and finally Puddington Lane. Described as a mosaic of freshwater wetland habitats, mixed farmland and

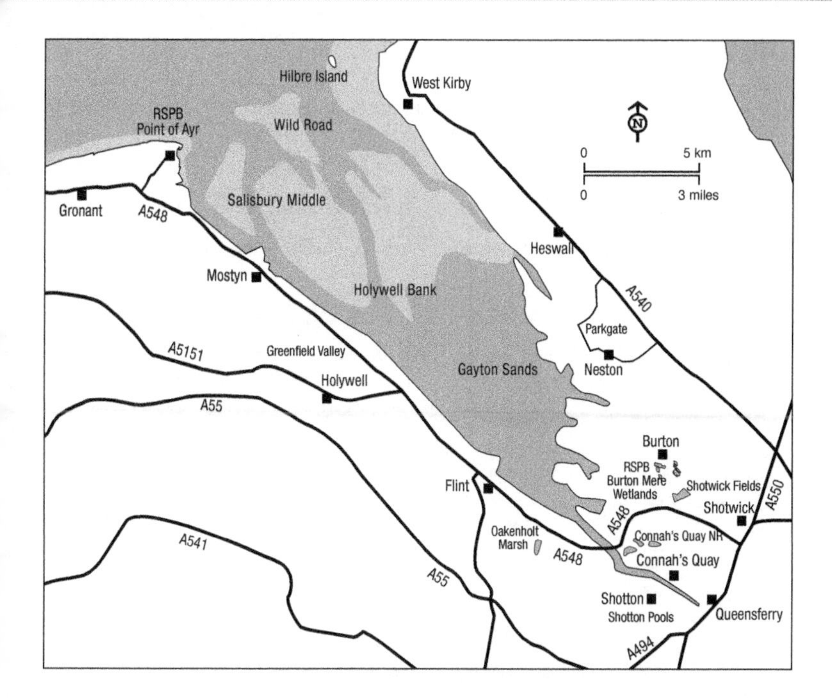

woodland, the reserve is bursting with wildlife. The reserve website offers up-to-date details of events and sightings: https://www.rspb.org.uk/reserves-and-events/reserves-a-z/dee-estuary-burton-mere-wetlands/

Connah's Quay Nature Reserve (SJ278713): A narrow strip of land about a mile in length beside the estuary at the south end the Uniper Power Station. There are five hides, including one that has two storeys. The reserve has a website, with details about membership of Deeside Naturalists' Society and access: http://www.deenats.org.uk/

Flint (SJ245735): The shore is accessible near Flint Castle, completed in 1286 and where Richard II was captured by Henry Bolingbroke, later Henry IV, even recorded in Shakespeare's play which includes the line 'For night-owls shriek where mounting larks do sing.' The battlements afford fine views over the estuary, while at SJ23587392 a footpath offers easy access almost without wellingtons, following the seawall north and providing a good vantage point for Bagillt Bank and Flint Marsh.

Ffynnongroyw (SJ142820): A footbridge over the railway at Ffynnongroyw provides both a vantage point and access to the foreshore. There is a lay-by for parking about 200 yards south of the bridge.

Greenfield Valley (SJ196775): Some 70 acres (28 ha) of woodlands, reservoirs, the ruins of the 12th-century Basingwerk Abbey and industrial history near Holywell. At Greenfield Valley there is ample parking and a visitor centre: https://greenfieldvalley.com/

Mostyn (SJ158803): On the south edge of Mostyn, a footpath from SJ163802 leads on to the estuary embankment and runs all the way to Greenfield.

Oakenholt Marsh (SJ266718): Southeast of Flint is part of the 11,651 acres (4,715 ha) managed by the RSPB on the Dee, embracing sites on both the Welsh and English sides of the estuary. There is no access to Oakenholt, though it can be viewed from a hide in the Deeside Naturalists' Society Connah's Quay Reserve.

Shotton Pools (SJ3068): Access only by permit, available from the steelworks.

Shotwick Fields (SJ3172): large boating lake, with surrounding fields.

CALENDAR

Resident: Cormorant, Grey Heron, Little Egret, Shelduck, Mallard, Shoveler, Oystercatcher, Ringed Plover, Lapwing, Snipe, Redshank, large gulls, Skylark, Meadow Pipit, Reed Bunting.

December–February: Great Crested Grebe, Mute Swan, Pink-footed Goose, Brent Goose, Wigeon, Gadwall, Teal, Mallard, Pintail, Common Scoter, Goldeneye, Red-breasted Merganser, Golden and Grey Plovers, Knot, Sanderling, Dunlin, Jack Snipe, Black-tailed and Bar-tailed Godwits, Curlew, Greenshank, Turnstone, Black-headed Gull, Hen Harrier, Marsh Harrier, Merlin, Peregrine, Short-eared Owl and Rock Pipit.

March–May: Winter visitors depart, summer visitors and passage migrants include Whimbrel, Common Sandpiper, Cuckoo, Yellow Wagtail and Wheatear. Breeding Little and Great White Egrets at Burton Mere RSPB. At Shotton: Swallows and Sand Martins usually over the pools by end of March, when Garganey occasionally call. Common Terns and Black-headed Gulls take up residence for breeding. From mid-April, summer visitors such as Grasshopper, Sedge and Reed Warblers, Whitethroat and Spotted Flycatcher may be seen.

June–July: Shelduck broods in crêches, but most have fledged and departed by end of period. Black-headed Gulls and Common Terns fledge from Shotton in early July, Green and Wood Sandpipers are possible by end of month.

August–November: Wildfowl and wader numbers begin to increase, some passage species such as Little Stint and Curlew and Green Sandpipers seen, terns remain in estuary until well into September. By October most wintering species have arrived.

80 POINT OF AYR AND GRONANT

OS Landranger Map 116
OS Explorer Map 265
Grid Ref: SJ124849
Website: https://www.rspb.org.uk/reserves-and-events/reserves-a-z/dee-estuary-point-of-ayr/

Habitat

The shingle spit with its guardian lighthouse at Point of Ayr marks the north-west side of the mouth of the Dee Estuary. The habitats here range from the sand dunes facing north into Liverpool Bay to the shingle spit, inside which extend the vast saltmarshes and mudflats of the estuary. Some 450 acres (182 ha) are now a reserve of the RSPB, by agreement with Dŵr Cymru (Welsh Water) and the Dee Wildfowlers' Club. To the west at Gronant, is the only remaining Little Tern colony in Wales. Close by, one of only two sites in Wales where Natterjack Toads (which became extinct in Wales in 1960 and were subsequently reintroduced) can be found, the other being Talacre Dunes almost at the Point itself.

Species

Little Terns, a threatened species at many colonies in Britain, once nested at various places along the North Wales coast, at Tywyn in Meirionnydd, at sites on the Llŷn Peninsula and Anglesey, in Denbighshire and in Flintshire east to Point of Ayr. Nowadays, disturbance in all its forms plays havoc with Little Tern nest sites and for the birds and most of the colonies the tale is one of decline. They do not require much room, but it must be away from treading feet, dogs and vehicles on beaches, while these birds also face many natural hazards too, such as Hedgehogs and Stoats, gulls and Carrion Crows – and, if they avoid all of these, just one extra-high tide can wash the nests away. There are now just two colonies in Wales, both in carefully wardened sites, about 200 pairs nest at Gronant and about 20 pairs at Point of Ayr. Breeding success is highly tide and predation related, but 206 young were fledged in 2019. Large numbers still gather at Point of Ayr in the late summer, when up to 500 have been noted, while at the same time the flock of Common and Arctic Terns may be 3,000 strong, in addition to which Sandwich Tern numbers may rise to 300 and there may also be the occasional Roseate and passing Black Tern.

Many wildfowl winter in the vicinity of Point of Ayr, Shelduck and Mallard being the main species, with small numbers of Teal, Wigeon, Pintail, Shoveler, Scaup, Eider, Common Scoter and Goldeneye. Red-throated and Great Northern Divers, Great Crested Grebes, Guillemots and Razorbills occur throughout the winter in the estuary mouth, or just out to sea where the water remains shallow.

The flocks of winter waders, part of the massive concentrations which make the Dee so important, can number up to 20,000 in midwinter. Oystercatcher, Dunlin and Redshank are most numerous, while others to be seen include Sanderling, Knot, Black-tailed and Bar-tailed Godwits, Curlew and Turnstone. The Point is a major high-water gathering place for all these birds and provides a spectacle for the birdwatcher perhaps seen nowhere else to this degree in Wales.

Seawatching from the Point of Ayr can be good given the right weather conditions, sustained northwesterly winds, when there can be spectacular movements of Fulmar, Manx Shearwater, Gannet and Kittiwake. Autumn gales, particularly in October, may push Leach's Petrel, Grey Phalarope, Pomarine and Long-tailed Skuas and Sabine's Gull into Liverpool Bay. These then pass the mouth of the Dee on their way back out into the Irish Sea.

The marshes and shoreline support a range of winter birds such as Hen Harrier, Merlin, Peregrine, Water Rail and Short-eared Owl. There is often a wintering flock of finches, anywhere between Gronant and Point of Ayr, including Brambling, Twite and occasional Shorelark, Lapland and Snow Buntings.

Scarce visitors to the point have included a Blue-winged Teal, American Golden Plover, Marsh and Buff-breasted Sandpipers, Wilson's Phalarope, Laughing Gull, Forster's Tern, Red-rumped Swallow, Tawny Pipit and Dusky Warbler.

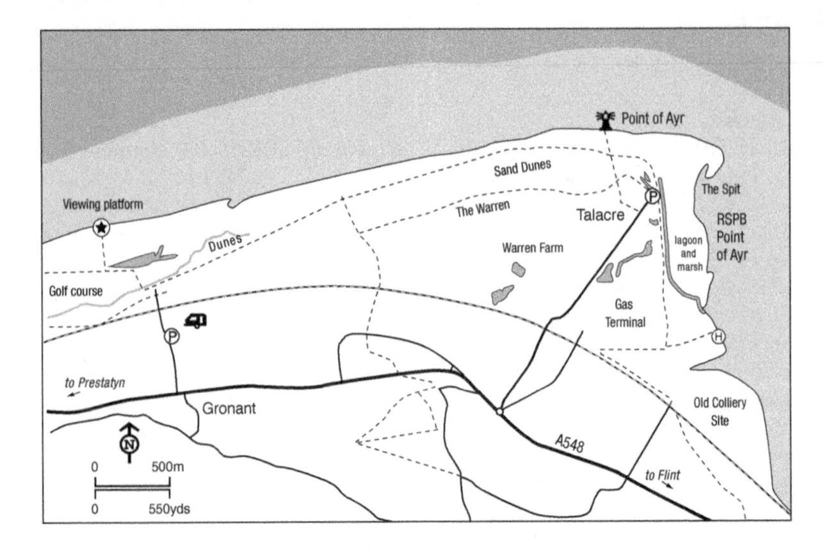

Timing

For seawatching, a visit at dawn, suitably equipped, is best. A visit to the estuary means careful checking of tide tables to ensure that one's arrival is around the vital high-water hours. For searching along the strandline for small passerines or the nearby dunes and marshes for birds of prey, the time of day is less critical.

Access

Leave the A548 Prestatyn to Flint road at SJ114834, by the unclassified road for Talacre and the coast, and park at the Lighthouse Inn. From here, one can walk west along the shore, backed by dunes, or east towards the point, or south for a

CALENDAR

Resident: Few breeding species but some wildfowl and waders seem to be present throughout the year, including Little Egret, Cormorant, Shelduck, Mallard, Oystercatcher, Ringed Plover, Curlew, Redshank, Black-headed, Lesser Black-backed and Herring Gulls.

December–February: Red-throated Diver, Great Crested Grebe, Brent Goose, Wigeon, Teal, Pintail, Shoveler, Common Scoter, Goldeneye, Red-breasted Merganser, Hen Harrier, Merlin, Peregrine, Golden and Grey Plovers, Lapwing, Knot, Sanderling, Dunlin, Snipe, Black-tailed and Bar-tailed Godwits, Turnstone, Short-eared Owl, Twite and occasional Lapland and Snow Buntings.

March–May: Winter visitors departing but wader passage continues well into May, with Whimbrel coming through from mid-April. First Sandwich Terns seen about the beginning of April, when passerine passage noted with birds such as Wheatear and Ring Ouzel among the dunes and Sand Martins and Swallows overhead.

June–July: Resident species present, parties of terns throughout period with numbers building up from mid-July. First returning waders occur.

August–November: Wader numbers rapidly increase during August, main influx of winter visitors from mid-October. Seabird passage offshore from mid-August until late September.

short distance beside the estuary to where there is a viewing screen at the edge of the gas terminal. Great care must be exercised so that the wader and tern roosts at high tide are not disturbed. Equal caution applies if you head west to see the Little Terns at Gronant.

81 NERCWYS MOUNTAIN

OS Landranger Map 116 & 117
OS Explorer Map 265
Grid Ref: SJ214589
Website:
https://naturalresources.wales/days-out/places-to-visit/north-east-wales/coed-nercwys/?lang=en

Habitat

Some 3 miles (4.8 km) south of the market town of Mold (Yr Wyddgrug), Nercwys Mountain, which rises to nearly 1,300 feet (400 m), is perhaps best known for its small but sustainable population of Nightjars, the large forestry plantation with open, clearfell areas proving ideal habitat. Denbighshire County

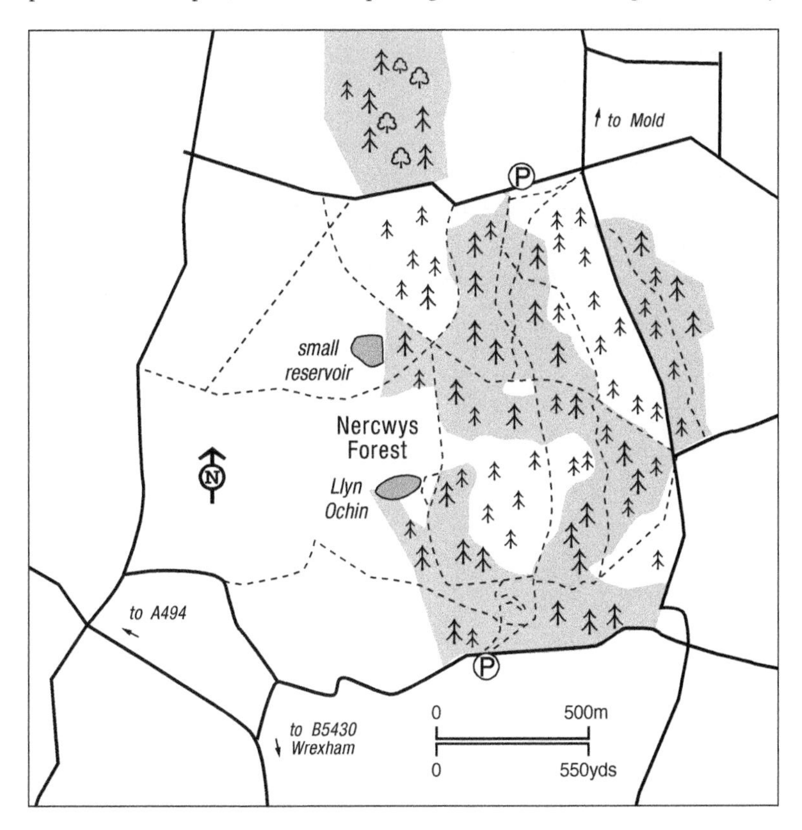

Council and Natural Resources Wales have developed several paths through the plantations, providing plenty of possibilities to observe the plentiful wildlife.

Species

Summer is best for birds with a chance of roding Woodcock, the 'cock of the woods', during early summer along with Nightjars and possible Long-eared Owls, all of which may be seen in the crepuscular hours. Somewhat easier to encounter are Redstarts, Blackcaps, Whitethroats, Chiffchaffs and Willow Warblers, while Firecrests have bred here on at least two occasions. A small reservoir close to the western boundary of the woodlands is always worth a peep, in late summer for passage waders, Common and Green Sandpipers and Greenshank have all been recorded, while in winter Mallard, Teal, Tufted Duck, Coot and Moorhen are present with good numbers of finches which include Crossbills, flocks of up to 40 having been recorded, and occasionally Bramblings in the woods.

Timing

In summer, late evening around dusk is best for Woodcock and Nightjars, though both may test your fortitude.

Access

Access is best from the A494 in Gwernymynydd and the plantation forest can be entered from either the north or south entrances; there is limited parking at both. The southern end is best as this is where the Nightjars are usually found.

CALENDAR

Resident: Kestrel, Sparrowhawk, Buzzard, Tawny Owl, Great Spotted Woodpecker, Great, Blue and Coal Tits, Goldcrest, Raven, Chaffinch, Siskin, Lesser Redpoll.

November–February: Fieldfare, Redwing with the possibility of Crossbill and Brambling.

March–June: Woodcock and Long-eared Owl. Arrival of summer visitors including Nightjar, Grasshopper Warbler, Whitethroat, Garden and Willow Warblers and Chiffchaff.

July–October: The summer migrants leave, but the woods come alive with wandering tit flocks, among which may be Long-tailed and Marsh Tits. Check the small reservoir for passage waders and wildfowl.

GWENT

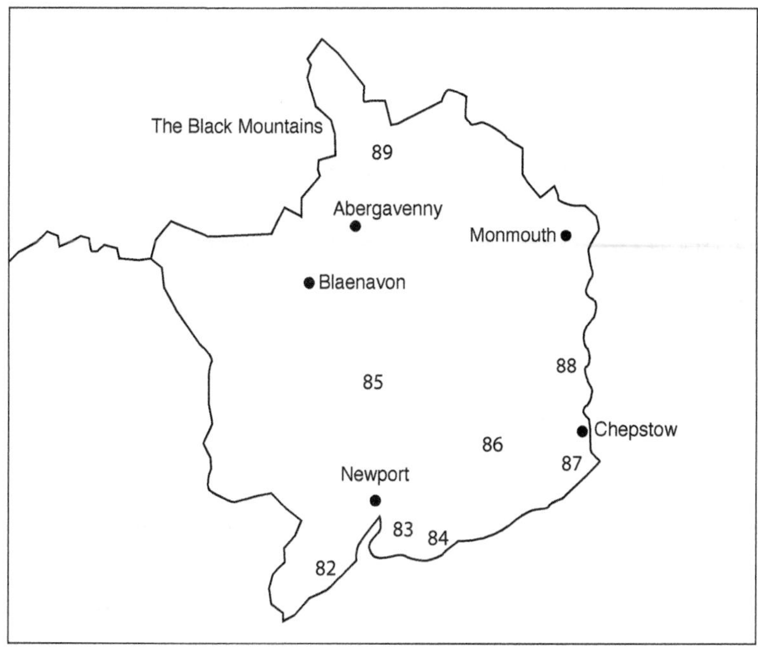

82 Peterstone Wentlooge
83 Newport Wetland Reserve
84 Goldcliff Lagoons and Goldcliff Point
85 Llandegfedd Reservoir
86 Wentwood
87 Blackrock and Sudbrook
88 Lower Wye Valley
89 Blorenge

Relative to its neighbours, Gwent is a small county and, being in the east of Wales, is more Anglicised. Bounded on the east by the border with England and the Wye Valley, with its bountiful deciduous woodlands and streams, Gwent is home to Goosanders, Buzzards, Goshawks, summering warblers, Redstarts, Pied Flycatchers, Dippers, Grey Wagtails and Kingfishers. In the north of the county are the Black Mountains which merge into the Brecon Beacons National Park – an area often overlooked by birdwatchers, but one that is home to Red Grouse, Ring Ouzel and other mountain species.

The funnel-shaped estuary of the Severn, unique in Great Britain, varies in width from 2 miles (3.2 km) just below the upper Severn Bridge to some 11 miles (17.6 km) at the Glamorgan boundary 20 miles (32 km) to the south-west: a vast area of intertidal mud and sandbanks. Behind the seawalls and the tidal sluices, which keep the Severn at bay, lie the Caldicot Levels, and west of the River Usk the levels of Peterstone. Together these are one of the largest remaining areas of unimproved wet grassland in lowland Britain, with Lapwings and the last breeding Yellow Wagtails in Wales. One of the fen remnants is Magor Nature Reserve, owned by Gwent Wildlife Trust, a 59-acre (24-ha) site in which are preserved some old, unimproved pastures, ditches, pools and scrub woodland. Close to the coast two National Nature Reserves – the Newport Wetlands Reserve and Goldcliff Lagoons – stand head and shoulders above the rest, created as part of the compensation for the Cardiff Bay development, at what was Uskmouth Power Station and at Goldcliff. The array of pools, scrapes and reedbeds have transformed this area into one of the top birdwatching sites in Wales.

82 PETERSTONE WENTLOOGE

OS Landranger 171
OS Explorer 152
Grid reference ST265795
Website: https://www.gwentwildlife.org/nature-reserves/peterstone-wentlooge-marshes-sssi

Habitat

Of all the places in Wales we would have liked to watch birds a thousand years ago, this is it! What gems might have been recorded when, what is now known as the Levels, was a huge marshland. About AD 200, the landscape began to change as slaves together with soldiers, probably from the garrison at Caerleon just 9 miles away, began to construct a great seawall. Over the ensuing centuries this has been improved, behind which a network of creeks, channels and wide ditches – the reens – developed eventually into the modern landscape of grazing pastures. This is one of the largest handcrafted landscapes in Great Britain and is much nibbled at by industrial facilities, landfill sites and a golf course.

Wentlooge Level, with a 4-mile (6.4 km) frontage between Sluice Farm and St Bride's Wentlooge, is distinguished by regularly planed blocks of long, narrow and generally straight-sided fields, separated by drainage channels, the latter now an important wildlife feature, especially when well fringed with reeds. Beyond the seawall, which provides a superb vantage point, the saltmarsh and mudflats of the mighty Severn Estuary extend at low tide almost half-a-mile from the shore. Gwent Wildlife Trust owns the fishing rights to a section of foreshore and has an agreement with the Wentlooge Wildfowling and Conservation Association regarding a no shoot zone at the mouth of Peterstone Gout – 'gout' from the Old English word for watercourse, channel or drain.

Species

Rare visitors here have included Glaucous, Mediterranean, Ring-billed and Sabine's Gulls. Avocets were once rare but now are nesting not so far away at Goldcliff Lagoons, while Stone-curlew, Kentish Plover, Greater Sand Plover, Pectoral, Broad-billed and Buff-breasted Sandpipers, Long-billed Dowitcher, Lesser Yellowlegs and Spotted Sandpiper offer a list to encourage every visitor to put that extra effort in when scanning the wader flocks of Ringed Plover, Dunlin, Knot, Curlew and Redshank or counting winter wildfowl including Shelduck, Mallard, Teal and Wigeon. The nearby fields and woods must be included in any summer visit with Whitethroats and Blackcaps in the hedgerows, Cetti's, Reed and Sedge Warblers in the reedbeds.

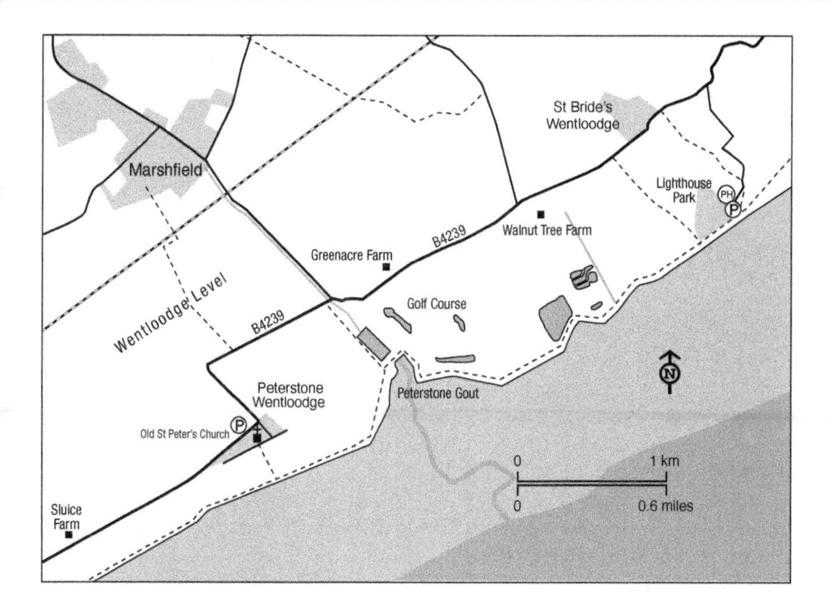

Timing

Arrive at least two hours before high-water, or if wishing to observe autumn migration, be settled in position by daybreak whatever the tidal situation.

Access

The Wales Coast Path follows the top of the massive seawall embankment which you may share with cattle and horses, and which may be reached from the limited parking across the road from Old St Peter's Church, Peterstone Wentlooge and close to the Six Bells public house. Taking the footpath next to the church, the delights of the shore just a short distance away are soon revealed with Peterstone Pill and Gout, which must not to be missed, just away to the left. Alternatively, park close to the bridge over the drainage channel at Sluice Farm. For St Bride's Wentlooge, leave the B4239 in the village for The Lighthouse pub, Shipwreck Cafe and a pay-and-display car park from where the Wales Coast Path is reached by steps up the embankment.

CALENDAR

Resident: Mute Swan, Canada Goose, Shelduck, Cormorant, Little Egret, Grey Heron, Oystercatcher, Ringed Plover, Lapwing, Curlew, Redshank, Black-headed, Lesser Black-backed, Herring and Great Black-backed Gulls, Skylark, Meadow Pipit, Cetti's Warbler, Linnet, Reed Bunting.
December–February: Wigeon, Teal, Pintail, Shoveler, Common Scoter, Merlin, Peregrine, Grey Plover, Knot, Dunlin, Snipe, Bar-tailed and Black-tailed Godwits, Turnstone, Short-eared Owl.

March–May: Sanderling, Bar-tailed Godwit, Whimbrel, Greenshank, Little Gull, Cuckoo, Lesser Whitethroats in the mature hedges, Reed and Sedge Warblers in the reedbeds.
June–July: Shelduck depart on moult migration, Greenshank, Ruff, Little Gull.
August–December: Little Stints, Curlew and Common Sandpipers on passage, while waders for the winter begin to arrive including Greenshank and Turnstone. Shelduck return in October, as do wintering duck. Prolonged stormy weather may result

in oceanic seabirds occurring in the Severn Estuary, so well worth a visit the stronger the winds the better. Summer visitors including Wheatears and Yellow Wagtails depart and winter thrushes arrive, and always a chance of a Kingfisher sweeping by along one of the reens or a Water Pipit at Sluice Farm.

83 NEWPORT WETLAND RESERVE

OS Landranger Map 171
OS Explorer Map 152
Grid Ref: ST331830
Website: https://www.rspb.org.uk/reserves-and-events/reserves-a-z/newport-wetlands/

Habitat

A significant expanse of the Gwent Levels is protected as a series of reserves, most notably the Newport Wetlands Reserve, co-managed by Natural Resources Wales and RSPB. This lies on an extensive part of the Caldicot levels to the southeast of the City of Newport and was created as mitigation for the loss of mudflats of Cardiff Bay, covering over 1,082 acres (438 ha) between Uskmouth in the west to Goldcliff in the east. The reserve is diverse, with deep-water pools in the west, remnants of ash settling pools from the now defunct Uskmouth Power Station, reedbeds, shallow scrapes and grazing meadow, all alongside the Severn foreshore.

Since construction, the area has quickly established itself as a top birdwatching locality, with American wader vagrants being recorded almost annually as well as a good supporting cast of other species scarce in this part of Wales. There are plenty of footpaths along the flood defences, at Uskmouth Ashponds (remnants of Uskmouth Power Station) at the western end and overlooking the Severn Estuary at Goldcliff (Site 84) at the eastern end. The Reserve is also an excellent place to see a wide variety of other wildlife, such as orchids, butterflies, dragonflies and Otters.

Species

The created reedbeds, saline lagoons and wet grasslands have already attracted a wealth of wetland birds, with significant numbers of overwintering Wigeon, Shoveler, Teal, Shelduck and Pintail. Bitterns are present throughout the year and may be heard booming during the breeding season, while Hen Harriers and Short-eared Owls are winter visitors. Up to 120,000 Starlings use the reedbeds to roost in winter. The scrub around the reedbeds has in excess of 30 pairs of Cetti's Warbler resident all year. Lapwing, Redshank, Oystercatcher and Ringed Plover breed on the wet grassland and saline lagoons. Little Egrets are resident all year and breed at Magor Marsh. Marsh Harriers first nested here in 2016 and have done so every year since. There is a small population of Bearded Tits, about six pairs breed each year, with youngsters often seen at other sites in South Wales during the autumnal dispersal.

Unusual visitors to the area in recent years have included Ring-necked Duck, Great White Egret, Cattle Egret, Spotted Crake, Pallid Swift, Red-rumped Swallow, Aquatic Warbler, Savi's Warbler, Woodchat Shrike and Bluethroat.

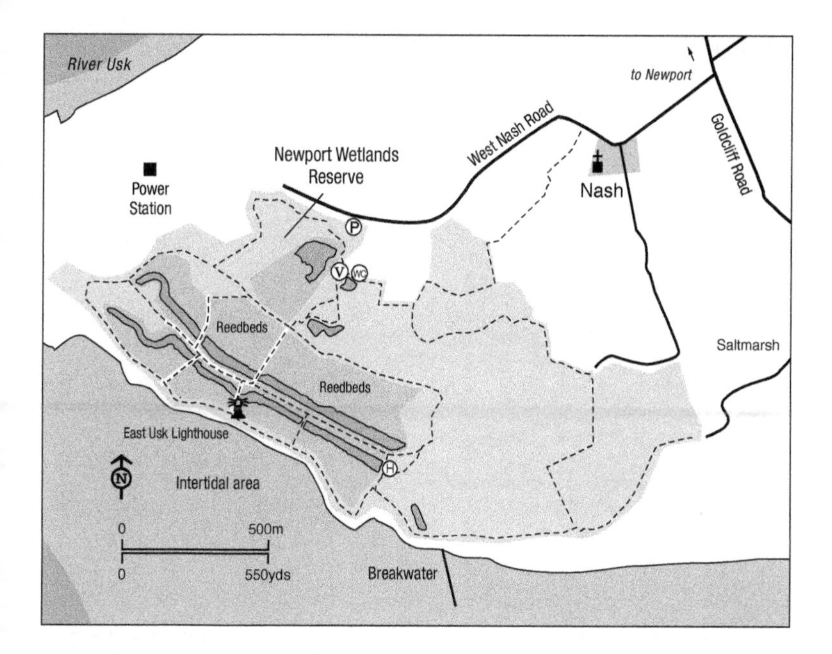

Timing

Anytime.

Access

Newport Wetlands Reserve car park is found at the western end of the site on West Nash Road, between Nash Village and the Uskmouth Power Station, grid reference: ST334834. From the M4 Junction 24, take the A48 to the Newport Retail Park, following the 'brown duck' signs, taking the first exit on the roundabout onto Queensway Meadows, then the third exit at the next roundabout onto Meadows Road, after 1.6 miles (2.5 km), turn right onto West Nash Road and continue to the reserve car park, about a mile. The car park is open from 9am until 5pm (or dusk in winter); RSPB members have free parking, there is a charge for non-members. There is a wheelchair-accessible toilet in the car park.

There are a number of footpaths around the reserve with a cycle track around part of the perimeter at Uskmouth, before it meanders alongside flooded fields at nearby Saltmarsh. From the car park, there are three waymarked trails varying in length from 2.74 km (1.7 miles) to 4.36 km (2.7 miles). The paths are surfaced but may be rough in places with some loose stones. The paths cross level ground, although there are a few slopes to climb the reedbeds which are 5m higher than the original land.

CALENDAR

Resident: Cormorant, Grey Heron, Bittern, Little Egret, Shelduck, Mallard, Marsh Harrier, Kestrel, Moorhen, Oystercatcher, Ringed Plover, Lapwing, Snipe, Redshank, Black-headed, Lesser Black-backed, Herring and Great Black-backed Gulls, Kingfisher, Skylark, Meadow Pipit, Pied Wagtail, Cetti's Warbler, Reed Bunting.
December–February: Wigeon, Teal, Pintail, Pochard, Tufted Duck, Scaup, Hen Harrier, Peregrine, Coot, Ringed and Grey Plovers, Knot, Curlew, Turnstone, Common Gull, Short-eared Owl, Rock Pipit.
March–May: Winter visitors depart but some species, especially the waders, stay well into spring (a few stragglers may even summer), while passage birds during April include Sanderling, Whimbrel and Common Sandpiper. Wheatears on the seawall from late March, Swallow passage commences soon after; by end of April, summer visitors such as Sedge and Reed Warblers, Whitethroat, and Lesser Whitethroat have arrived.
June–July: All summer residents present. Wader numbers show a steady increase on estuary as July proceeds, while most Shelducks depart on moult migration.
August–November: Wader numbers continue to increase and the first wintering ducks appear during September, by which time most summer visitors have left. Movements of Skylarks, Meadow Pipits, autumn thrushes, Starlings, finches and Woodpigeon from mid-October.

84 GOLDCLIFF LAGOONS AND GOLDCLIFF POINT

OS Landranger 171
OS Explorer 152
Grid Reference ST369829
Website: https://www.thewildlifeoculus.com/p/goldcliff-map.html

Habitat

Named after a limestone cliff remarked on in Roman times, above a bed of yellow mica which glittered in the sun, great energies were devoted to reclaiming the fertile lands along the Severn shore, extending eastwards from the River Usk. In one of the worst natural disasters affecting the coast of Great Britain, a storm surge on 30 January 1607 caused much destruction and loss of life each side of the Bristol Channel. A glance in St Mary Magdalene's Church, Goldcliff, reveals a plaque installed soon after the storm surge, recording the level that the water reached and the effect on the parish: 22 people were drowned and the damage was valued at more than £5,000 (nearly £700,000 in today's money).

As with Newport Wetlands (Site 83), the creation of the lagoons here was part of the compensation for destruction of the Taff Ely Estuary and the development of Cardiff Bay, and was declared a National Nature Reserve in 2008. Three spacious hides overlook a mosaic of lagoons with both fresh water and saline habitats, wet grassland, reedbeds and saltmarsh, while beyond the seawall is the Severn Estuary – with the world's second-highest tidal range and as always the subject of plans for a barrage. Close by the Lagoons, Goldcliff Point is renowned for seawatching, as one guidebook says: 'If you want to see seabirds in Gwent, this is the place to try. As always, the aftermath of a period of strong winds in late summer and autumn is likely to produce dividends for the dedicated watcher.'

Species

Pride of place without any doubt must go to the Avocets, a pair of which were observed displaying in 2002. They returned to nest the following year, the first in Wales, fledging four young. Numbers increased over the rest of that decade to about 50 breeding pairs fledging around the same number of young each year. Did Avocets nest here 2,000 years ago? The remains of one among 1st-century Roman food refuse at the Caerleon Roman Fortress Baths just 5 miles (8 km) from Goldcliff are rather intriguing.

The first hide offers good views over Monk's Lagoon, not only of Avocets but also for the wealth of other waders present, including Oystercatcher, Lapwing, Ringed Plover, Little Ringed Plover and Redshank, all of which nest. Grey Plover, Knot, Whimbrel, Curlew, Black-tailed and Bar-tailed Godwits, Curlew Sandpiper, Dunlin, Snipe, Common Sandpiper, Green Sandpiper, Greenshank, all occur on passage or remain for much of the non-breeding months. Shelduck are present throughout the year together with Mallard, Teal, Gadwall and Shoveler, joined in winter by Wigeon and Pintail.

The shore hide provides a splendid shelter on a blustery day and really comes into its own from immediately after high-water when returning waders, some you may already have enjoyed on the lagoons, swoop in to feed. Among scarce visitors have been a Long-billed Dowitcher in March 2005 and Black-winged Pratincole in June 2001, while the first three records of the White-rumped Sandpiper for the county were all at Goldcliff in 1995, 1999 and 2003. Other gems have included Glossy Ibis, Kentish Plover, Temminck's Stint, Baird's, Broad-billed, Pectoral and Buff-breasted Sandpipers, Red-necked Phalarope. The first Hudsonian Whimbrel for Wales and a second for Great Britain was on

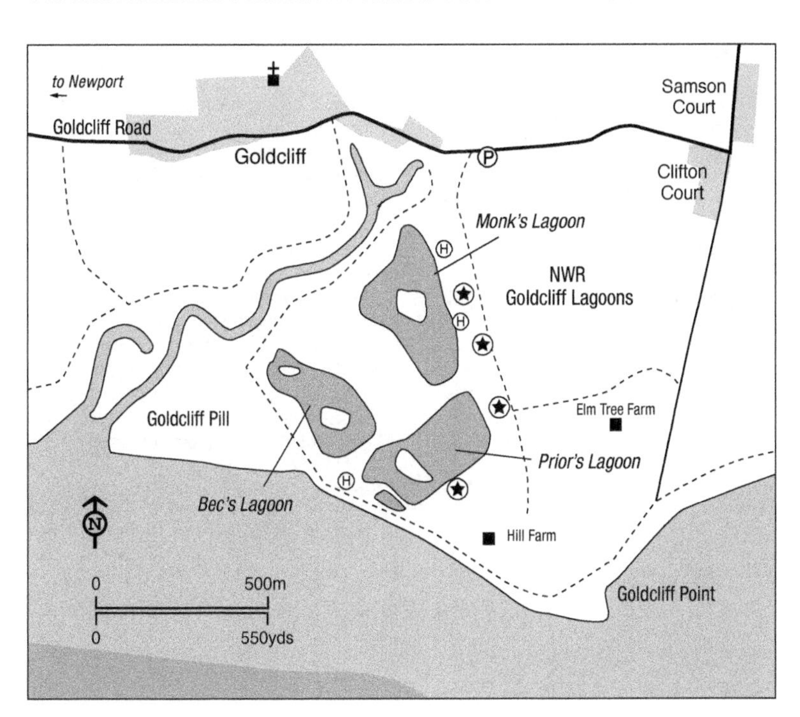

one at the lagoons for two days in early May 2000, while two years later, almost to the day and presumed to be the same bird, one spent another two days at Goldcliff; these remain the only sightings for Wales.

Stormy weather in the autumn means seawatching opportunities for the hardy at Goldstone Point. Fulmars, Gannets, Kittiwakes, Manx Shearwaters and Storm Petrels from the Pembrokeshire colonies are most likely. Arctic and Great Skuas together with passage Arctic and Common Terns from farther afield are likely to add that extra excitement for the few hours of watching, while there can be rewards like Leach's Petrel, Pomarine and Long-tailed Skuas, Sabine's and Little Gulls, Little and Black Terns, Guillemots and Razorbills.

Timing

Mornings are best light-wise for the lagoons, while for the estuary, check the tide tables and begin viewing from high-water or an hour or so before.

Access

Rather limited roadside parking in Goldcliff or immediately to the east, taking care not to impinge on the residents. From just east of the village a path leads to the nature reserve, continuing below a high embankment upon which the first hides and viewing platforms are situated, to reach the shoreline and a section of the Wales Coast Path. For the Pill take the coast road passing eastwards through Goldcliff to where, between Clifton and Samson Court, the road makes a right-angle turn left, immediately on the right a minor road leads direct to the shore from which it is protected by the seawall. Simply choose the most comfortable position on the seawall towards Goldcliff Point itself.

CALENDAR

Resident: Cormorant, Grey Heron, Shelduck, Mallard, Teal, Gadwall, Shoveler, Tufted Duck, Moorhen, Oystercatcher, Avocet, Ringed Plover, Lapwing, Curlew, Redshank, Reed Bunting.

December–February: Wigeon numbers increase, with up to 3,000 shared between Goldcliff and Newport Wetlands.

March–May: Winter visitors depart while the first Wheatear on the seawall will soon be followed by more and in April and May passages of Swallows and martins. At one time Yellow Wagtails featured strongly on this coast in mid-April, now numbers are very small, while never overlook the possibility of a Blue-headed Wagtail.

June–July: Most Shelduck depart on moult migration. First passage waders by late July.

August–November: Shelduck numbers increase from September onwards as arriving waders reach a peak, particularly Curlew, Dunlin and Redshank and for the diligent observer sightings of Little Stint, Sanderling, Curlew Sandpiper, Ruff and Spotted Redshank. Seawatching at Goldcliff Point can often bring dividends during stormy weather. Later passages of winter thrushes and Starlings heading for the opposite side of the Bristol Channel and always the possibility, usually as single birds, of Snow Buntings on the coastal embankment during October and November.

85 LLANDEGFEDD RESERVOIR

OS Landranger Map 171
OS Explorer map 152
Grid Ref: ST326993
Website: http://www.llandegfedd.co.uk/

Habitat

Situated just east of Pontypool, and separated from the valley of the Afon Lwyd to the west and the Usk to the east by ridges rising to about 500 feet (150 m), Llandegfedd Reservoir covers nearly 400 acres (162 ha) and was completed in 1963 to supply water for Cardiff. The Sôr Brook flows from the reservoir and joins the Usk at Caerleon, nearly 5 miles (8 km) to the south. The land surrounding the reservoir is mainly agricultural, with woodland on the steeper slopes. The lake has a multi-activity centre for fishing, water sports, walking and nature, run by Welsh Water.

Species

Llandegfedd Reservoir quickly established itself as the most important open freshwater site for birds in south-east Wales. Divers are generally very scarce, with Great Northern being the most frequent; Great Crested Grebes are present throughout the year, several pairs nest, while in late summer numbers rose at one time to over 100 birds, more recently no more than half this. Slavonian and Black-necked Grebes are occasional visitors, with there being a number of spring occurrences of Black-necked Grebes in their superb summer plumage. Cormorants are present throughout the year with late-summer gatherings of up to 80 or so, but numbers then quickly decline to a winter maximum of about 20. The heronry generally contains approximately 15 nests, with birds always to be seen on the reservoir shores with the addition of Little Egrets, mainly from autumn onwards.

It seems amazing that prior to 1960 the only record of the Bewick's Swan in Monmouthshire was one found dead near Newport in 1909. From 1960 it quickly became a regular winter visitor, with no fewer than 20 in December 1963 at Llandegfedd which began to be used as a roosting site for those feeding in the Nedern Valley. Numbers of Bewick's Swans wintering in Great Britain flourished during the 1970s as birds relocated from Holland. Then more change: from the 1990s numbers declined so that the Bewick's Swan in Wales is now an irregular winter visitor in vanishingly small numbers, with just two present from 25 January to 7 February 2019.

Winter ducks include several hundred Wigeon, by far the most numerous, followed by smaller numbers of Mallard and Teal, the peak count of the latter being 940 in 1986, more recently about 500, these in December. Pochards have declined while Tufted Ducks occur in almost all months, their numbers a mere shadow of the several hundred 30 years ago. Small parties of Goldeneyes visit the reservoir throughout the winter, sometimes remaining until late May, as do Goosanders, often with birds flighting in at dusk, presumably from the nearby rivers. Red-necked Grebe, Smew and Long-tailed Duck have been recorded in the past but are not regular.

The Mandarin Duck, which was not admitted to the British List until 1971, first attempted to breed in Wales in Denbighshire in 1974 (although it was unsuccessful) and Glamorgan in 1985, with the first sighting in Gwent two years later at Llandegfedd Reservoir. Since then, records of Mandarin in eastern Wales have become too numerous to merit individual mention in county reports.

The Hobby has nested in Gwent since 1966 and now breeds at a minimum of 22 sites in the county being regularly seen over the reservoir, as are Osprey on their way to and from their breeding grounds. Other birds of prey have included Goshawk, Sparrowhawk, Buzzard, Kestrel, Merlin and Peregrine.

Waders recorded on passage include Oystercatcher, Ringed Plovers, Dunlin, Common and Green Sandpipers and Turnstone. Other waders occur less frequently, these include Little Ringed Plover, Wood Sandpiper, Little Stint and Avocet. Large numbers of Black-headed, Common, Lesser Black-backed and Herring Gulls visit the reservoir to roost throughout the winter months. Other less usual species recorded here include Kittiwake, Glaucous and Iceland Gulls. Terns pass through in small numbers in most autumns, so 36 Black Terns on 21 August 1970 and 20 on 8 September the same year was remarkable. Equally remarkable were the first two records of the White-winged Black Tern in Gwent, both at Llandegfedd, at the end of May 1991 and 24 September 2000.

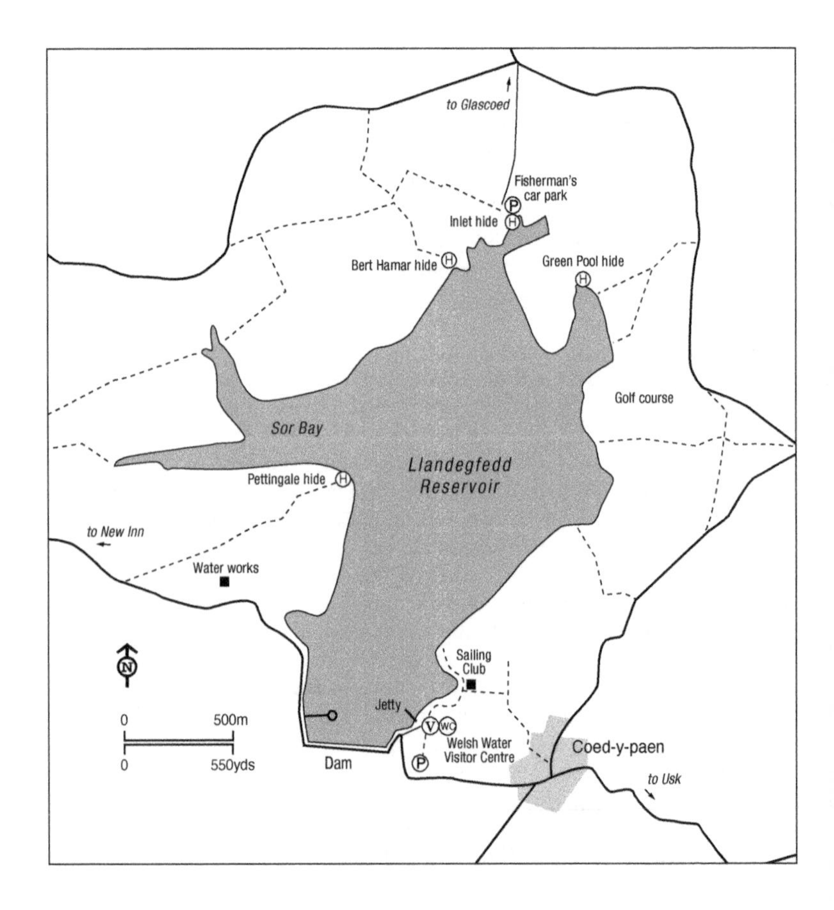

Timing

Not too critical. Sailing is popular here, so there is always a chance of disturbance.

Access

Views of the lake are possible from Welsh Water's visitor centre on the eastern side of the reservoir or from the Fisherman's Car Park (open during the summer season 8:30am – 5:30pm daily) at the northern end. There are three hides situated at the northern end and one near Sluvad Water Treatment Works, south of Sôr Bay, at Pettingale Point.

Access is restricted to certain areas of the reservoir during the winter months to protect the wildfowl from disturbance. Between 1 October and 1 March the visiting birdwatcher requires a permit, which can be purchased from the visitor centre (£8 in 2021). For entry to the northern end during the winter months, visitors will need a key for the car park gate (£6 in 2021, obtainable from the visitor centre). Enjoy a long circular walk of just over 6 miles (10 km) through the countryside and woodland surrounding the reservoir commencing from the visitor centre car park, which then heads along the western side towards Glascoed, before returning on the eastern side. Some moderate climbing is involved, with the walk reaching a peak height of over 450ft (150 m). From the high points there are great views over the reservoir.

CALENDAR

Resident: Great Crested Grebe, Cormorant, Grey Heron, Mute Swan, Mallard, Buzzard, Red Kite, Sparrowhawk, Coot, Water Rail, Grey and Pied Wagtails, woodland passerines including Goldcrest and Lesser Redpoll.

December–February: Wigeon, Teal, Pochard, Tufted Duck, Goldeneye, Goosander, Lapwing, Snipe, Black-headed, Common, Lesser Black-backed and Herring Gulls.

March–May: Winter visitors depart, though some, especially Tufted Duck, remain until May. Summer passerines from early April onwards, including Nightjars, when Swallows and Sand Martins start hawking over the water, being joined in May by Swifts.

June–July: Summer visitors including Hobbies. Depending on water levels, returning waders such as Green Sandpiper and Little Ringed Plover may call.

August–November: Tern and wader passage in August and September, from then on winter visitors begin to arrive. Fieldfare and Redwings from early October, while there is always a chance of Siskins and Redpolls in the Alders beside the reservoir and Woodcock in the woodlands.

86 WENTWOOD

OS Landranger Map 171 & 172
OS Explorer Map 152 and OL14
Grid Ref: ST427949
Websites:
https://www.woodlandtrust.org.uk/visiting-woods/woods/wentwood/
https://naturalresources.wales/days-out/places-to-visit/south-east-wales/wentwood/?lang=en

Habitat

Wentwood is the largest ancient woodland in Wales, a woodland known as Coit Gwent when first mentioned in the Book of Llandaff in the 12th century, when it stretched 15 miles (24 km) from the outskirts of Newport to the Wye Valley. Although now much reduced in size, Wentwood still extends over some 2,500 acres (1,000 ha), and so is quite easy to get lost in! Despite domination by conifers, relics of the ancient woodland remain, together with evidence of earlier activities such as sunken lanes, boundary banks and charcoal hearths. In 2006, Coed Cadw (the Woodland Trust in Wales) purchased about a third of the woodland, the rest being owned by Natural Resources Wales. Present management is aimed at reducing in stages the coniferous forest and replanting with native broadleaved species.

Species

Wentwood offers an interesting mixed mosaic habitat and a variety of bird species which include Buzzard, Sparrowhawk, Goshawk, Woodcock (Wentwood together with the Wye Valley its breeding stronghold in Gwent), Nuthatch, Treecreeper, Crossbill, Siskin and Lesser Redpoll. Nightjars have long been known – Wentwood has been described as their main location in the county though fortunes have changed across the years. Up to eight pairs in the 1970s had risen to 28 pairs by 2004 then down to six pairs in 2017. Searching at dusk in the clearfelled areas can bring rewards. Other summer visitors include Cuckoo, Tree Pipit, Whinchat, Wood Warbler, Garden Warbler, Pied and Spotted Flycatchers. Wintering Great Grey Shrikes are occasionally reported in the

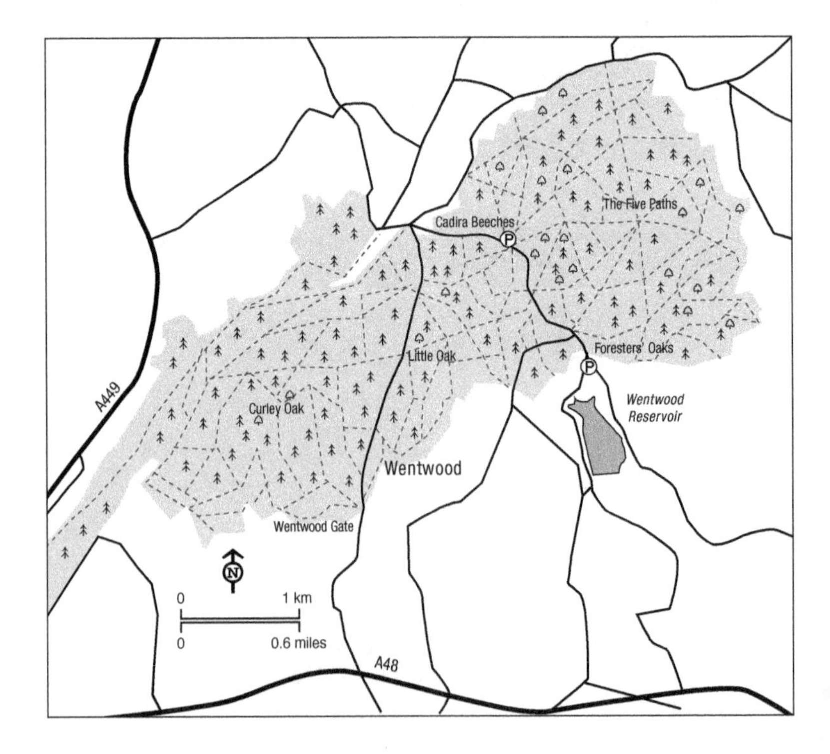

clearings, while Firecrests, which used to breed here before the turn of the millennium are still occasionally located by dedicated observers. A male Iberian Chiffchaff, the twenty-second for Great Britain, was discovered at Cadira Beeches where it sang from 10 May until 18 June 2010.

Wentwood Reservoir is always worth a look, the roads each side providing viewing opportunities. Great Crested and Little Grebes have nested in some years and are normally present in winter, along with Mallard, Gadwall, Mandarin Duck, Tufted Duck, Pochard, Coot and Moorhen. It has also held in the past Great Northern Diver, Slavonian Grebe, Long-tailed Duck, Ring-necked Duck and Smew.

Timing

All year.

Access

Good access by way of tracks, paths and bridleways which you can visit on foot, bicycle or horseback. The main entrances are located at the two car parks: Foresters' Oaks and Cadira Beeches, on the road to Usk. There are a number of other entrances, including Wentworth Gate which is closest to bus routes and the ancient Curley Oak. The roads each side of the reservoir have only limited parking opportunities.

CALENDAR

Resident: Coot, Red Kite, Buzzard, Sparrowhawk, Goshawk, Great Spotted Woodpecker, Goldcrest, Great, Blue, Marsh and Coal Tits, Nuthatch, Treecreeper, Grey Wagtail, Jay, Chaffinch, Siskin, Lesser Redpoll and Crossbill.

November–February: Great Crested and Little Grebes, Mallard, Mandarin Duck, Wigeon, Gadwall, Tufted Duck, Pochard, Coot, Fieldfare, Redwing, Brambling.

March–June: Woodcock, Nightjar, Whinchat, Tree Pipit, Wood, Willow and Garden Warbler and Spotted Flycatcher.

July–October: Roving tit flocks.

87 BLACKROCK AND SUDBROOK

OS Landranger 171/172
OS Explorer Outdoor Leisure 14
Grid Ref: ST515880

Habitat

Two thousand years ago, at this point, there was a ferry used by those travelling to Venta Silurum, modern-day Caerwent. A rail tunnel was completed in 1886 and here, where the Severn Estuary stretches just over 2 miles (3.2 km) to the English shore at Aust Warth, the second Severn Crossing, the Prince of Wales Bridge opened in 1996. This stretch marks the end of the River Severn and the start of the Severn Estuary. Three miles offshore and surrounded by sandbanks known as the Welsh Grounds is tiny Denny Island, home to some 55 pairs of Cormorants and the only breeding location of the Great Black-backed Gull in the county, with up to 30 pairs. Looming in the future, as it has been for decades, is the possibility of a tidal barrage. If it is built the loss of the intertidal area on this, one of the most important estuaries in Europe, a Special Protection Area, would have a huge impact on the almost 100,000 migratory and wintering birds the use these habitats, and see the end of lave net fishing here, the last of its kind carried out in Wales.

Species

Peregrines use the underside of the Severn Bridge as a vantage point and hunt over the saltmarsh and adjacent fields in winter, habitats that on occasion are alive with Skylarks, Meadow Pipits, Linnets, Reed Buntings and in winter Fieldfares and Redwings, occasionally a Merlin or Short-eared Owl. Sometimes there might even be an encounter with a Snow Bunting in late autumn or early winter. The drainage ditches fringed with reeds provide homes for summer-visiting Reed and Sedge Warblers and wintering Water Rails.

The shoreline from Sudbrook north, even at low water, unlike most of the estuary, is no great distance from the seawall and coast path, so affording much closer views of Shelduck and wintering Wigeon, Mallard and Teal, while waders like Oystercatcher, Curlew and Redshank are prominent. Careful scrutiny of the gull flocks may be rewarded by the presence of Mediterranean Gulls. To the west of Sudbrook, the intertidal area rapidly expands; the same waders are present but unless one has taken account of the time of low water are likely to be too far for observation from the coast path. Seawatching in autumn from the old fort at Sudbrook can prove exciting during stormy weather, though there may also be surprises at other times like a flock of about a thousand Manx Shearwaters one early June day.

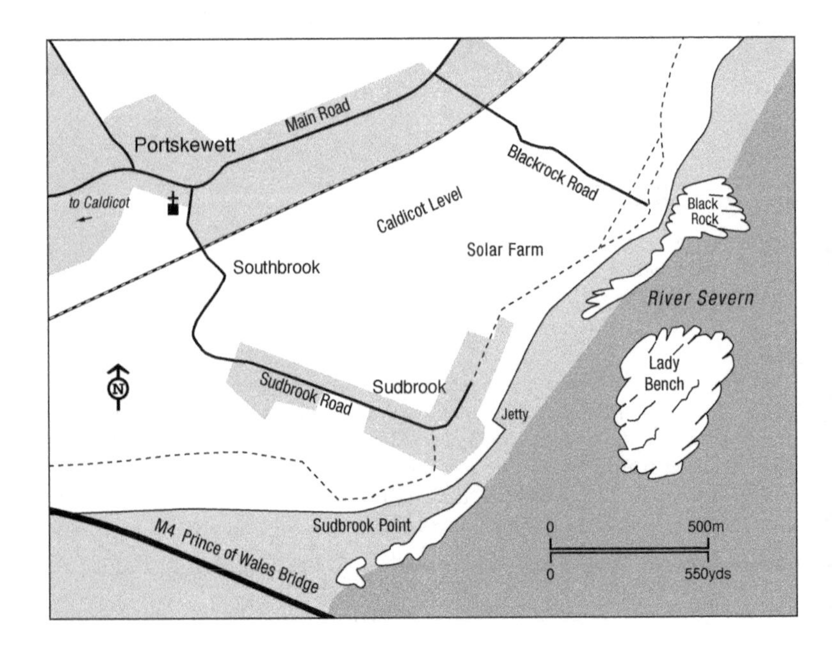

Timing

As with all large estuaries, a check of the tide table before you set forth is essential and if you can choose a day with high-water in early afternoon – then the light at both Sudbrook and more so at Black Rock will be behind you.

Access

From the east join the A48 on the western outskirts of Chepstow for 4 miles (6.4 km); having crossed the M4 turn left on the B4245 then left again for Portskewett. After just over a mile (1.6 km) the road swings sharp right, shortly afterwards turn left for Black Rock. From the west, leave the M4 at junction 23A taking the B4245 passing through Magor for Caldicott. On the eastern edge of the town turn right for Portskewett, continuing along Main Road before turning right into Black Rock Road. Park at the picnic site at the end of the road. The Wales Coast Path offers easy access in both directions along the bank above the estuary shore, while the Iron Age fort at Sudbrook is an excellent vantage point.

CALENDAR

Resident: One can almost describe the Shelduck as resident as many are only absent on their moult migration for a few weeks in July and August. Cormorant, Grey Heron, Little Egret, Oystercatcher, Curlew, Redshank, Great Black-backed and Herring Gulls, Peregrine, Meadow Pipit, while Rock Pipits are known to nest on Denny Island and occasionally occur along the mainland shoreline, also Pied Wagtail, Raven, Reed Bunting.

December–February: Wigeon, Teal, Ringed Plover, Grey Plover, Lapwing, Snipe, Jack Snipe, Dunlin, Redshank, Turnstone, Black-headed, Common and Mediterranean Gulls.

March–May: Bar-tailed Godwit, Whimbrel, both Pomarine Skua and Little Gull have occurred on spring passage.

June–July: Shelduck ducklings which having fledged make a rapid departure with the adults on a moult migration, the majority heading for the Wadden Sea – a World Heritage Site off the coasts of Denmark, Germany and the Netherlands, the largest tidal flats system in the world – where for a period they will be flightless.

August–December: Waders on migration with many remaining throughout the winter. Shelduck return from October onwards having completed their moult. Arctic and Great Skuas possible during stormy weather, Summer migrants depart, with Swallows and Martins prominent when setting off across the Bristol Channel, winter visitors from mid-October onwards.

88 LOWER WYE VALLEY

OS Landranger Map 162
OS Explorer Map OL14

Habitat

The rivers Severn and Wye rise almost together on the slopes of Plynlimon, little more than 2 miles (3.2 km) apart, but each plunges forth in different directions, their journeys of 220 miles (352 km) and 130 miles (208 km) respectively, embracing separate catchments, only for their estuarine waters to eventually mingle where Wales meets England near the ancient town of

Chepstow. The Wye, from not far below the equally ancient town of Monmouth, marks the boundary with England, to the east of which extends the Forest of Dean, while on the steep slopes on the Welsh side of the river during its tortuous course there are gorge woodlands of great importance for conservation. These include several that have been designated as National Nature Reserves, while the woodlands and the bat caves with their Greater and Lesser Horseshoe Bats have been notified as Special Areas of Conservation (SACs). Dormice and Yellow-necked Mice dwell in the woodlands while Otters frequent the river, and presumably include in their diet the rare Allis Shad, a member of the herring family also known as the May fish, as this is when they enter the Wye to spawn.

The Eagle's Nest, one of the best viewpoints in the Wye Valley, was constructed in 1828 for the Duke of Beaufort and is well worth the climb as the Cotswolds and the Mendips are visible on a clear day. No visit to the valley is complete without at least a brief halt at Tintern, the magnificent monastery founded by Cistercian monks in the 11th century.

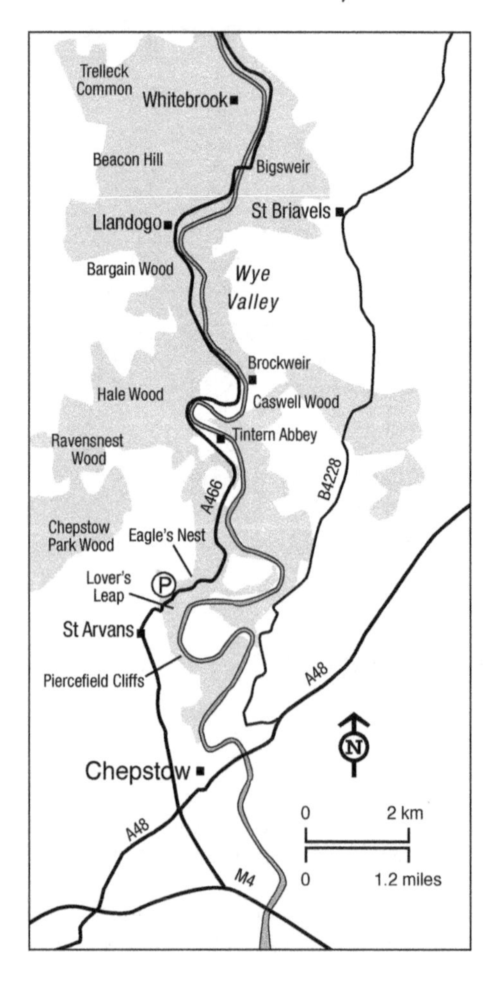

Species

The harbinger of spring, certainly as far as the river is concerned, must be Common Sandpiper. These birds appear here on passage before moving to waterways on higher ground further inland. Dippers and Kingfishers tend to be winter visitors, while there are always Grey Wagtails to be seen. There is a heronry at Piercefield close by where the river makes its final bend before reaching Chepstow, while Little Egrets never seem far away. A pair or two of Shelduck usually nest, while Mallard with broods are regularly seen along the whole river. Goosander and Cormorants fish on the river – the latter, sometimes 100 strong, roost on the cliffs at Piercefield, their nearest breeding colony is on Denny Island some 10 miles (16 km) distant in the Severn Estuary. The only breeding seabirds are the few pairs of Herring Gulls on the cliffs at Chepstow and on the castle itself. Sand Martins nest at several locations and are often the first summer migrant to arrive in the valley. Birds of prey in the valley woodland include Buzzard, Sparrowhawk, Tawny and Little Owls. Hobby, Peregrine and Red Kite numbers have increased, with as many as a dozen pairs of each now breeding in the county.

The Wye Valley is undoubtedly a key area for Hawfinches, especially woodland where there are good numbers of Wild Cherry trees, an excellent food source as the stones are easily opened by the stout bill of the Hawfinch to gain access to the kernel. Not all the woodland is broadleaved, so other species such as Goldcrests and Crossbills can sometimes be located in the conifer plantations, those at Trellech being one of the strongholds for the latter. Great Spotted Woodpeckers, Nuthatches and Treecreepers are frequently encountered, Green and Lesser Spotted Woodpeckers less so. Chiffchaffs are usually the first of the woodland summer visitors to arrive, quickly being followed by Willow Warblers, later Redstarts, Garden Warbler, Blackcap, Wood Warbler, Spotted and Pied Flycatchers.

Timing

Morning the best, the earlier the better in spring and early summer for the dawn chorus.

Access

The A466 traverses the valley with several minor roads leading westwards from it through the woodland. There are numerous footpaths and forest walks providing access to key sections of the woodlands, which tower on both the Welsh and English slopes above the Wye.

CALENDAR

Resident: Cormorant, Grey Heron, Mute Swan, Mallard, Goshawk, Sparrowhawk, Red Kite, Buzzard, Kestrel, Peregrine, Pheasant, Herring Gull, Little and Tawny Owls, Green, Great Spotted and Lesser Spotted Woodpeckers, Grey and Pied Wagtails, Wren, Dunnock, Robin, Blackbird, Song and Mistle Thrush, the tit family, Nuthatch, Treecreeper, Jay, Raven, Crossbill, Hawfinch.

December–February: Winter visitors such as Fieldfares, Redwings and Bramblings, Siskins and Lesser Redpolls in riverside alders.

March–May: Still some winter birds about, but Ravens have already commenced nesting before the first Sand Martins and Chiffchaffs arrive. By mid-April the woodland birds are well into the nesting season, the last migrants to arrive being Swifts and Spotted Flycatchers.

June–July: Breeding season now well advanced, some birds such as the tits which have until the early part of June been so conspicuous feeding young now seem to disappear, domestic duties finished. Mallard broods on the river, where there is always the chance of early passage Green and Common Sandpipers.

August–November: Departure of summer visitors, with Swifts screaming about the villages one day and gone the next, in early August. Blackcaps and Chiffchaffs the last of the warblers to depart, though both can overwinter. Big movements of finches and thrushes, not forgetting Starlings during October and November.

89 BLORENGE

OS Landranger Map 161
OS Explorer Map OL13
Grid Ref: SO269118
Website: https://www.breconbeacons.org/

Habitat

The Blorenge, 1,840 feet (560 m), is an area of upland heather moorland, with areas of Bilberry and patches of Purple Moor-grass. It represents an easterly bastion of the Brecon Beacons National Park immediately north of Blaenavon. The north-eastern ridge of the Blorenge affords commanding views over the Usk Valley, the town of Abergavenny and the nearby hills of the Sugar Loaf and Skirrid Fawr westwards to Mynydd Llangattock. Today it is difficult to imagine that these quiet uplands were once a key part of the industrial prosperity of Wales. Pen-fford-goch Pond, known locally as Keeper's Pond, once fed steam engines at a large forge and can be a good place to see several species of dragonfly, while elsewhere there are the remains of quarries, mills and tramroads, and the Monmouthshire and Brecon Canal. The Beech woodlands on the steep lower slopes are one of the most westerly native populations in Great Britain.

Species

Typical species that can be seen on the moorland include Red Grouse – one route is known as the Grouse Walk – which are best seen early mornings in spring and summer and here the most southerly natural population in Great Britain with 56 pairs in 2007 but just five territories in 2018. Wheatear, Meadow Pipit and Skylark together with Stonechat, Tree Pipit, Whinchat and Yellowhammer on the hillsides. There is a small pool, known as the Punchbowl, on the eastern side of the mountain which occasionally has Mallard. Coed-y-Prior Common and the woodland high above the Usk valley is also worth visiting, with Redstart, Pied Flycatcher, Wood Warbler, Marsh Tit likely species.

Once again there a timely reminder here that all things, or pretty well all things, are possible when it comes to birdwatching, as a male Marmora's Warbler spent nearly two weeks singing on the Blorenge in June 2010, this just the sixth for Great Britain since the first, also on upland, at Midhope Moor Yorkshire, in the summer of 1982.

Timing

Spring and summer.

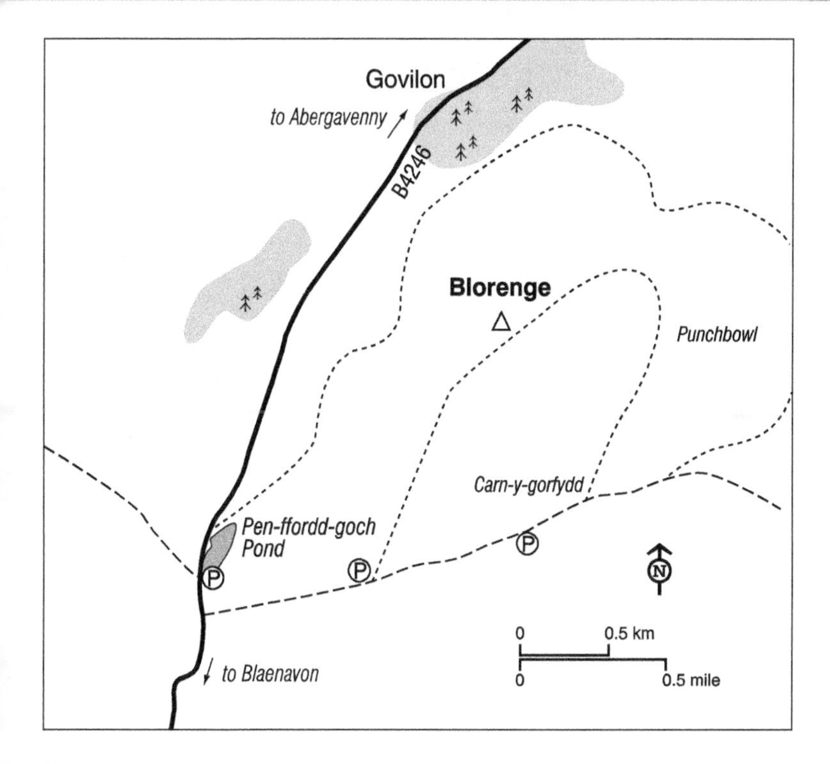

Access

Take the B4246 from Blaenavon to Abergavenny road. Turn right onto a minor road which takes you alongside the television masts. Park in the nearby Foxhunter's car park (please ensure that no valuables are left or show in the car). There is a track that then takes you towards the summit of the Blorenge and to the nearby cairn. The track continues to the edge of the mountain which provides spectacular views of the Usk valley and the town of Abergavenny. From this location there are often a large number of hang-gliders which fly out over the Usk valley to Abergavenny nearby.

CALENDAR

Resident: Red Grouse, Buzzard, Peregrine, Sparrowhawk, Meadow Pipit, Skylark, Stonechat, Marsh Tit, Raven and Yellowhammer.

November–February: Hen Harrier, perhaps from breeding locations further north in Wales.

March–June: Cuckoo, Wheatear, Whinchat, Tree Pipit, Redstart, Grasshopper, Wood and Willow Warblers, Whitethroat, Chiffchaff and Pied Flycatcher, Linnet. Ring Ouzel on passage.

July–October: Skylarks and Meadow Pipits cease singing and summer visitors depart, so that by October few birds are immediately obvious; however, an early morning visit may well reveal Starlings, finches, pipits and thrushes flying south along the edge of the uplands overlooking the Usk valley.

MEIRIONNYDD

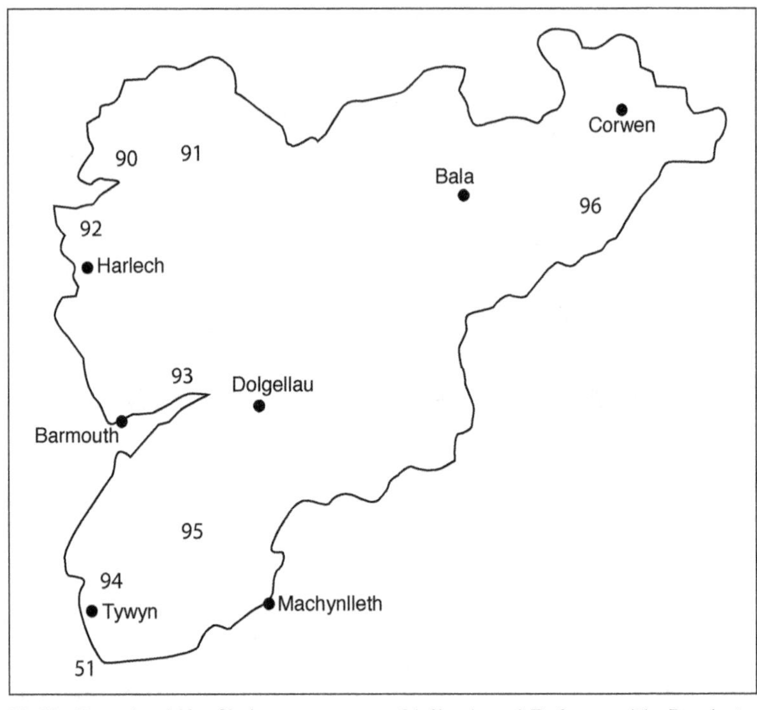

90 Afon Dwyryd and Afon Glaslyn
91 Vale of Ffestiniog
92 Morfa Harlech and Morfa Dyffryn
93 Mawddach Estuary
94 Aberdysynni, Tonfanau and the Broadwater
95 Craig yr Aderyn
96 Llyn Tegid

Meirionnydd is a mainly mountainous county, much of it within the boundaries of the Snowdonia National Park, Britain's second-largest national park at 884.9 sq miles (2,142 sq km). Aran Fawddwy is the highest mountain in Meirionnydd at 2,970 ft (905 m), then Cadair Idris at 2,930 ft (893 m). The upland areas are very underwatched from a bird's point of view, but visitors should be able to find Buzzard, Red Kite, Peregrine, Ring Ouzel (although much scarcer than it used to be), Wheatear, Stonechat, Whinchat, Meadow Pipit, Skylark and Raven. Red and Black Grouse, Hen Harrier, Short-eared Owl and Merlin are also present. There is an approximate total in the county of 50,162 acres (20,300 ha) of upland moorland, largely comprising dry and wet heath. The oak woodlands throughout the county are alive in late spring and early summer with the song of Redstart, Wood Warbler, Chiffchaff, Willow Warbler and Pied Flycatcher. Meirionnydd has roughly 81,790 acres (33,100 ha) of woodland, 60,800 acres (24,600 ha) of conifer plantation and 16,800 acres (6,800 ha) of semi-natural broadleaved woodland, much of this being

Atlantic wet oak woodland. The western coastal plain is mainly lowland and its sea edge is mostly sand dunes and very low sea cliffs, bisected by several sheltered narrow wood-sided estuaries. Its main land characteristics are the sand dunes and slacks, raised bog, saltings, coastal grazing marsh and semi-improved grassland on the whole grazed with sheep. The plain gradually slopes up through woodland and then onto the ffridd (between 600 and 1,200 ft, 180–360 m) and finally the uplands proper.

This stretch of coast forms the north-eastern part of Cardigan Bay and is nationally important for wintering sea-duck, divers and grebes, usually off the mouth of the Dyfi or off the Dwyryd and Glaslyn. Between these two are the Broadwater and Aberdysynni area (good for Eider) the Artro and the Mawddach estuaries. Bird Rock, or Craig yr Aderyn, is a unique inland nesting site for Cormorants and an important roosting site for Chough – up to 50 have been recorded here. Meirionnydd has at least 100 upland lakes, varying greatly in size, including Llyn Tegid the largest man-made lake in Wales.

90 AFON DWYRYD AND AFON GLASLYN

OS Landranger Map 124
OS Explorer OL18
Grid Ref: SH570365 and SH555365
Website: https://www.glaslynwildlife.co.uk/live/

Habitat

The Dwyryd estuary, northern boundary of Morfa Harlech NNR, together with its neighbour the Glaslyn estuary (including the area immediately north of Porthmadog Cob which carries the A497, most of which lies in Caernarvonshire), is one the most important estuaries on the coast of West Wales. As an aside from birdwatching, you can always visit the Italian-style village of Portmeirion. The woodland here is good for birds, while the peninsula itself is an excellent vantage point for scanning the estuary. There is also the Ffestiniog and West Highland steam railway, which runs from Blaenau Ffestiniog to Porthmadog via Porthmadog Cob across the estuary and then onto Caernarfon. Ospreys have bred in this area since 2004, with viewing facilities originally provided by RSPB and more recently by Bywyd Gwyllt Glaslyn Wildlife, at Pont Croesor, near Prenteg, Porthmadog. The website for this site contains further details about access and information on the Ospreys.

Species

Ospreys were confirmed to be breeding in Wales in 2004 when two pairs bred, one of which was at Afon Glaslyn, the male having been ringed as a chick near Aviemore, Scotland in 1998. Sadly the nest was damaged during a gale at the end of June and both nestlings were lost. The following year the same adults returned and successfully reared two chicks, with this pair rearing a further 24 young until 2014. The male was not seen again, but with a new mate, the female has reared a further 16 young up to 2021. Arriving in late March from wintering grounds in West Africa, the Ospreys are present until early September. They fish out in the estuary, often off Portmeirion, and are regularly seen passing over Porthmadog on the way back to the nest – just 5 miles distant (8 km) – with their catches.

Cormorants are present for much of the year, with numbers sometimes rising as high as 300 in early autumn. Look especially for the birds that roost on the Llandecwyn pylon towards the head of the Dwyryd estuary (these pylons will be removed in the future when the powerlines are routed under the estuary). With very few exceptions, Wales is not renowned for its wintering geese; however, Canada Geese and Greylags are virtually resident on the Dwyryd and Glaslyn estuaries and on the marshes upstream of Porthmadog Cob. Up to 90 of the latter have been recorded in midwinter, while birds have also nested here. Canada Geese breed on the marshes, where peak numbers of about 200–300 birds can occur in late summer. A record count of 1,186 was recorded on the January 2014 Wetland Bird Survey (WeBS). Up to 3,000 wildfowl winter, the main species being Shelduck, Wigeon, Teal, Mallard and Pintail. Scaup are now scarce but used to occur in relatively large numbers, for example 160 in January 1997; the recent decline is part of a general trend for this species across northwest Europe. There are small numbers of Shoveler, Tufted Duck, Goosander and occasional Long-tailed Duck and Goldeneye. The Wigeon flock is worthy of a detailed look, as in the past American Wigeon have also been present here, or off Porthmadog Cob, as in 2000 and 2007.

A regular flock of wintering Whooper Swans, up to 70 in number, the largest flock in North Wales, tends to frequent the fields near Pont Croesor (SH593413). Red-breasted Mergansers were first proved nesting in Meirionnydd on the Dwyryd in 1957, where about 3–5 pairs now nest, with birds recorded throughout the year. Although Great Crested Grebes occur in larger numbers, up to 100 on occasions, they are present only from November to March, though they can be found breeding on the upland lakes. Look out for Slavonian Grebes during the same period: the northern section of Cardigan Bay and its estuaries have become an important wintering area for this species, though numbers have dropped considerably over the last 15 or so years. Little Grebes appear in August and overwinter here, with the last birds normally departing in April.

The main waders are Oystercatcher, Lapwing, Dunlin, Curlew, Redshank, Grey Plover, Knot, Ringed Plover, Sanderling, Bar-tailed and Black-tailed Godwits and Greenshank. Wherever you are on the coast of Wales it is always worth carefully searching through gull flocks. Black-headed Gulls are generally the most numerous gull on the two estuaries, with numbers reaching almost 1,000 in midwinter, but as many as 3,000 Herring Gulls were present in March 1997. Mediterranean Gulls are making increasingly frequent visits, though still in very small numbers, while Little Gulls occur most years on spring passage and sometimes in winter. Peregrine regularly hunt across the estuary and saltings, while Hen Harrier, Merlin and sometimes Short-eared Owls are present in winter. Rock Pipit and Black Redstart can sometimes be seen along the seaward Cob wall during the winter months, especially at the Boston Lodge (eastern end of the Cob).

Among rare birds have been a Long-billed Dowitcher in October 1993, which conveniently crossed the county boundary and thus appears on the Caernarfonshire list too, as did an adult Laughing Gull from November 2005 to April 2006. A Ross's Gull was not so considerate and remained on the Caernarfonshire side of the boundary in 1995. More recently there was a Green-winged Teal in 2007, a Cattle Egret in the Llyn Bach area in November

2009, a Bonaparte's Gull off the Cob in May 2011 and a Glossy Ibis present in mid-September 2020.

Timing

As for all estuary work consult the tide tables, those indispensable guides for all coastal birdwatchers. The woodland walks around Portmeirion are best enjoyed in the early morning, and you may have them to yourself unless you chance to meet other readers of this book. Take care if venturing out onto the sand/mudflats and saltings as the tide can come in very quickly, and human presence also can disturb the birds. Estuaries can be busy with birds at any time of year but tend to be quieter in midsummer.

Access

For the Ospreys follow the signs from Porthmadog to the visitor facilities at Pont Croesor. The observation hide is quite distant from the nest, to reduce disturbance, but the high-powered telescopes and CCTV in the observation hide give excellent views.

On the north side, leave the A487 at Minfford (SH596384) for Portmeirion, entrance fee payable, which provides access both to the woodland and the estuary itself. There is also a small car park by the edge of the Dwyryd estuary if one carries on along the minor road for half a mile past the Portmeirion entrance. For Porthmadog Cob and the Glaslyn, one can park near the old tollgate (Boston Lodge end) of Porthmadog Cob in two parking areas, with resulting good views of the Glaslyn estuary: either from the new Lon Las cycleway, the hide at the Boston Lodge end or from the raised walkway on the Cob. Parking

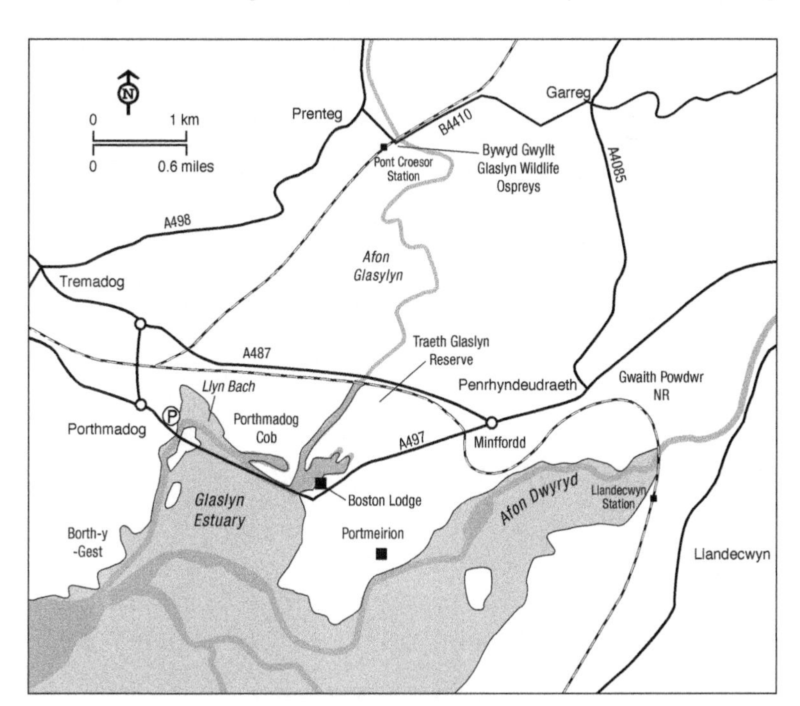

is also available in the town itself for the west end of the Cob and also where paths lead around Llyn Bach (a tidal pool), the harbour and the west shore of the Cob. Views across the Glaslyn and mouth of the Dwyryd can also be found in the Borth y Gest and Morfa Bychan areas. Park on the beach at Black Rock Sands and walk east to view the outer parts of the estuary (Traeth Bach). The Caernarfonshire/Meirionnydd county boundary is in the middle of the river channel here.

On the southern side, the best of several possible vantage points is reached by leaving the A496 in Llanfihangel y Traethau/Ynys (SH598352), continuing until a car park is reached, beyond which it is only a short walk to the shore of the Dwyryd. Spectacular views of Portmeirion and Ynys Gifftan can be had from here. The minor road into this site can flood on some major high tides, thereby prolonging the birdwatching experience. Several footpaths lead off this minor road and provide access to stretches of the outer estuary further west. Access can also be gained from Morfa Harlech forestry plantation and to the Morfa Harlech NNR itself (Site 82). Views can also be gained from the Llandecwyn and Talsarnau areas using various footpaths; these two places give good views of the upper saltings where large flocks of Canada Geese can be present.

CALENDAR

Resident: Cormorant, Grey Heron, Greylag and Canada Geese, occasional Barnacle Goose, Shelduck, Red-breasted Merganser, Oystercatcher, Lapwing, Dunlin, Curlew, Redshank, Black-headed, Lesser Black-backed and Herring Gulls and most woodland birds.

December–February: Red-throated and Great Northern Divers, Little, Great Crested and Slavonian Grebes, Mute and Whooper Swans, Wigeon, Teal, Pintail, Shoveler, Scaup, Common Scoter, Goldeneye, Ringed Plover, Sanderling, Snipe, Black-tailed and Bar-tailed Godwits, Green Sandpiper, Greenshank, Fieldfare, Redwing.

March–May: Wheatear and Pied/White Wagtails can often be found on the upper Dwyryd saltings toward the end of March. The first returning Osprey that nest locally usually appear in the third week of March. Some winter visitors may remain until May, while passage waders such as Whimbrel occur. Swallows and martins arrive, as do summer migrants to the woodlands in the area.

June–July: Shelduck and Red-breasted Merganser broods if one is lucky. First, returning waders put in an appearance and gull numbers increase. The nearby nesting Osprey fledge in late July or early Aug. Nightjar breed at Gwaith Powdwr and the Rhyd areas.

August–November: Wader passage reaches its height and by end of period; winter wildfowl, including divers and grebes, beginning to establish themselves.

91 VALE OF FFESTINIOG

OS Landranger Map 124
OS Explorer OL18
Grid Ref: SH652413
Websites:
https://naturalresources.wales/about-us/what-we-do/welsh-government-woodland-estate/ffestiniog-frp/?lang=en
https://www.woodlandtrust.org.uk/visiting-woods/woods/coed-cymerau-isaf/

Habitat

The beautiful Vale of Ffestiniog effectively divides the main massif of Snowdonia from the Rhinog range to the south. The vale contains one of the highest concentrations of Western Atlantic Oak woodlands in Britain, those on the north side are particularly important. Eight nature reserves are managed by Natural Resources Wales, though five, because of the rugged terrain, are not for the faint hearted; others fortunately are more accessible, including a major section of some 168 acres (68 ha) which forms Coedydd Maentwrog National Nature Reserve, extending for nearly 1.5 miles (2.4 km) along the valley side. Most is made up of Sessile Oak, well over 100 years of age, with Birch, Rowan, Sycamore and Ash with a low mixed understory and also some Bracken areas. When the reserve was acquired, Rhododendron had come to dominate one section, but this has now been removed to allow native species to re-establish, work that is ongoing across the whole area. Rhododendron is a huge problem in many areas and no less so in Snowdonia. Woodland Trust, National Trust, Snowdonia National Park and others have done huge amounts of this type of work in the area over the last 30 years.Large-scale felling of conifer plantations is also occurring, either to replant in kind or to plant or allow the natural regeneration of broadleaved trees. This creates a mosaic of age structures and diversity within woodland, which many bird and plant species prefer. Coed Hafod y Llyn, part of the reserve, has been described as a Welsh rainforest, which offers a hint as to the need to be well prepared during unsettled weather.

At the eastern end there are some wet-heath areas between drier ridges. A deep rocky gorge and several small streams and flushes cross the reserve, and these add further diversity and help with an impressive plant list. There is an extensive lichen flora on the trees, including several scarce species in the *Parmelia* and *Lobaria* groups. Flowering plants include Primrose, Selfheal, Lesser Celandine, Corn Mint, Fairy Flax and Lady's Bedstraw. Red Squirrels have not been seen since 1971, and alas Grey Squirrel is now abundant. Other mammals present include Fox, Badger, Weasel, Wood Mouse and Bank Vole, while American Mink are present. Polecat records are increasing in the area and, who knows, maybe Pine Marten will recolonise at some point in the future. Otters are occasionally seen in the river valley below. A small herd of Wild Goats, sometimes up to 20 strong, uses the reserve from time to time; they are descended from domesticated animals abandoned several centuries ago when herding became uneconomic. Coed y Bleiddiau, the 'Forest of the Wolves' is where the last Wolf in Wales was slain early in the 16th century. The Coed Cymerau National Nature Reserve and other woodlands including the Woodland Trust property Coed Cymerau Isaf are well worth visiting, some of the waterfalls on the Afon Goedol being especially impressive.

Species

Hole-nesting birds have been further attracted to the woodlands by the provision of nest boxes, the most numerous of the occupants being Pied Flycatcher. A main food source for this handsome summer migrant and indeed for other insectivorous species is the larvae of the Mottled Umber, a moth which on occasions reaches plague proportions, stripping the oaks of up to 95% of their foliage. Other species include Redstart, Wood Warbler and several members of the tit family. All three woodpeckers are present, though Lesser Spotted is now becoming very scarce, as well as Nuthatch and Treecreeper. Birds of prey include Goshawk, Sparrowhawk, Buzzard, Peregrine, Kestrel and Tawny Owl. Jays are common in the woodland, and there is usually a pair or two of Ravens. Nightjar can still be found in the Vale, usually around Gwaith Powdwr Nature Reserve (Penrhyndeudraeth) and at Rhyd. A grand view of the Dwyryd/ Glaslyn estuaries and some of Morfa Harlech (Sites 90 and 93) and the setting sun can be gained from the highest section of Gwaith Powdwr reserve if arriving at the Nightjar site pre-sunset and pre-churring. Well worth it just for the view and a chance of hunting Barn Owl to conclude the day.

The small lake of Llyn Mair can attract Little Grebe, Grey Heron, Mute and Whooper Swans, Canada Goose, Mallard, Pochard, Tufted Ducks, Teal, Goldeneye, Moorhen, Coot and Black-headed Gull. Red-breasted Merganser, Goosander, Common Sandpiper, Grey Wagtail and Dipper occur along the river. If you follow the footpaths out of the woodland onto the open higher ground you should encounter Buzzard, Kestrel, Peregrine, Tree and Meadow Pipits, Whinchat and Raven. Also possibly Goshawk and maybe a chance encounter with an overflying Osprey from the nearby Glaslyn nest (Site 90). A Great Grey Shrike was recorded nearby in a naturally regenerating clearfell forestry area in November 2018. Breeding Sedge Warbler, Reed Bunting and Sand Martin can be found in summer down in the valley bottom alongside the Afon Dwyryd – access from Tan-y-Bwlch where there are footpaths at SH657405 or SH648402 and a very pleasant summer's evening walk. Hobby is also occasionally seen here.

Timing

Visits are best made in May or June for woodland-breeding species, no particular time being best unless you wish to hear the dawn chorus. Llyn Mair can get busy people-wise, especially in summer, so is better early in the day. The area is good at any time of year, especially for woodland species, or as already suggested, for a pleasant evening's walk.

Access

The A487 together with the B4410 pass through the valley, Llyn Mair is easily viewed from the latter. A nature trail leaflet is produced by Snowdonia National Park, Ffestiniog Railway, National Trust, Natural Resources Wales, Tilhill Forestry and Woodland Trust plus two private landowners, who also do all the land management. There is a whole network of trails in this area. There is also a leaflet, 'Tan y Bwlch woodland walk', detailing the paths from Tan-y-Bwlch steam railway station to Plas Tan y Bwlch, available from both those sites. One trail commences at SH655412, and passes through the western end of Coedydd Maentwrog. For the main block of woodland there is a public footpath from the entrance at SH658409. There are also many tracks through the forestry blocks to the east of this site. Further east, a footpath leaves the A496 from a car park at SH693431 and takes one through Coed Cymerau and on to the open country around Tanygrisiau Reservoir. Attention should be drawn to the Ffestiniog Light Railway, which provides spectacular views of the valley and the surrounding hills, even if birdwatching opportunities may be limited. Parking is possible at Llyn Mair itself just off the B4410 or half a mile to the west in a car park on the edge of forestry or in Tan-y-Bwlch station itself. Café and toilets are available here.

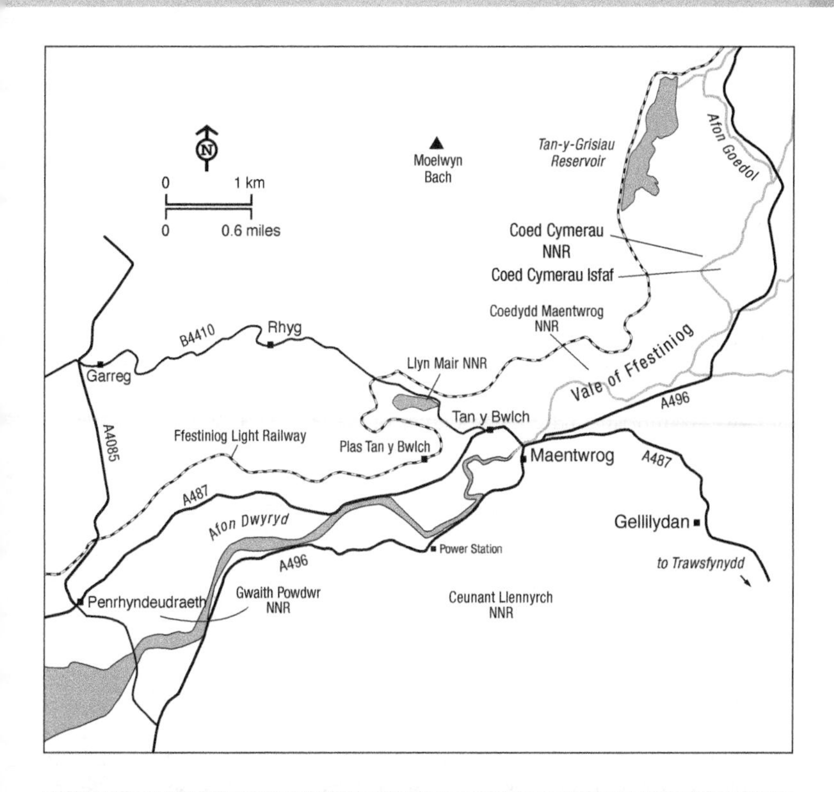

CALENDAR

Resident: Grey Heron, Mute Swan, Mallard, Goshawk, Sparrowhawk, Buzzard, Kestrel, Pheasant, Moorhen, Coot, Woodpigeon, Stock Dove, Tawny Owl, all three woodpeckers (though Lesser Spotted is scarce now), Grey Wagtail, Goldcrest, tits, Nuthatch, Treecreeper, Jay, Jackdaw, Raven, Chaffinch, Lesser Redpoll, Siskin, Bullfinch. Crossbill some years. Hawfinch are possibly present in small numbers.

December–February: Little Grebe, Whooper Swan, Teal, Pochard, Tufted Duck, Goldeneye, Fieldfare, Redwing, Siskin.

March–May: Winter visitors depart, though some ducks may remain well into April, summer visitors arrive and include Cuckoo, Tree Pipit, Redstart, Whinchat, Garden Warbler, Blackcap, Wood Warbler, Chiffchaff, Willow Warbler, Spotted and Pied Flycatchers.

June–July: All residents and summer visitors present.

August–November: Summer visitors have largely departed by early September, though chance of both Blackcap and Chiffchaff remaining to October and possibly even overwintering. Winter wildfowl start coming to Llyn Mair from late September, followed by Fieldfare and Redwing and possibly Brambling moving to the valley during late October. Woodcock can arrive in November and in harsh winters into January and February to escape the much colder weather in Russia and Eastern Europe.

92 MORFA HARLECH AND MORFA DYFFRYN

OS Landranger Map 124
OS Explorer OL18
Grid Ref: SH565337 and SH555257
Websites:
https://naturalresources.wales/days-out/places-to-visit/north-west-wales/morfa-harlech-national-nature-reserve/?lang=en
https://www.first-nature.com/waleswildlife/n-nnr-morfaharlech.php

Habitat

Most spectacular of all the Welsh castles is that of Harlech, built by the master craftsmen of Edward I between 1283 and 1289 on a 200-foot-high (61-m) spur of Lower Cambrian sandstone, an outlier of the Harlech Dome. Such a superb site, such fortifications, that in 1294 a mere 37 men defended the castle against the entire Welsh army. The song 'Men of Harlech', much loved by Welsh rugby supporters and regimental bands, is said to have been inspired by the bravery of the castle defenders commanded by Constable Dafydd ap Ieuan, during a prolonged siege by the English, between 1461 and 1468, the longest known siege in the history of the British Isles. At that time the sea reached to the base of the cliff on which the castle stands. Subsequently, the sea retreated and in its place we find a long sandy shore backed by a vast dune system extending north to the Glaslyn estuary.

One can obtain few finer views of coastal Wales than from the now peaceful battlements, whose only sentinels are Feral Pigeons and Jackdaws: north to the sandhills and warrens of Morfa Harlech and beyond to Traeth Bach and the confluence of the Afon Dwyryd and Afon Glaslyn estuaries (see Site 90) or south towards Mochras (Shell Island) and the equally great sandhills of Morfa Dyffryn. Both sites now include National Nature Reserves (NNRs) managed by Natural Resources Wales.

Morfa Harlech NNR covers some 2,184 acres (884 ha), where wave action from the south has continually pushed shingle, silts and sands northwards, so that now a vast expanse of flatland has developed, fronting the original coast which ran to the north-east. Here is one of the most important actively growing dune systems in Britain, one of only a handful in Wales; such is the growth in some winters that sections of boardwalk have to be removed in autumn, otherwise by spring they could be buried under tons of sand. Between the dunes and high ground is a marshy area with prolific Sharp-flowered Rush, while the wetter areas contain Bogbean, Common Pondweed and Yellow Flag. The dunes have formed into a series of parallel ridges which curve towards the estuary. Those close to the shore are dominated by Sand Couch, then, as one moves inland, by Marram Grass and on the older dunes by a rich lichen and bryophyte flora with higher plants such as Portland Spurge, Lady's Bedstraw, Burnet Rose, Restharrow, Wild Thyme and some scrub and gorse clumps. Tucked among the dunes are botanically rich slacks with Creeping Willow, Variegated Horsetail, Silverweed, Marsh Helleborine, all three varieties of Early Marsh-orchid, Pyramidal and Bee Orchids, Dune Pansy and Sharp Rush, the latter first recorded in Britain here in 1639.

On 23 September 1988 Mike Alexander, warden of Morfa Harlech NNR, discovered a very special dead turtle on the beach: a male Leatherback Turtle, the largest ever found at 9ft 5in long and almost the same width, weighing 144 stone. This majestic animal had roamed the seas for over a hundred years before becoming trapped in fishing nets and drowned. The turtle was taken to the National Museum of Wales where it has been the centrepiece of one of the galleries.

Sandwiched between the two NNRs is the small Artro estuary, fringed with a reedbed at the western edge. The whole site is well worth a look and can be covered in an hour or two. Sheltering this is the point of Mochras (Shell Island) which holds a large campsite. The areas of scrub and reedbed on the edge of Mochras held Water Rail and Cetti's Warbler in October 2019, and Great White Egret seem to prefer this quiet corner of the estuary. Beyond this to the south is Morfa Dyffryn, a 499 acre (202 ha) NNR. The dunes have much the same structure as Morfa Harlech but are taller and wider, and once again the slacks (often pools in winter) offer the highlights, with plants such as Sea Centaury, Sea Spurge, Green-flowered Helleborine and Bee and Pyramidal Orchids.

Species

The marshes and dunes at Morfa Harlech are of interest, with Shelduck, Oystercatcher, Ringed Plover, Lapwing, Curlew (now almost gone), Redshank, Skylark, Stonechat and Linnet all nesting. Small numbers of waders, including Sanderling, frequent the beach while on passage. Red-throated Divers are regularly seen offshore, together with the occasional Great Northern Diver, while Great Crested Grebe are winter visitors in small numbers, occasionally joined by Slavonian Grebe. Some 2,000 Common Scoter regularly winter offshore while the total numbers in this northern sector of Cardigan Bay can exceed 6,000 birds. Search carefully and you can be rewarded by views of Velvet Scoter (20 in February 2021 in the inner Artro harbour) while Surf Scoter has also been recorded on a number of occasions. Short-eared Owls and Hen Harriers regularly winter on the marshes. The Artro saltmarshes hold several pairs of Redshank in the breeding season and the sand dunes at Mochras harbour held 110 Sand Martin nest-holes in 2019. Although not so noted for its birds, and with only a small number of species breeding, Morfa Dyffryn can prove of interest, with Eider, Common Scoter and occasional Scaup and Long-tailed Duck seen offshore in winter. Surf Scoter has occurred here too.

In winter, Hen Harrier, Merlin, Peregrine and Short-eared Owl hunt over the sandhills when the flocks of Skylarks, Meadow Pipits and finches such as Chaffinch, Brambling, Goldfinch and Linnet deserve the observer's attention for they may include scarce species such as Snow Bunting or even potentially Twite – and if not they are still a joy to watch in their own right. Llanbedr Airfield (no general access) is also worth a good scan from the fenceline by the dunes. There are often Wheatear here in autumn, with flocks of Golden Plover and Lapwing present in winter. One or two Shorelark occur irregularly on the upper strand line, the last being at Traeth Bennar in November 2007.

Rarities have included a Sharp-tailed Sandpiper at Morfa Harlech on 14–15 October 1973, a first/second for Wales as one turned up at Shotwick, Flintshire on the same date. A Pectoral Sandpiper at Mochras in May 1997 was the first county record and in 2002 an adult Ivory Gull was seen at this site and also at Porthmadog and Black Rock Sands (Caernarfonshire). Two Bee-eaters were seen just below Harlech Castle on 30 April 2010. Once again, this is just a flavour of the rarities to show that anything is possible.

Timing

When visiting the estuaries, consult the tide table as always. For the coastal hinterland time of day is less important, though best in the early morning, especially at Morfa Dyffryn, during the holiday season. Do watch the tide times for accessing Mochras as the tidal causeway can flood. Parking on the Artro sandflats is not advised as the tide can come in quickly here. Seawatching from

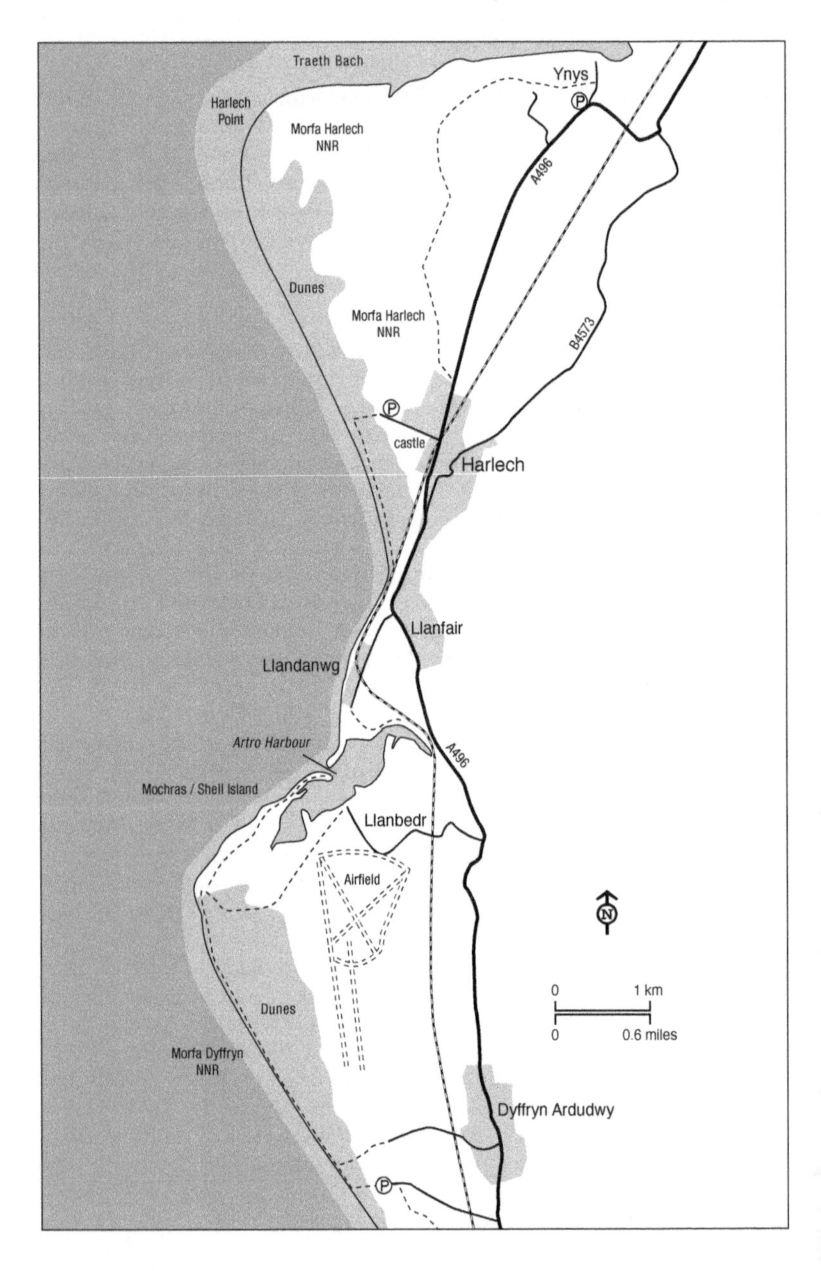

the dunes is possible in early autumn, though windblown sand can be a problem getting into optics. Probably best from Mochras itself on north-westerly or westerly winds.

Access

Morfa Harlech: One can approach from Harlech by way of the car park at SH574317, where there is access to the shore and dunes. A footpath also leaves the A496 at SH581320, while further east others leave the A496 at SH599355, a track and then a footpath leads to Traeth Bach or branch left into Morfa Harlech forestry, while at SH610362 a footpath leads to Ynys Gifftan on the eastern boundary of the reserve, though care should be taken on high tides. Morfa Harlech can also be walked to from the Ynys car park SH599357 around the edge of Traeth Bach and into the reserve itself. Best stick to the very edge of Traeth Bach saltmarsh rather than walk out onto the sands as it could flush many of the birds. Once again be wary of the tide situation.

Morfa Dyffryn: Head for Mochras (Shell Island), leaving the A496 at SH585267 in Llanbedr; good views out to sea may be obtained from Mochras and from Llandanwg to the north. For the nature reserve, leave the road at SH568271 and continue south on foot over the raised footway alongside the airfield or down the Mochras raised road causeway. Alternatively, leave the A496 at SH586223 south of Dyffryn Ardudwy and drive to the beach car park; from here one can walk north along Benar Beach or turn west off the beach path just after the car park and into the dunes via several paths. The scrub around this car park and running up west of the track is worth a good look, especially in autumn, as are the trees and hedges in the Dyffryn Seaside caravan park which is accessible via two footpaths.

CALENDAR

Resident: Cormorant, Buzzard, Red Kite, Grey Heron, Mute Swan, Shelduck, Mallard, Red-breasted Merganser, Kestrel, Oystercatcher, Ringed Plover, Lapwing, Curlew, Redshank, Black-headed, Herring and Great Black-backed Gulls, Skylark, Stonechat, Meadow Pipit, Pied Wagtail, Reed Bunting.

December–February: Red-throated and Great Northern Divers, Great Crested Grebe, Whooper Swan, Wigeon, Teal, Pintail, Tufted Duck, Scaup, Eider, Common Scoter, Goldeneye, Hen Harrier, Merlin, Peregrine, Short-eared Owl, while Woodcock frequent the dune slacks and scrub.

March–May: Wheatear, Swallow and Sand Martin from late March, Cuckoo and smaller migrants such as Grasshopper and Sedge Warblers, Whitethroat, Wheatear, Spotted Flycatcher during April, also wader passage at the same time including Whimbrel, Ringed Plover and Common Sandpiper.

June–July: Breeding species still present. Manx Shearwater occasionally seen offshore, also some terns. First autumn waders appear on the Artro estuary, Morfa Harlech shoreline or at Harlech Point.

August–November: Wader passage throughout early part of period, with a chance of scarce species such as Little Stint, Curlew Sandpiper, Ruff, Bar and Black-tailed Godwit and Spotted Redshank. Winter ducks commence arriving from late September, and a chance of seeing Whooper Swans from end October to mid-November. Do not forget the sandhills for winter predators and possibility of birds such as Twite and Snow Bunting. Any scrub is worth spending time on at this time of year, Benar Beach car park west (scrub) held a Yellow-browed Warbler in October 2018. Winter thrush arrival in November.

93 MAWDDACH ESTUARY

OS Landranger Map 124
OS Explorer OL18 (shows half of estuary only) and 23
Grid Ref: SH695185
Websites:
https://www.rspb.org.uk/reserves-and-events/reserves-a-z/mawddach-valley-arthog-bog/
https://www.rspb.org.uk/reserves-and-events/reserves-a-z/mawddach-valley-coed-garth-gell/
https://www.snowdonia.gov.wales/visiting/walking/leisure-walks/abergwynant-woods

Habitat

Although not so renowned for its birds as some other estuaries, the Mawddach deserves inclusion for, in the words of William Condry, one of Wales's greatest naturalists, 'The Mawddach estuary does not need me to sing its praises for it is justly celebrated as one the most beautiful places on earth.'

The Afon Mawddach, some 22 miles (35.2 km) in length, is the longest river wholly within Snowdonia National Park. Rising on high ground to the north-east, it plunges through the Coed-y-Brenin forest, gathering force with the addition of numerous tributaries before it reaches the estuary and its wide sandflats surrounded by steep wooded hills. It is then a further 7 miles (11.2 km) before it gains the sea at Barmouth, squeezing through a narrow exit past Fairbourne spit extending north from the southern shore. Both Allis Shad and Twaite Shad occur in the river, members of the Herring family often known as May fish as this is the time they swim upstream to spawn. Both species are now extremely scarce in Great Britain, pollution being one cause for their decline, while the construction of weirs and other barriers on watercourses has meant the loss of ancestral spawning areas. The Mawddach contains one of the largest areas of saltmarsh in Cardigan Bay, where scarce plants such as the Dwarf Spike-rush and Lax-flowered Sea-lavender, and common plants such as Sea Aster abound.

A little inland on the south side is the RSPB reserve Arthog Bog, an estuarine raised mire, one of only two in Wales, the other being on the south side of the Dyfi estuary. This the most northerly place in the world for Wavy St John's-wort, a beautiful plant much sought after by botanists in the rushy pastures and damp heaths of West Wales and south-west England, while Brown Beak-sedge, another scarce species, can also be found at this one of its few locations in Wales. The bog is, alas, a mere vestige of its former glory, for vast quantities of peat were extracted in days gone by to be burnt on the fires of Barmouth, Dolgellau and nearby villages. A couple of the footpaths crossing the bog show the 'raised' profile well if one views east of Arthog RSPB section.

The RSPB holding in the Mawddach Valley extends for some 1,288 acres (521 ha) much of it fine woodland, with Coed Garth Gell, 114 acres (46 ha), being open at all times. The reserve is bounded on one side by a dramatic gorge and best visited in May and June; Sessile Oak and Birch are the commonest trees within the reserve, which is interspersed with heathy glades.

Another site in the valley worth a look is Coed Abergwynant, accessed from the Mawddach Trail on the south side. These woods hold Redstart, Wood Warbler and Pied Flycatcher, as well as the other usual woodland species.

Species

Shelducks and Red-breasted Mergansers breed on or close to the estuary, and one of the sights during June and July is the broods of ducklings escorted by adult birds. Mandarin Ducks have arrived with one or two pairs now frequenting the river near Dolgellau, but they do not seem to be present every year. Common Sandpipers breed on the upper estuary and along the tributary rivers, and Oystercatchers in the wet fields at several spots on the southern side of the estuary. Only small numbers of wildfowl and waders frequent the estuary, but most species can be seen and in due season include Oystercatcher, Ringed and Golden Plovers, Lapwing, Dunlin, Bar-tailed Godwit, Curlew and Redshank, with occasionally such birds as Curlew Sandpiper and Little Stint. Small numbers of terns move into the estuary in late summer as they travel south from colonies in Anglesey and Lancashire. Little Egrets are now commonly seen and probably breed, while Ospreys can occur on feeding forays from nesting sites on nearby estuaries and on passage. Sedge Warbler, Reed Warbler and Reed Bunting all occur in suitable habitats close to the estuary, while Siskin and Lesser Redpoll are also seen, especially in the riverside alders in winter. Most regular species occur in the surrounding woodland and on the hills above, including Buzzard, Red Kite, all three woodpeckers (though Lesser is now very scarce), Tree Pipit, Whinchat, Redstart, Wood Warbler, Pied Flycatcher, Willow Tit (possibly now gone), Long-tailed Tit, Nuthatch and Treecreeper. Hawfinch now regularly breed in the area and Crossbill have also bred in recent years. Best areas where most estuary birds can be feeding or at high tide roost are Fairbourne Spit end and saltings, SH615140, Cutiau Marsh SH637171 where there is a high-tide roost in *Spartina*, Garth Isaf saltings SH650160, Arthog saltings SH642152, Glandwr marsh SH648175, Bontddu area SH676186, Penmaenpool fields SH685186, and east of Penmaenpool Toll Bridge, in fields, marsh and reedbed SH705185. On first glance, the estuary can appear quite birdless but most birds are at the eight sites above as these are the optimum places for feeding. Other good sites can be the main channel at the mouth of the estuary, the rocks below the high-tide mark off Barmouth and rocks off Fairbourne. Rarities have included a Great Reed Warbler at Penmaenpool in July 1978, and a Rustic Bunting at Arthog in 1991 remains the only county record for this species, while Lapland Buntings have been seen on several occasions at Fairbourne.

Timing

For visits to the estuary, tide tables should be consulted. The woodlands can be enjoyed at all times of day, though a spring visit is best made in the early morning, when birdsong is at its peak.

Access

The Snowdonia National Park operates an information centre (formerly run by RSPB) imaginatively situated in a long-disused signal box beside the toll bridge at Penmaenpool (SH695185) and open in the summer months. The upper floor provides an excellent vantage point for this part of the upper estuary. Using the Mawddach Trail, originally an old railway line, there is a walk along a section of the southern shore from Penmaenpool to Morfa Mawddach, from where you can cross to Barmouth by footbridge. The cycleway east from Penmaenpool allows one to look at another part of the estuary together with a

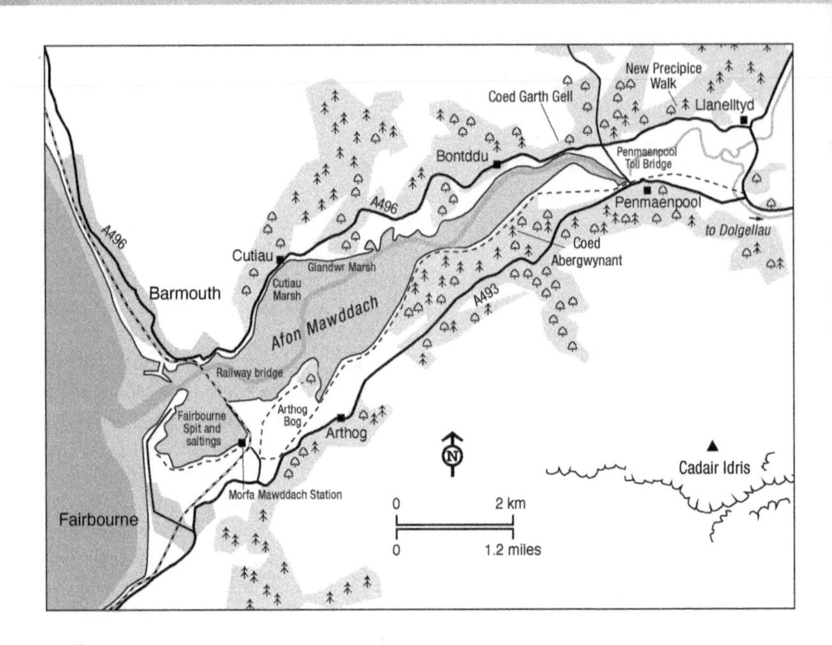

small area of reedbeds and willow scrub. This 9-mile (14.5 km)-long cycleway runs from Dolgellau to Barmouth. The northern estuary shore is not so accessible save on the lower section at the approaches to Barmouth. Here, about 2 miles (3.2 km) east of the town is Barmouth, or Cutiau Marsh, easily seen from the A496 where there are several lay-bys. There is also access to the shore in the Bontddu area. On the south side of the estuary, the minor road which follows the light railway along the isthmus north from Fairbourne offers views over Morfa Mawddach and the sea if one walks through the narrow sand dunes. A small passenger ferry runs from the tip of Fairbourne spit to Barmouth in the summer months. This is a relatively unexplored area for its birds and merits concentrated attention. Footpaths from the A493 at Arthog cross the bog to the estuary shore, while a circular route commencing close to Morfa Mawddach Station (SH628142), where there is parking, takes one through the RSPB reserve. Access to the RSPB reserve at Coed Garth Gell is by public footpath from SH687191, just west of the north side of the Penmaenpool toll bridge, while excellent general views may be obtained from the New Precipice Walk, high above the upper reaches of the estuary, at SH705200. Various parking places are available in Dolgellau itself, just south of Dolgellau, Penmaenpool, Arthog, Morfa Mawddach, the Fairbourne area and in Barmouth.

CALENDAR

Resident: Cormorant, Canada Goose, Grey Heron, Little Egret, Mute Swan, Shelduck, Mallard, Red-breasted Merganser, Goshawk, Sparrowhawk, Buzzard, Red Kite, Kestrel, Oystercatcher, Ringed Plover, Herring and Great Black-backed Gulls, Tawny Owl, woodpeckers, Grey Wagtail, tit family, Nuthatch, Treecreeper, Jay, Raven, Hawfinch, Chaffinch, Goldfinch, Siskin. Although not resident, Great White Egret can now be seen at any time of year.

December–February: Great Crested and Little Grebes, Wigeon, Teal, Goldeneye, Pintail, Golden Plover, Lapwing, Dunlin,

Snipe, Water Rail, Curlew, Greenshank, Turnstone, Black-headed and Common Gulls, Kingfisher, Dipper, a chance of overwintering Chiffchaff at Arthog.

March–May: Winter visitors depart. Whimbrel, Common Sandpiper, Sandwich Tern, Tree Pipit, Redstart, Sedge, Garden and Wood Warblers, Blackcap, Chiffchaff, Spotted and Pied Flycatchers. Cetti's Warbler (not common in Meirionnydd) can occasionally be heard in the willows.

June–July: All resident and summer visitors present. Ringed Plover breed in the area. First passage waders return, chance of terns particularly at estuary mouth.
August–November: Summer visitors depart, wader numbers peak during mid-September and winter ducks begin to arrive shortly afterwards. Woodcock arrive while tits band together, often joined by Goldcrests and Treecreepers. Winter visitors such as Fieldfare and Redwing from mid-October, Siskins and Lesser Redpolls in the valley.

94 ABER DYSYNNI, TONFANAU AND THE BROADWATER

OS Landranger Map 135
OS Explorer OL23
Grid Ref: SH580025

Habitat

This site is a few miles down the valley from Craig yr Aderyn (Site 95). Where the Afon Dysynni meets the coastal grazing marsh it forms a large 'pool', known as the Broadwater (SH580025), and from here the river wends its way through sand/mudflats and eventually to the sea at Aber Dysynni. The edge of the Afon Dysynni near the Broadwater is fringed with a large reedbed leading eventually to mudflats and a few higher islands often not covered by the tide. As the estuary has a narrow outlet, the water is often slow in leaving the area, allowing the mud to be exposed for less time.

Tonfanau area is possibly the prime migration watch point in Meirionnydd. This small headland north of the Broadwater is ideal as a migrant stopover point with its tree shelter belts, scrub, semi-improved grassland, small reed-sided pools and wetter grass areas with shingle ridge beyond. From 1938 until 1966, the land immediately north of Broadwater was occupied by a Royal Artillery base and mostly used by anti-aircraft batteries firing at targets towed by aircraft from the RAF station between Broadwater and Tywyn.

Beyond Tonfanau are low mud sea-cliffs in which Sand Martin nest and at low tide large areas of exposed rock on which many waders and Eider can rest. The Aber Dysynni shingle ridge stretches all the way to Tywyn, wider near the river mouth with two or three gravel storm berms with interesting halophyte (salt-tolerant) vegetation and saline pools. Some of this ridge area is managed to prevent inland flooding. The wet sandy rough fields to the south of the Broadwater are worth a look botanically as well as for birds. In winter large flocks of Golden Plover and Lapwing can be present in these fields and Penllyn Marsh, south of Tywyn, is an area of coastal wet grassland and warrants a visit if in the area.

Species

Eiders were first recorded off Harlech in 1951 and by the early 1970s were present throughout the year. In 1973 a raft of 50 was discovered off Tonfanau, with numbers over the next 10 years rising to about 90 before declining by 1991 to 31. Numbers have continued to fluctuate, with 100 here in March 2009 and a record count of 181 in November 2018. Two nests found in 1998 about 4 miles (6.4 km) north of Tonfanau was the first record of breeding in Meirionnydd, while five downy ducklings in June 2007 were the first evidence of breeding at Tonfanau. A female King Eider spent a winter in 1986 and another was here between August 2017 and May 2019; the bird was often at Ynyslas in Ceredigion but would come across the Dyfi to the Aberdyfi area and also along the coast off Aber Dysynni with the Eider flock.

The demise of Little Terns in Meirionnydd meant that by the 1950s the colony on the shore at Aberdysynni, known since 1894, was the only one remaining. In 1974 the RSPB set up a protection scheme with numbers rising to 36 pairs by 1977; however, eggs and young were lost to predators including Kestrels, Stoats and even an Oystercatcher! Numbers rapidly plummeted, the last attempts at breeding being in 1987 and 1988, while two or three birds seen occasionally were the last, leaving the Flintshire colonies (see Site 80) – the only ones in Wales.

The estuary may play host to small numbers of ducks and waders during the winter months or on passage, including Teal, Wigeon, Dunlin, Redshank, Curlew and Ringed Plover. There are plenty of gulls and terns, along with Bottle-nosed Dolphin, which can be seen feeding in this outlet area as well. It is well worth giving the sea a good look at any time of year. Rock Pipit breed along the coast to the west of Aber Dysynni.

During summer evenings Manx Shearwaters from their nesting colonies on Bardsey, Ramsey, Skokholm and Skomer. In late summer these wheeling masses may contain other shearwater species as they feed on shoals of Whitebait. This phenomenon can be seen all the way down to Aberystwyth with counts of 40,000 off the Borth and Ynys Las areas in Ceredigion, though the numbers do not appear to be as big off Aber Dysynni. Some larger counts can also be made off Aberdyfi if one stays within the confines of Meirionnydd. Large numbers of Sandwich Terns congregate here on their autumn migration. Being halfway along Cardigan Bay and within the hollow of the bay, seawatching is not all that productive but with winds in the westerly quarter – and some persistence – both Arctic and Great Skuas may turn up, as well Balearic Shearwater. Seawatching can be carried out from the top of the low cliff at Tonfanau but if you prefer more height to gain a view further out, a short walk up the coast to Foel Llanfendigaid SH567055 and view from partway up the slope, behind some conveniently situated rocks supplying shelter from the wind. A winter count of 791 Red-throated Divers was made in December 1996 off Aber Dysynni, though counts are much lower than this now.

Lapland Bunting and Jack Snipe have been seen on the shingle bank at Aber Dysynni. This site had Black-headed Bunting in June 2000. Tonfanau area receives the expected migrants while in October 2020 Firecrest, Yellow-browed Warbler, Richard's Pipit and Barred Warbler were all seen on the same day. Other rarities have included a Surf Scoter at Aber Dysynni in October 2011 and on the Broadwater, an American Golden Plover on 30 October 2016, two White-rumped Sandpipers on the 5–7 November 2011 and a Temminck's Stint on 10 March 2004. The whole area is under-watched and anything is possible – a chance for the reader to make a name for themselves!

Timing

Winter and passage periods. Tonfanau can be especially productive in late August to early November.

Access

From the A493 coast road from Dolgellau to Tywyn, turn left at Rhoslefain (SH577057) and park in the Tonfanau area. This gives you access to Tonfanau and beach over the railway at Tonfanau Halt from where a short walk down the cycleway over the 'Arnhem' footbridge takes you to the Broadwater. The roadside tree strips at Tonfanau are well worth spending some time over especially in autumn, though bear in mind access is not possible as this area is private. To access Tonfanau and Broadwater from the east side take the minor road at Pont Dysynni (SH598038). Another access point to view the Afon Dysynni would be through Ynysmaengwyn (SH601021) via the footpath to the river's edge. There is also a minor dead-end road that leads up from Tywyn to the east side of the Afon Dysynni and Tonfanau can be accessed over the 'Arnhem' footbridge. There are plenty of footpaths from Tywyn to the Broadwater area, or up the shingle ridge from Tywyn. A large bend in the Afon Dysynni can be viewed from the A493 near Bryncrug at SH600037. The Aber Dysynni shingle ridge

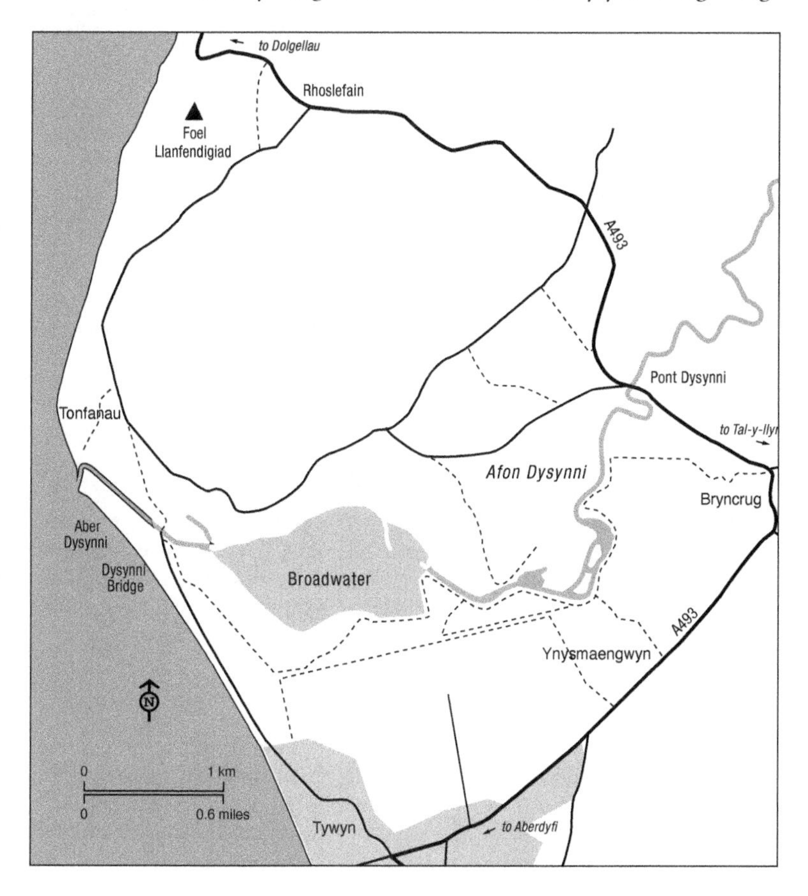

can be accessed along the ridge from Tywyn or under the railway bridge at low tide near the 'Arnhem' bridge over the Dysynni. There are no other railway crossing points to gain access to the Aber Dysynni ridge east of the river. The north side of the Broadwater shore is private and not generally accessible.

CALENDAR

Resident: Mute Swan, Mallard, Eider, Grey Heron, Cormorant (often in large numbers wing drying on the Broadwater), Buzzard, Red Kite, Rock Pipit (uncommon in the county).
November–February: Shelduck, Wigeon, Teal, Goldeneye, Red-breasted Merganser, Ringed Plover, Dunlin, Curlew, Whimbrel, Grey Plover, Turnstone, Greenshank, Redshank and Water Rail on Tonfanau Pool and in Afon Dysynni reedbeds.
March–June: Wader passage, Manx Shearwaters, Gannets and Fulmar offshore, passing Sandwich Terns. Shelduck breed on the Broadwater, a few Teal and Wigeon may still be present until May. Reed and Sedge Warbler breed in reedbeds and ditch scrub. Osprey are often seen fishing over the river, estuary or river mouth. Ringed Plover breed on the shingle areas (beware of disturbance to this species).
July–October: Large numbers of Sandwich Tern may turn up in late July or early August, together with occasional Arctic, Common and Little Terns. Also offshore are Manx Shearwater, Gannet, occasional skuas, auks and Fulmar. Passage waders include Ringed Plover, Dunlin, Curlew, Whimbrel, Black-tailed and Bar-tailed Godwits, Greenshank and Redshank. Little Stint and Curlew Sandpiper are also occasionally seen. Migrant passerines such as Wheatear, Tree Pipit, Skylark and Whinchat.

95 CRAIG YR ADERYN

OS Landranger Map 124
OS Explorer OL23
Grid Ref: SH642066
Website: https://www.snowdonia-active.org/wp-content/uploads/2019/08/Craig-yr-Aderyn-eng-s.pdf

Habitat

The Cormorant is sufficiently widespread on the Welsh coast and as a visitor to inland waters that it does not normally spark great interest in the birdwatcher. One colony, however, merits mention – indeed a section of its own – for that at Craig yr Aderyn, 'Bird Rock', near Tywyn in south Meirionnydd is unique. Prehistoric seas once flowed up the broad inlet of what is now the flat farmland bordering the Afon Dysynni, to the very foot of the rock which towers nearly 850 feet (258 m) above the valley. Nowadays, Cardigan Bay and the sea is some 5 miles (8 km) distant, but still the Cormorants come to this ancestral crag, as they have probably done from the very earliest of times.

Species

55 pairs of Cormorants together with 12 pairs of Herring and seven of Lesser Black-backed Gulls nest on Craig yr Aderyn. The colony has probably never been much larger, despite local tradition, as numbers are restricted by the availability of suitable ledges. Edward Lhuyd first described the colony in 1695, when he referred to 'corvorants, rock pigeons and hawks that breed on it',

while that other great Welsh naturalist, Thomas Pennant (1726–1798), noted Craig yr Aderyn as 'The Rock of Birds' with Cormorants, Rock Doves and hawks nesting upon it. William Catherall, in his *History of Wales* written in 1828, provides an evocative early description both of the rock and its seabird inhabitants:

> Craig Aderyn is a most picturesque and lofty rock, so called from the numerous birds which nightly retire among its crevices: the noise they make at nightfall is most hideously dissonant, and as the scenery around is extremely wild and romantic, the ideas engendered by such a clamour in the gloom of the evening in so dismal and desolate a spot are not the most soothing or agreeable. Towards twilight some large aquatic fowls from the neighbouring marsh may be seen majestically wending their way to this the place of nocturnal rest.

Do not be put off by this, in places sombre, account. Craig yr Aderyn is a remarkable place, and you will be rewarded with spectacular views of Cormorants here flighting to and fro some 500 feet above your head. Look out also for Choughs, for all of the pairs nesting in Meirionnydd do so at inland sites and this is one; indeed, here they nest at one of the few natural inland sites in Wales, most pairs preferring quarries and mineshafts (in fact, a decline is taking place at these inland quarry breeding sites). Up to 50 roost on the crag and there is a good chance of birds of prey: Buzzard, Red Kite, Kestrel and even

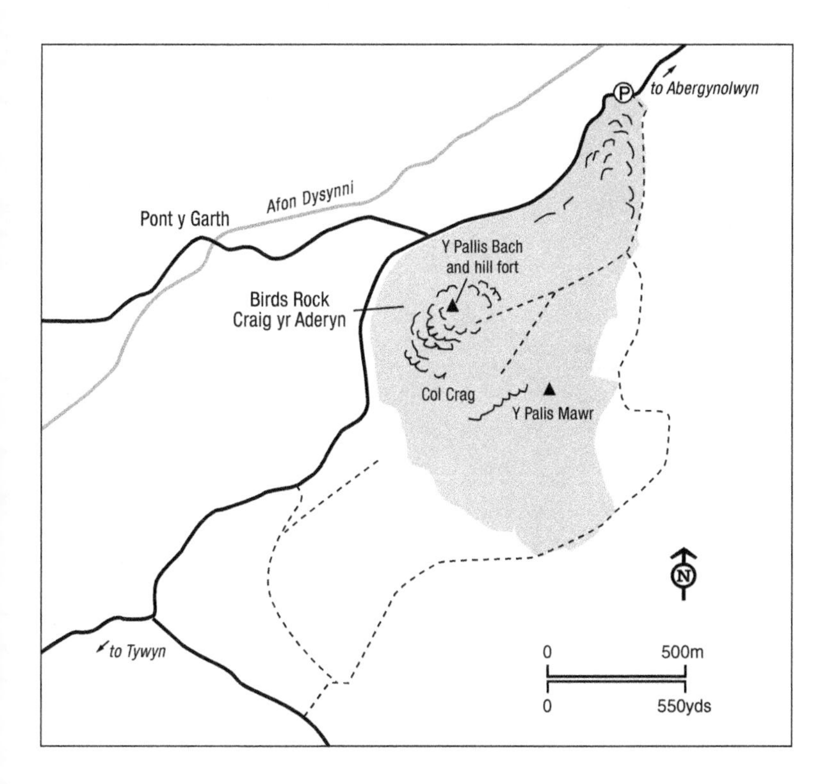

Peregrine. Cliff climbing is popular at Craig yr Aderyn but prohibited on the nesting cliffs from 1 March to 21 July, though climbers are advised not to use this section at any time of year.

Timing

No particular time. Outside the breeding season, however, the late afternoon and early evening, when the Cormorants return to roost, may reveal most activity.

Access

Leave the A493 Tywyn to Dolgellau road at Bryncrug (SH609033) and follow the unclassified road along the east side of the Dysynni Valley. The road eventually passes directly below Craig yr Aderyn, from where views may be obtained looking up at the colony. A more whole view of the colony can be gained from the footpath (SH636071) that runs alongside the river off the minor road that leads to Pont y Garth. For the more energetic, a footpath takes one to the top of the crag and thus an aerial view of passing Cormorants. Many of the woodlands in the area have footpaths, including a new planting along the Craig yr Aderyn to Llanegryn minor road which is well worth a look.

CALENDAR

Resident: Wren, Cormorant, Buzzard, Red Kite, Kestrel, Peregrine, Skylark, Meadow Pipit, Magpie, Chough, Jackdaw, Raven, Carrion Crow, Chaffinch, Mistle Thrush.

December–February: Fieldfare, Redwing, Brambling.

March–May: Summer migrants arrive, including Swallow, Wheatear, Redstart, Whinchat, Willow Warbler, Whitethroat. Herring and Lesser Black-backed Gulls also nest on the cliff.

June–July: All residents and summer visitors present. Wood Warbler and Pied Flycatcher are present in nearby woodlands.

August–November: Summer visitors depart. Swallows among the last, with some in the area until early October, overlapping with the arrival of winter thrushes and the influx of finches. Dippers possibly on the river.

96 LLYN TEGID

OS Landranger Map 125
OS Explorer OL23
Grid Ref: SH906330

Habitat

Nearly 4 miles (6.4 km) long and almost three quarters of a mile wide at its widest point, this is the largest inland water in Wales – if one discounts the fact that the levels and river flow can be controlled and altered artificially for various reasons, but mainly to prevent flooding down river. The lake is 414 ha in size and is 43 metres deep at its deepest point. The Afon Dyfrdwy (Dee) is 70 miles (112 km) long, flowing in at the southern end of the lake from its source in the nearby Dduallt bog in the hills at Penaran, and out through the Bala end and on to Llangollen, eventually meandering across the Cheshire plain to below Chester and finally exiting by the Wirral to become the famous Dee Estuary. The lake nestles between the Arans, Arenig and Berwyn mountains in the faultline valley running north-east to south-west from Bala all the way past Craig yr Aderyn (Site 95) and through Tal-y-llyn and eventually reaching the coast at Aber Dysynni, Tonfanau and Broadwater (Site 94).

Llyn Tegid is relatively sheltered, save when exposed to the south-west winds when the size of waves in winter can be somewhat alarming. It has a lowland feel and is quite unlike the other hundred or so mainly upland acidic lakes in Meirionnydd. The north end by the small and bustling market town of Bala is reed fringed with some Alders and Downy Birch and mostly rough pasture. The southern end is more open, with some exposed mud and shingle on lower water levels and grassland beyond. Most of the bird species and numbers occur in this southern quarter of the lake. Although one of the best lakes in Wales for birds, when compared with other lakes in England and Scotland that support huge numbers of waterfowl is only considered moderate in quality having a relatively low number of species breeding, but is much busier in winter.

Llyn Tegid holds many species of fish, the most famous being the Gwyniad, a species endemic to the bottom depths of this lake and rarely caught. The Gwyniad is regarded as a freshwater Herring but actually belongs to the Salmon family. Some authorities consider this species to be the same as the Schelly of Lakeland and the Powan of Scotland. In the past, it was regularly eaten in the local hostelries as it was much cheaper than meat, but the practice died out in the mid-1800s, probably owing to its poor flavour, 'fish of an insipid taste' as the Welsh naturalist Thomas Pennant (1726–1798) described it. Other fish are Perch, Pike, Trout, Salmon, Grayling, Eels, Lampreys, Roach, Gudgeon, Bullheads, Stone Loaches and Minnows, with Silver Bream and Rudd added in the mid-1900s. Otters are frequently observed along the shores, while Pine Marten have been seen in the surrounding area as well as increasing numbers of Polecat, their presence often only revealed by road casualties. Red Squirrels used to be present in nearby forestry areas and may still be here in small numbers.

Species

Goosander regularly breed in the area and broods can be seen in summer. Common Sandpiper breed around the lake edge and up the feeder streams, as do Grey Wagtail and Dipper. Grey Heron breed in the area. Many woodland species, such as Garden and Wood Warblers, Redstart and Pied Flycatcher breed in the woodlands around the lake edge and surrounding area. Common Sandpipers breed along the tributaries, while occasionally Oystercatcher and Little Ringed Plover nest on the shingle areas.

Wintering duck are mainly found at the northern and southern ends of the lake. Great Northern Diver (if present) are usually at the northern end. Teal can often be present in the lakeside vegetation at both north and south ends. The Canada Goose flock can often be on the west side of the lake in lakeside fields viewable from Llangower Point or in fields at the southern end of the lake. Goldeneye and Goosander frequent the Dyfrdwy outlet or on the river upstream of this or at the south end in Glan-llyn bay. Large Brambling flocks have occurred in recent years at Glan-llyn, feeding on beechmast.

A male Ring-necked Duck was first seen in autumn 2018 and returned the following two autumns. Great Northern Diver, rare inland, are seen in most

winters, with sometimes two present. There are older records of Red and Black-throated Divers but none recently.

Timing

Waterfowl winter on the lake with many species arriving in the autumn. The best time is early in the morning before too much disturbance, especially in summer when the lake edge can become busy with walkers and fishermen and the water busy with small boats and canoeists. The two ends are often the busiest for water traffic. Fortunately, the lake is fairly undisturbed in winter.

Access

There are various access points and car-parking spots along the east side of the lake beside the minor road from Bala to Llanuwchllyn and it is this side from where the best views of the birds can be gained. A miniature steam railway runs along the east shore in summer, autumn and also the Santa Special. The west side of the lake has the A494 (Corwen to Dolgellau) running alongside it and there are a few access points with paths down to the water's edge, but is much more restricted on this side. Viewing from this side can also be poor in the morning as you may be looking into the sun a lot of the time.

CALENDAR

Resident: Grey Heron, Mallard, Goosander, Mute Swan, Canada Goose, Great Crested Grebe, Red Kite, Buzzard, Goshawk, Sparrowhawk, Peregrine, Tawny Owl, Great Spotted Woodpecker, Green Woodpecker, Grey and Pied Wagtails, Dipper, Mistle Thrush, Goldcrest, tit family, Raven, Reed Bunting.

December–February: Coot, Wigeon, Goldeneye, occasional Pochard, Teal, occasional Scaup and Great Northern Diver.

March–May: Common Sandpiper, Cuckoo, Swift, Swallow, House and Sand Martin, Redstart, Wood Warbler, Chiffchaff, Spotted Flycatcher, Pied Flycatcher, Sedge Warbler, occasional Reed and Cetti's Warblers.

June–July: All residents and summer visitors present. Often large flocks of Swift feeding over the lake or town (where they nest) Mandarin occasionally seen – this latter species breeds further down the Afon Dyfrdwy valley in the Corwen area.

August–November: Great White Egret now fairly regularly seen but Little Egret remain rare here. Wintering duck species return including Teal, Goldeneye, Wigeon, Tufted Duck, occasional Scaup, Pochard. Kingfisher, Jack Snipe, Cetti's Warbler.

MONTGOMERYSHIRE (TREFALYDWYN)

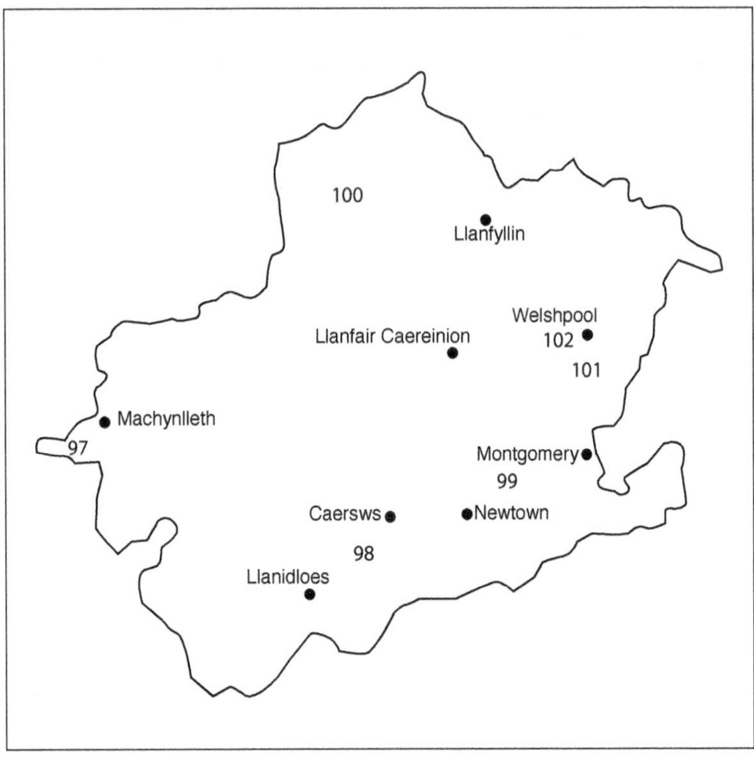

97 Cors Dyfi and Dyfi Osprey Project
98 Llandinam Gravel Pits
99 Pwll Penarth
100 Lake Vyrnwy
101 Dolydd Hafren
102 Llyn Coed-y-Dinas

Montgomery is land-locked, bar the small patch at the base of the Dyfi estuary in the far west of the county. This area (Cors Dyfi and the site of the Dyfi Osprey Project) has brought many species to the county, following the Dyfi upstream from RSPB Ynyshir (see Site 52). The third-largest county in Wales and the only one to extend the full width of the country reaches from the Dyfi estuary to the Shropshire plain. In the north is an arc of mountain and moorland, the highest point Moel Sych rising to 2,700 feet (827 m). Most of the river valleys run eastwards including that of the meandering River Severn, here are key sites such as Dolydd Hafren and Llandinam where Goosander, Kingfisher, Sand Martins, Grey Wagtails and Dipper may be found.

The development of the Welshpool bypass from 1993 resulted in a new lake, Llyn Coed-y-Dinas, now a Montgomery Wildlife Trust nature reserve with good access and a hide which quickly became a magnet for wildfowl and passing waders. The uplands are important for breeding Red and Black Grouse, Hen Harrier, Merlin, occasional Short-eared Owl and waders such as Curlew, Snipe, Dunlin and Golden Plover.

In the north of the county is the artificial Lake Vyrnwy and its surrounding deciduous woodlands and conifer plantations. The management of this lake, its woodlands and the upland moors has been taken over by the RSPB to form an extensive tract of mid-Wales. Here, together with Severn Trent Water (which owns the site), they are attempting to reverse the decline in moorland species such as Red Grouse, Hen Harrier and breeding waders. So far, their management has been very successful, with Black Grouse numbers rising dramatically.

97 CORS DYFI AND DYFI OSPREY PROJECT

OS Landranger Map 135
OS Explorer Map OL23
Grid Ref: SN703984
Websites:
www.dyfiospreyproject.com
https://www.montwt.co.uk/nature-reserves/cors-dyfi

Habitat

At the inland end of the Dyfi Estuary is the county's only section of salt water. Here the Montgomeryshire Wildlife Trust have a small nature reserve (17 ha), the Cors Dyfi Reserve. The reserve comprises wet meadows, reedbeds, bog, scrub and wet woodland and is the site of the Dyfi Osprey Project. Ospreys have summered in the area for several years and first nested in 2011, on a nest-platform beside the Aberystwyth to Shrewsbury railway, with the adults fishing in the estuary. In 2021 a new visitor centre was opened, with an observation hide providing distant views and CCTV giving live footage from the nest. Beavers were introduced into a 7-acre enclosure on the reserve in April 2021 as part of the Welsh Beaver Project, which since 2005 has been investigating the feasibility of bringing Beavers back to Wales, this work led by North Wales Wildlife Trust on behalf of all five Wildlife Trusts in Wales as part of the Living Landscapes strategy. Beavers are known as nature's engineers: they make changes to their habitats, which in turn create diverse wetlands for other species to thrive.

Species

The main ornithological interest here is the pair of breeding Ospreys. In the autumn of 2007, an Osprey nesting platform was erected in response to an increasing number of Osprey sightings, a male taking up residence the following spring being joined by a female later in the year. A male maintained a presence in 2009 and 2010 but failed to attract the attention of occasional passing females. There was success in 2011 when the male attracted the attention of a female hatched at Rutland Water, the pair successfully raising three chicks – the first since 'fishey-hawkes' were reported nesting on the Dyfi in

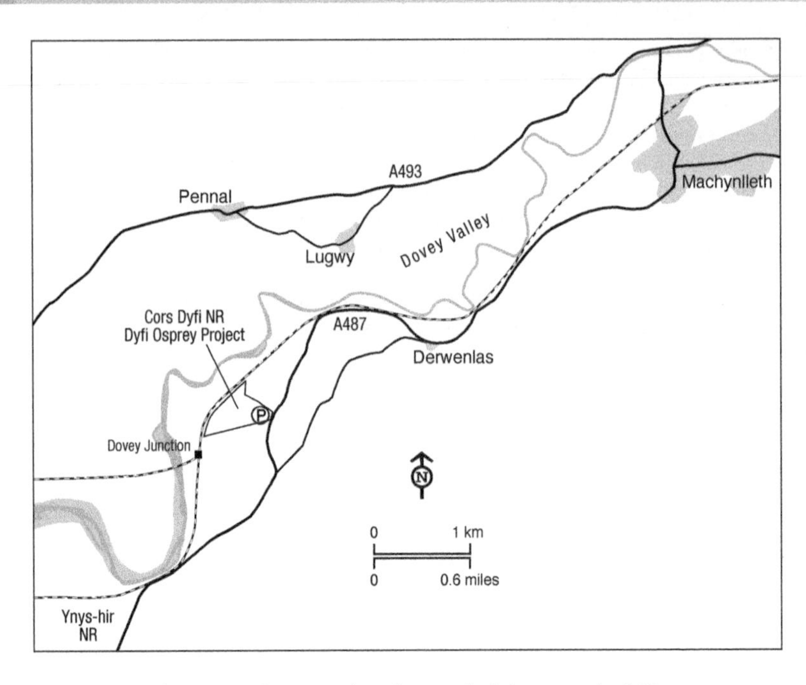

1604! Since then a pair has nested each year, fledging a total of 21 young up to and including 2020.

The nature reserve also has breeding Nightjars, Whitethroat, Grasshopper, Reed and Sedge Warblers, Tree Pipit and Stonechat. In 2020, Long-eared Owls nested, three juveniles being observed while the adults took a dislike to the Ospreys, which they repeatedly attacked at their nest.

The recent addition of nest-boxes in the copse has attracted Pied Flycatchers, while bird feeders attract good numbers of Lesser Redpoll, Siskin, Chaffinch along with the occasional Willow Tit, Great Spotted Woodpecker and Nuthatch. During the winter months Hen Harrier, Merlin and Bittern have been recorded. Rare visitors have included Crane, Great White Egret, Cattle Egret, Firecrest and a Nightingale, recorded singing by the Osprey CCTV in spring 2020.

CALENDAR

Resident: Red Kite, Buzzard, Sparrowhawk, Lesser Redpoll, Siskin, Chaffinch, Reed Bunting.

December–February: Occasional Hen Harrier, Merlin, distant geese and wildfowl.

March–May: Return of Ospreys late March onwards, spring migrants in the form of Chiffchaff, Willow, Sedge, Reed and Grasshopper Warblers, Whitethroat, Nightjar.

June–July: Osprey, Nightjar, Chiffchaff, Willow, Reed, Sedge and Grasshopper Warblers.

August–November: Ospreys and summer migrants depart. Arrival of winter thrushes, Brambling.

Timing

Reserve is open daily.

Access

Entrance to reserve and car park is well signed, on the NW side of the A487 road, three miles south-west of Machynlleth at SN703984. Charges apply. Facilities include two bird hides, visitor centre, with a small shop, refreshments and accessible toilets. The site is fully accessible to wheelchair users.

98 LLANDINAM GRAVEL PITS

OS Landranger Map 136
OS Explorer Map 214
Grid Ref: SO022876
Website: https://www.montwt.co.uk/nature-reserves/llandinam-gravels

Habitat

Beside the fast-flowing River Severn at Llandinam, the Llandinam Gravels Reserve of Montgomery Wildlife Trust comprises river shingle banks backed

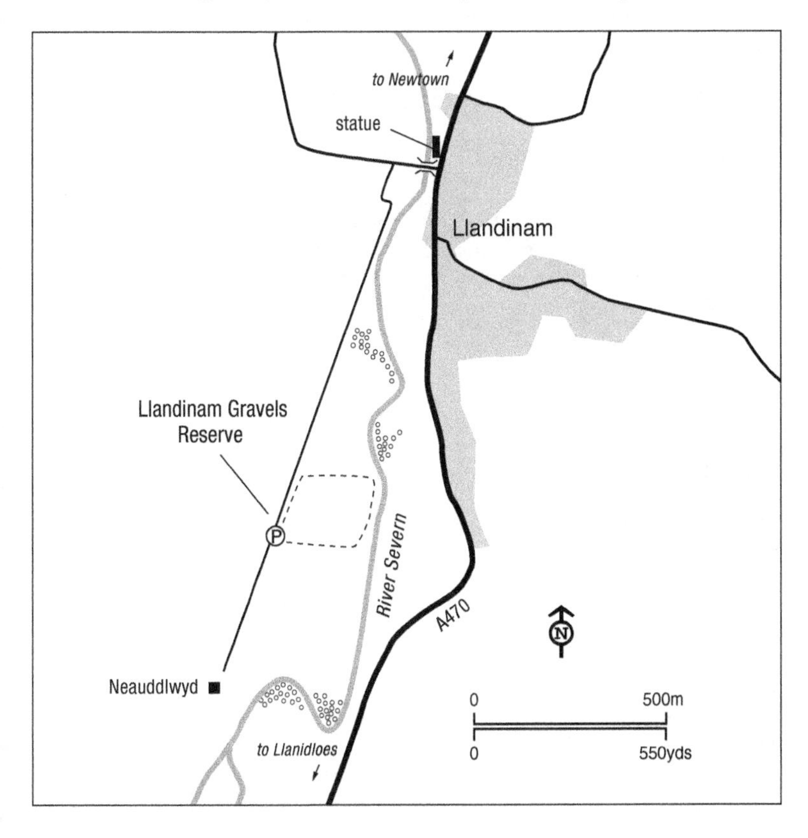

by the water-meadows. There is no point in pushing against such a strong force as the River Severn, so the Trust is working with the river to create a harmonious balance of habitat erosion and creation. The ever-changing banks of river shingle provide ideal nesting areas for Common Sandpipers and Little Ringed Plovers and a home for a rich variety of invertebrates, while Otters regularly pass this way.

Species

Little Ringed Plover and Common Sandpipers arrive in spring and breed, as do the occasional Oystercatcher, a species which as been slowly expanding its inland breeding range in Wales. Other species include Grey Heron, Little and the occasional Great White Egrets, Dipper, Grey Wagtail, Sedge Warbler and Reed Bunting.

Timing

Spring and autumn passage periods, summer for breeding birds.

Access

Turn off the A470 at Llandinam by the statue of David Davies, an industrialist and politician who died in 1890, cross the river and turn left onto a track that follows the dismantled railway. The reserve car-park is at SO022876 from where a footpath of three-quarters of a mile provides a circular walk of the nature reserve.

CALENDAR

Resident: Grey Heron, Little Egret, Kingfisher, Buzzard, Red Kite, Pied Wagtail, Grey Wagtail, Meadow Pipit, Dipper and Reed Bunting.
November–February: Occasional Great White Egret, Goosander, Snipe, Fieldfare and Redwing.
March–June: Returning species include Little Ringed Plover, Common Sandpiper, Sand Martin, Swallow, Swift, House Martin, Willow Warbler and Chiffchaff. Waders may include Oystercatcher, Curlew and Redshank.
July–October: Green Sandpipers may pass through, while winter thrushes arrive in early October.

99 PWLL PENARTH

OS Landranger Map 136
OS Explorer Map 215
Grid Ref: SO138927
Website: https://www.montwt.co.uk/nature-reserves/pwll-penarth

Habitat

Pwll Penarth is one of the string of sites along the River Severn which are managed for nature conservation by the Montgomeryshire Wildlife Trust (like Dolydd Hafren and Llyn Coed-y-Dinas below, Sites 101 and 102 respectively). Here the redundant settling beds of an old-fashioned sewage farm have been remodelled to create two valuable habitats. It is bounded on the south and east

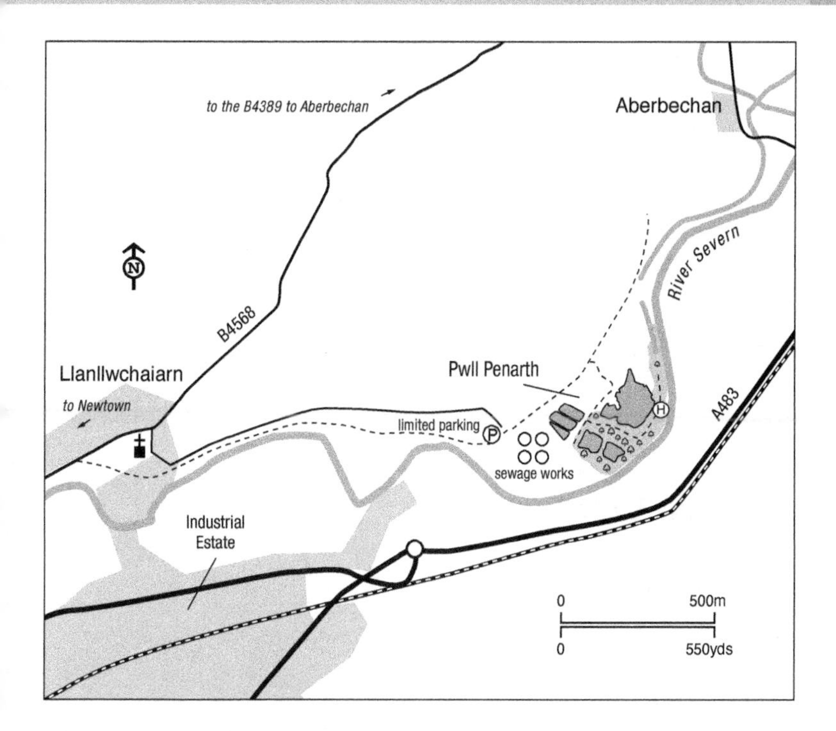

by a bend of the Severn and a canal feeder, while the northern boundary is the now dry bed of the Montgomeryshire Canal. A 27.5-acre (11-ha) lake has been landscaped to provide deep waters and shallows, and islands have been constructed.

Species

Little Grebe, Kingfisher, Cetti's Warbler, Reed Bunting are often present joined in summer by Reed and Sedge Warblers. Water Rail and small numbers of wildfowl such as Teal occur in winter. Mandarin Duck has been recorded. Rarities like Scaup and Garganey have been seen, while passage waders include Curlew and Green Sandpiper. The adjacent weir on the River Severn frequently attracts Grey Wagtail and Little Egret, while the leaping Salmon in autumn can be spectacular.

Timing

The reserve is open daily.

Access

The reserve is just off the B4568 from Newtown to Llanllwchaiarn, where a lane (next to the church) turns south and continues for about a mile (1.6 km) to Newtown sewage works, to which there is no access. Limited parking from where a substantial footpath leads for just over a mile through woodland to the reserve entrance. From here, one can make a complete circuit of the reserve with its two birdwatching hides, one at ground level the other raised high.

CALENDAR

Resident: Cetti's Warbler (heard more than seen!), Grey Heron, Mute Swan, Canada Goose, Mallard, Moorhen, Coot, Kingfisher, Grey and Pied Wagtails, Wren, Dunnock, Robin, Blackbird, Song and Mistle Thrushes, Magpie, Chaffinch, Goldfinch, Bullfinch, Reed Bunting.

December–February: Cormorant, Wigeon, Teal, Tufted Duck, Snipe, Curlew, Fieldfare, Redwing, Brambling, Siskin, Redpoll.

March–May: Departure of winter visitors, though a few ducks may remain until April depending on water levels. Summer visitors arrive, including Common Sandpiper, Sand Martin, Blackcap, Chiffchaff and Willow Warbler.

June–July: Hirundines feeding over the lake, Green Sandpiper possible, Kingfisher.

August–November: Departure of summer visitors and arrival of winter wildfowl. Usually good numbers of Fieldfares and Redwings from late October, when Siskin and Redpoll feed in the riverside alders.

100 LAKE VYRNWY

OS Landranger Map 125
OS Explorer Map 239
Grid Ref: SJ000210
Websites: https://www.rspb.org.uk/reserves-and-events/reserves-a-z/lake-vyrnwy/

Habitat

Lake Vyrnwy is the largest artificial lake in Wales, being nearly 5 miles (8 km) long and with a perimeter of 11 miles (17.6 km). The RSPB manages a total of 24,239 acres (10,080 ha), some owned, some leased from Hafan Ddyfrdwy (the Welsh branch of Severn Trent Water), around the lake and the surrounding uplands, making this the third-largest nature reserve in the Society's extensive landholding.

Large areas of conifers surround the lake, intermingled with patches of deciduous wood and scrub. Climb out of the valley, and you encounter the real gem of Vyrnwy: the extensive heather and grass moorlands at the south-western end of the Berwyn Range. Few mammal species occur on these uplands, though there can be large numbers of Short-tailed Voles and Pygmy Shrews, while Moles are found on even the highest ground. Sadly, there are no longer any Red Squirrels in the woods but there are Polecats and a large population of Badgers, while bats including Daubenton's hunt insects at dusk over the lake. Woodland plants include Oak Fern and Beech Fern, Wood-sorrel and Enchanter's Nightshade, while in flushes on the heather uplands you can find the insectivorous Round-leaved Sundew and Common Butterwort. RSPB Lake Vyrnwy farm is the largest organic farm in England and Wales and supports farming activities that benefit farmland wildlife, including such birds as Curlews. These vary from growing appropriate crops to managing river corridors, fencing woodlands and restoring walls and hedges.

Species

Pride of place among the small numbers of wildfowl which frequent Lake Vyrnwy must go to Goosander. In the early 1950s, a pair was present during the breeding season and could well have nested, though in this then very under-recorded area it is not surprising that it was 1968 before this was suspected and 1970 before a nest was located in a hollow tree close to the lake and a brood of ducklings seen, the first for Wales. Since then, Goosanders have greatly extended their range in Wales and increased with at least 685 pairs in 2016. From January onwards, Goosanders engage in their courtship displays. The preferred nest-site is a spacious tree-hole, and to ensure that sufficient are available specially designed nest boxes have been provided. One of the midsummer sights of Lake Vyrnwy is broods of ducklings closely escorted by a female. The males depart during May, probably on a moult migration which takes them to Finnmark.

Other breeding waterbirds include several pairs of Great Crested Grebe, Mallard and Mandarin Duck which now number some five or six pairs (the population in Wales now estimated to at least 500 pairs). There has been no confirmed nesting at Lake Vyrnwy of Red-breasted Mergansers for many years and this species is now only occasionally seen. In winter there are Wigeon, Tufted Duck and Goldeneye, while occasionally Common Scoters are seen, especially in late summer – a sign of overland migration to winter quarters off south-west Wales or in Cardigan Bay. On one occasion a drake Surf Scoter spent a day on the lake. Whooper Swans regularly visit the lake in winter, up to 20 having been recorded, while Coots are rare visitors.

Common Sandpipers, Kingfishers, Grey Wagtails and Dippers all nest beside the lake or on the numerous feeder streams tumbling down from the surrounding hills. When the receding water of late summer leaves a suitable muddy margin at the north-western end of the lake, waders stop briefly on passage: these have included Oystercatcher, Ringed Plover, Spotted Redshank, Greenshank and Green Sandpiper. Gulls are a rare sight, the wardens become excited when one is seen.

The woodlands hold as rich a variety of birds as one can find anywhere in Wales. The RSPB has carried out much woodland management to diversify the habitat, and also provided many nest boxes for a range of species from Goosander to Willow Tit. Red Kite, Sparrowhawk, Goshawk and Buzzard nest in the woodland, and there are usually several pairs of Tawny Owls and possibly Long-eared Owls. Other breeding species include Great Spotted Woodpecker, Redstart, Blackcap, Wood Warbler, Pied Flycatcher, Siskin and Lesser Redpoll, while Crossbills occur regularly. Ospreys have occasionally been recorded over the lake, while the high hills surrounding Lake Vyrnwy attract several other raptors, including Hen Harrier and Merlin, the abundant Meadow Pipits being the main prey of the latter, while the Hobby is a common sight in summer, maybe even nesting on the nature reserve. Peregrines and Kestrels regularly hunt these uplands, and Short-eared Owls have bred. At the moorland edge, where patches of Bracken, scrub and smaller trees occur, look out for Tree Pipit, Whinchat, Stonechat, Grasshopper Warbler and Whitethroat, all summer visitors to the area, for even Stonechats move to lower ground at the end of the breeding season. Skylarks and Meadow Pipits are numerous across the hills, and where rocky areas occur look out for Wheatear and Ring Ouzel. Red and Black Grouse breed in small numbers and so do a few Curlew.

Timing

Any time of the day, but it is really worth the effort to arrive early in the morning as you then seem to have the whole of the lakeside to yourself. Early

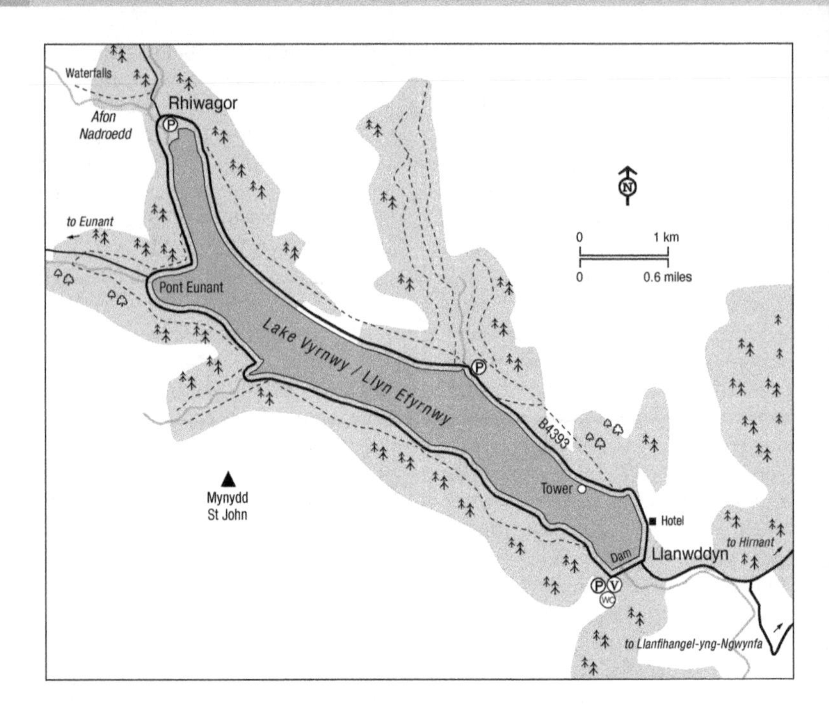

morning is certainly preferable on the hills, as bird activity will be at its peak at this time. If it is Woodcock you hope to see, then a dawn or dusk arrival is essential.

Access

If coming from the east, pass through Llanfyllin, take the B4393 to Llanwddyn near the south-east corner of the lake. The road continues on a shore-hugging circuit. If coming from Dolgellau, leave the A458 at SJ010111 near Llangadfan and follow the B4395 and then the B4393 to reach the lake. Two quite superb minor roads leave the lake near its northern extremity and climb out of the valley and across the moors. The first leaves Pont Eunant (SH963224) and rises to some 1,600 feet (488 m) before dropping steeply into the upper reaches of the Dyfi. The other leaves the lake at SH964242, passes through the plantations bordering the Afon Nadroedd, and then across several miles of high ground before descending to Bala. There is an RSPB reception and information centre at SJ016191, near the southern extremity of the lake, with a hide suitable for wheelchairs beside the car park and three other hides elsewhere on the reserve.

CALENDAR

Resident: Black Grouse, Great Crested Grebe, Grey Heron, Teal, Mallard, Goosander, Sparrowhawk, Goshawk, Buzzard, Red Kite, Kestrel, Peregrine, Red Grouse, Pheasant, Moorhen, Tawny and Short-eared Owls, Great Spotted Woodpecker, Grey and Pied Wagtails, Dipper, Goldcrest, Firecrest, Carrion Crow, Raven, Chaffinch, Greenfinch, Siskin, Redpoll, Crossbill.

December–February: Whooper Swan, Wigeon, Pochard, Tufted Duck, Goldeneye.

March–May: Hen Harrier, Merlin, Golden Plover, Lapwing, Snipe, Woodcock, Curlew, Common Sandpiper, Cuckoo, Kingfisher,

Redstart, Whinchat, Stonechat, Wheatear, Garden Warbler, Blackcap, Wood Warbler, Chiffchaff, Willow Warbler, Spotted and Pied Flycatchers.
June–July: All summer residents present, a chance of Goosander ducklings and passage waders at lakeside. Occasional Common Scoter.
August–November: Departure of summer visitors, Common Scoter and terns possible, while waders continue to visit if the water level is low. First winter ducks from late September, and from mid-October winter visitors such as Fieldfare and Redwing arrive.

101 DOLYDD HAFREN

OS Landranger Map 126
OS Explorer Map 216
Grid Ref: SJ205004
Website: https://www.montwt.co.uk/nature-reserves/dolydd-hafren

Habitat

Dolydd Hafren is the largest area of unmodified river floodplain in Montgomeryshire. It comprises a mosaic of oxbows, shingles, wet meadows, reedbeds, scrub and willow and is now a nature reserve of the Montgomeryshire Wildlife Trust where the River Severn meets two tributaries: the Camlad, which flows from the Forest of Clun, and the Rhiw, which flows from the west. The valley is wide and flat, and the Severn has changed its course more than once to leave a series of oxbow loops with resulting pools. On the 113-acre (42-ha) nature reserve, further pools have been excavated by the Wildlife Trust and Natural Resources Wales to enhance the habitat and provide even greater opportunities for wildlife. This includes Otters, which have returned to this part of the Severn valley after many years of decline, while the Brown Hare still thrives here.

Among the notable plants is the inconspicuous Mudwort, an ephemeral plant, erratic in its occurrence and with a rather limited distribution in Great Britain, even more so in Wales. It likes the exposed mud at the edge of pools, especially where this is disturbed by feeding waterfowl. Flowering Rush, with its tall stems and pink flowers, is much more noticeable. In Wales, this species is restricted to Anglesey, some mainly coastal localities in South Wales and a few sites along the English border, of which this is one.

Species

Oystercatchers and Little Ringed Plover have colonised the Severn and there are usually a couple of pairs of each at Dolydd Hafren, along with Common Sandpipers. Sadly, both Redshank and Lapwing no longer breed in the area, although they are sometimes seen along the river. Snipe and especially Curlew winter here: the latter has reached up to 600 on occasions in the past but numbers now rarely exceed 100. Little Egrets are regular along the river and there have been several records of Great White Egret. Little Grebes winter on the river, Grey Herons are usually to be seen, as are Mute Swans which can number 20 or so in winter, when there can be hundreds of Canada and Greylag Geese present. Mallard breed, and in the winter reach about 100 strong, with

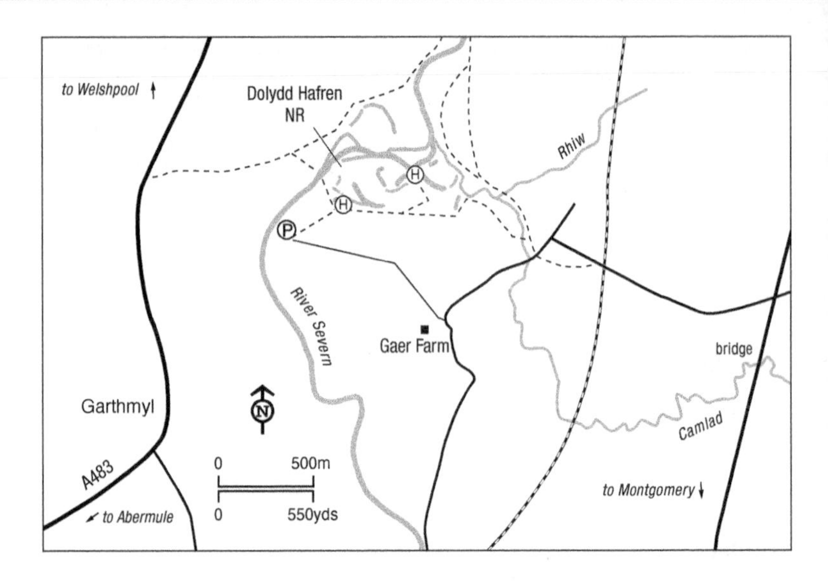

smaller numbers of Wigeon and Teal and the occasional Mandarin Duck, indeed one nested in 2021. Also Pintail, Shoveler and Shelduck, while Goosanders are more common. Moorhen and Coot nest and Kingfishers are frequently seen. Reed Buntings are resident, while Reed Warblers began to nest in 1998 and have since increased to around a dozen pairs. The hedgerows planted as screening have greatly increased the numbers of small birds including Sedge and Garden Warblers and Blackcaps. Tree Sparrows have been seen in the breeding season, but Yellowhammer, once the most common bunting in Wales, is now just a winter visitor. Peregrines are seen daily and, in the summer, so are Hobby, which nest not far away over the border with Shropshire. Ospreys are seen annually on passage and bred in 2004 just off the reserve but alas have never returned.

Timing

As one of the hides is on the eastern boundary you should aim for morning visits, with the sun behind you.

CALENDAR

Resident: Grey Heron, Little and Great Egrets, Mute Swan, Canada and Greylag Geese, Mallard, Goosander, Buzzard, Kestrel, Peregrine, Moorhen, Coot, Lapwing, Kingfisher, Meadow Pipit, Pied Wagtail, Blackbird, Jackdaw, Linnet, Reed Bunting.
December–February: Little Grebe, Wigeon, Teal, Pintail, Merlin, Water Rail, Curlew, Black-headed Gull, Stonechat.
March–May: Hobby, Oystercatcher, Little Ringed Plover, Redshank, Common Sandpiper, Sand Martin, Sedge, Reed and Garden Warblers, Blackcap. Green Sandpiper have been recorded in 11 months of the year, with a maximum 16 in September.
June–July: Most breeding species have young.
August–November: Best chance of waders on passage, winter visitors including Fieldfares and Redwings begin to arrive as last summer visitors depart.

Access

The reserve is open daily. Situated at the junction of three Ordnance Survey maps – 126, 136 and 137 – follow the B4388 north from Montgomery, and shortly after crossing the Camlad turn west and, once over the railway, take the first turn south back over the Camlad and so to Gaer Farm and a track to the reserve where there is a small car park. Screened by vegetation, a footpath follows the southern boundary to reach two hides which are both up steps.

102 LLYN COED-Y-DINAS

OS Landranger Map 126
OS Explorer Map 216
Grid Ref: SJ222052
Website: https://www.montwt.co.uk/nature-reserves/llyn-coed-y-dinas

Habitat

This is the story of the Welshpool bypass, or rather the potential benefit of a bypass, where its construction, which commenced in 1993, across the Severn meadows immediately to the east of the town, required large quantities of gravel excavated from close at hand. The resulting hole, extending for some 13 acres (5.26 ha) of the 20-acre (8 ha) site, proved a marvellous opportunity for the Montgomeryshire Wildlife Trust. With the assistance of Natural Resources Wales, the Trust was able to develop an excellent wetland nature reserve. Shallows and deep water, islands and reedbeds, the reeds and other appropriate aquatic plants were introduced, a hay meadow created, trees planted and a dipping pond provided – thus transforming within a few years a previously low-interest area into one of the most exciting birdwatching sites in the county.

Species

Great Crested Grebes quickly came to nest, while Little Grebes are winter visitors. Many Cormorants can be seen here and have recently begun breeding. Mute Swans, Mallard, Tufted Ducks nest as do Canada Geese. Black-headed Gulls are much in evidence, in 2019 roughly 200 pairs fledged about 300 young, making this the largest inland colony in Wales. Winter wildfowl on the lake may include Greylag Goose, Wigeon, Teal, Mallard, Shoveler, Tufted Duck, Pochard, Goldeneye and Goosander, while Whooper Swan make occasional winter visits. Grey Heron and Little Egret are usually present, they nest nearby in Welshpool. Great White Egret, Lapwing, Snipe and Oystercatcher are regular visitors. Hirundines (Sand and House Martins and Swallows) and Swifts are often abundant in summer. Breeding passerines include Sedge, Reed and Garden Warblers and Reed Buntings. Among the more unusual visitors have been Gadwall, Spotted Crake, Little and Mediterranean Gulls, Dunlin, Ruff, Redshank, Turnstone, Curlew Sandpiper, Wood Sandpiper, Little Stint, Black, Arctic and Common Terns, Bearded Tit and Bittern.

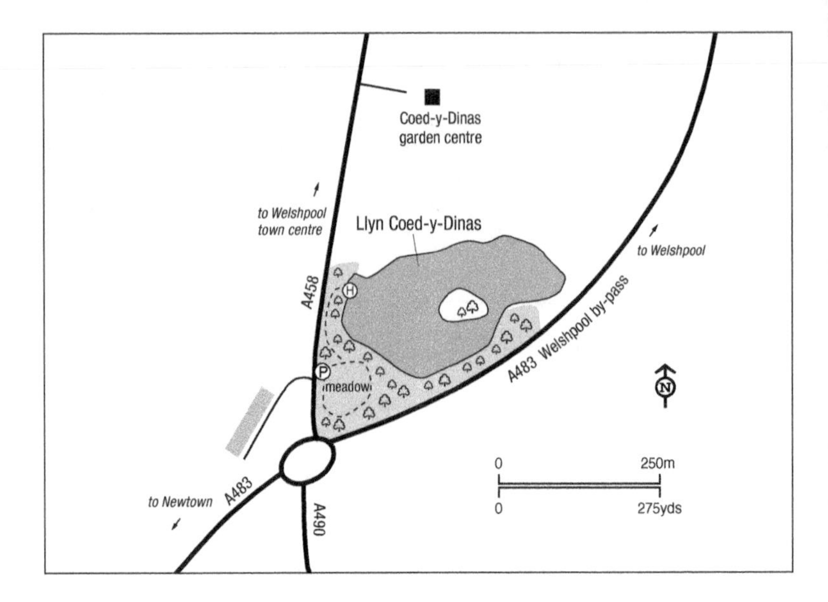

Timing

The reserve is open daily. An afternoon visit, especially on bright days, is best, then you will have the sun behind you.

Access

Leave the A483 Welshpool to Newtown road at the western end of the Welshpool bypass and head towards Welshpool, after a short distance the reserve car park is on the right. From here a path, suitable for wheelchair use, leads to the hide, while another route provides a circuit of the hay meadow and leads to the dipping pond.

CALENDAR

Resident: Cormorant, Grey Heron, Little Egret, Mute Swan, Canada Goose, Mallard, Sparrowhawk, Buzzard, Moorhen, Coot, Black-headed Gull, Grey and Pied Wagtails.
December–February: Wigeon, Shoveler, Pintail, Gadwall, Goldeneye, Water Rail, Lapwing, Snipe, Fieldfare, Redwing, Brambling, Siskin. There is also a large winter roost of Starlings.
March–May: Winter visitors depart. Passage of summer migrants, with Sand Martins the first to arrive, others include Common Sandpiper and Little Ringed Plover.

June–July: Breeding waterfowl have young, the bustling colony of Black-headed Gulls is in full swing, none more so than when the young birds make their first flights. Always a chance of waders on passage, while Hobbies often hunt over the pool.
August–November: Summer migrants departing, though Swallows in evidence until well into September. Passage waders still likely. From early October onwards winter visitors arrive.

PEMBROKESHIRESHIRE (SIR PENFRO)

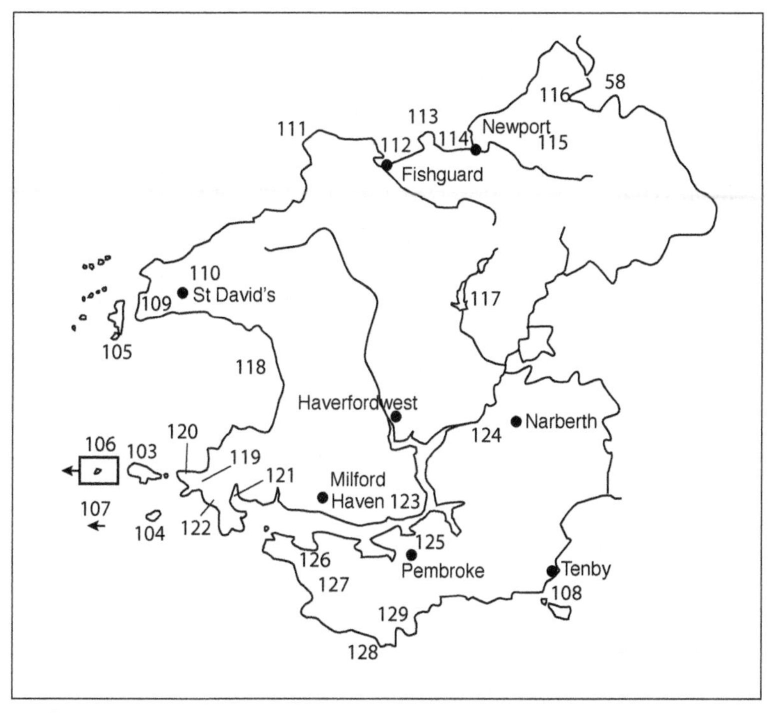

103 Skomer Island
104 Skokholm Island
105 Ramsey Island
106 Grassholm Island
107 The Smalls and the Celtic Deep
108 Caldey and St Margaret's Island
109 St David's Head and Valleys
110 St David's Commons
111 Strumble Head
112 Fishguard Harbour and Goodwick Moor
113 Dinas Island
114 Nevern Estuary
115 Pengelli Forest
116 Teifi Marshes (Welsh Wildlife Centre) and Gorge Woodlands
117 Llys-y-frân Reservoir
118 St Bride's Bay
119 Marloes Mere
120 Martin's Haven and the Deer Park (Wooltack Point)
121 The Gann Estuary & Pickleridge Pools
122 Dale Airfield
123 Milford Haven Waterway
124 Blackpool Mill and Minwear Wood
125 Pembroke Mill Ponds
126 Angle Bay
127 Castlemartin Corse
128 Castlemartin–Stackpole Coast
129 Bosherston Lakes

Pembrokeshire or "Little England beyond Wales" as it is sometimes called, is quite a diverse county, with stunning coastline, upland moors (Preseli Mountains) and wooded valleys. The coastline and a swathe of uplands constitutes the National Park, a 186-mile (299-km) long-distance footpath, from Amroth in the south to St Dogmaels in the north. Peregrine, Chough, Stonechat and Raven breed along nearly its whole length. The top birdwatching sites on the coast are Stackpole–Bosherston in the south, with its cliffs and auk colonies, St Bride's Bay in the middle, St David's Head on the far west and Strumble Head in the north, where keen watchers have studied movements for the last 50 years.

Offshore are the Pembrokeshire islands of Grassholm and Ramsey (RSPB), Skokholm and Skomer (Wildlife Trust of South & West Wales) and Caldey and St Margaret's. Skokholm's claim to fame was as the first Bird Observatory, opened in August 1933, and the ensuing studies on Manx Shearwater navigation. Skomer is a National Nature Reserve as are the waters surrounding it – one of the few National Marine Reserves in the United Kingdom.

In the centre of the county is the Cleddau Estuary, nationally important for wintering wildfowl and waders, with the best sites to view being Angle Bay, Dale and Landshipping. At the top of the eastern arm is Blackpool Mill, not really a good site for viewing the estuary but the river is an ideal spot for Kingfisher and Dipper, and its woods may hold Wood Warbler, Redstart and, until only a few years ago, Pied Flycatcher. Further inland, the intensive silage-driven agriculture doesn't provide much for wildlife but the large reservoir at Llys-y-frân attracts a big winter gull roost, a few ducks and the occasional rare visitor.

103 SKOMER

OS Landranger 157
OS Explorer OL36
Grid Ref: SM725093
Websites:
https://www.welshwildlife.org/skomer-skokholm/skomer/
http://www.pembrokeshiremarinesac.org.uk/

Habitat

Dominating the southern shores of St Bride's Bay, Skomer, at 781 acres (316 hectares), is the largest true island in Wales (Anglesey and Holy Island no longer islands as both are reached by road and rail). A National Nature Reserve since 1959, 13 years before that, in 1946, Skomer was the scene of the establishment for seven months of a field centre, one of the first in Great Britain, by the West Wales Field Society, now the Wildlife Trust of South & West Wales (see Buxton & Lockley (1950) *Island of Skomer*). Following an appeal by the Society, the island was purchased in 1958 and declared a National Nature Reserve with staff and research workers in occupation for about nine months of the year.

You walk with history when visiting Skomer. Excavations in recent years provide evidence of a sub-Megalithic occupation around 6,000 years ago, followed by Bronze Age then Iron Age farmers. This was a period of flourishing, as attested by the remains of numerous hut sites, well-preserved field boundaries and other features, contributing to one of the finest relict prehistoric agricultural landscapes anywhere in Great Britain. Vikings, their first raid in Pembrokeshire being in AD 850, named the island Skalmey the 'cleft' or 'split' island – one glance at the map reveals why. The first written reference is that in the valuation of William Marshal, 2nd Earl of Pembroke, who died in 1231. Most exciting of all is the first, in 1326, of many written records referring to Rabbit-catching and, in 1387, to the 'farm of birds' to the value of 6s 8d. In medieval times, a chapel was constructed on the isthmus of The Neck.

Occupation continued across the centuries. The farmhouse built around 1830 at the very centre of the island, has been in ruins since a great storm in 1954; the impressive outbuildings opposite have been extensively renovated as accommodation for visitors and research workers together with a small information centre.

The first naturalist to visit Skomer was the Reverend Murray Mathew in May 1884, who reported on the 'vast multitudes' of Manx Shearwaters and Puffins. He was followed by Robert Drane, a Cardiff pharmacist who provided an entertaining account of his journey in *A Pilgrimage to Golgotha*, where he describes the predation of Manx Shearwaters by Great Black-backed Gulls. It was Drane who first recognised that the Bank Vole on Skomer differs from its mainland counterpart, thus the Skomer Vole.

The Skomer Vole resides in areas of Bracken while Wood Mouse, despite its name and the absence of trees thrives, on open grassland right to the cliff edge. Both Common and Pygmy Shrews are present and Rabbits abound, while at least eight species of bat have been recorded, though whether these breed or hibernate on Skomer is unknown. Common Frogs, Common Toads, Palmate Newts and Slow-worms are present, but no snakes.

Grey Seals are seen throughout the year, with about 230 pups born annually, most between August and December, on open beaches and in the sea caves, one named Seal Hole. The haul-out of adults and immatures on the main North Haven beach is an impressive sight. Harbour Porpoise occur offshore throughout the year, while Common and Risso's Dolphins are regularly sighted.

The waters surrounding Skomer Island are designated as a Marine Conservation Zone (Skomer MCZ) currently the only MCZ in Wales. Following establishment as a voluntary Marine Nature Reserve in 1976, it was designated a Marine Nature Reserve proper in 1990, and this changed to a Marine Conservation Zone in 2014. The wider sea area is also part of the Pembrokeshire Marine Special Area of Conservation (SAC).

In spring vast swathes of Skomer are covered in Bluebells to be followed by Red Campion while the cliff slopes support Sea Campion, Thrift and Sea Mayweed. Later, much of the island is dominated by Bracken, which sadly has taken over areas where Heather once flourished. There are no big trees but Willows flourish beside the North Pond and in the valley below, while areas of Bramble also provide welcome shelter for both breeding and migrating birds.

Species

Pride of place must be given to the Manx Shearwater, one of the greatest travellers in the bird world and the most important species on Skomer. More than 300,000 pairs nest here – more than half the global population of this extraordinary seabird. Surveys across the years have shown an increase in numbers, partly a genuine increase partly the more refined methods being used. Still, this species is difficult to observe as it nests in burrows and is only active on the surface after dark; indeed, the darker the night the better.

Throw in a light drizzle as well and their evocative calls reach a wild crescendo amid a whirring of wings as birds scamper to take off while others hurtle in to land.

A night or two on Skomer with the Manx Shearwaters offers an unforgettable experience. Yet in daylight, there is virtually no sign of them other than the occasional remains of those which fall prey to Great Black-backed Gulls, or distant views of a few over the tide race to the west of the island. Using GPS loggers, we now know that during the breeding season Manx Shearwaters from Skomer can forage far out into the Western Approaches a long way west of Ireland, while others travel to the northernmost reaches of the Irish Sea.

Especially exciting are the weeks from late August and through September when young Manx Shearwaters now deserted by their parents are departing after some 10 weeks in their burrow. Emerging after dark they exercise their wings, perhaps for several nights, before heading for the nearest clifftop from which to launch on their first flight. Immediately they embark on a direct journey to the seas off Brazil, Uruguay and Argentina. The return journey in late winter takes them in a huge sweep past the Caribbean and the east coast of North America before a trans-Atlantic crossing. How many miles will they have travelled between the Pembrokeshire islands and the South Atlantic during a lifetime which may be half a century?

Although early work on Manx Shearwater migration and movements by R.M. Lockley in the 1930s and in the 1950s by G.V.T. Mathews was carried out largely on Skokholm, from 1976 onwards the emphasis has been on Skomer. Here, micro-cameras in burrows transmit pictures of nesting Manx Shearwaters to the information room on the island and also to the Lockley Lodge Information Centre at Martin's Haven. However, there is no substitute for spending a night or two on a Pembrokeshire island and sharing for a few precious hours the Manx Shearwaters' world. This is a world that was also shared in 1981 and 1982 by a Little Shearwater (also known as a Macaronesian Shearwater), for which the nearest colonies are in Madeira and the Azores.

The Puffin colony on Skomer, with approximately 40,000 birds in 2022, is the largest in southern Britain – though as with Skokholm perhaps it is just a shadow of its past glory, with Mathew in 1884 stating that 'there is scarcely a yard of ground free of them', while the 50,000 pairs of 1946 had dwindled to about 5,000 in 1960. From the early 1990s onwards there has been a steady increase, and the Skomer Puffins are more consistently successful that any other studied colony in Great Britain.

Puffins return in March, a single egg is laid in the nesting burrow in late April, with chicks from early June, when the adults start bringing fish ashore. This entails seven weeks of hard labour, not only in catching the fish but also avoiding the attentions of Jackdaws. Herring, Great and Lesser Black-backed Gulls all lie in wait for a possible meal too. Peak numbers of Puffins will be encountered towards the end of July, when the breeding population is supplemented by non-breeders prospecting for nesting burrows. A cloudy, even a rainy day will help ensure maximum numbers are ashore from late morning onwards. Numbers rapidly tail away during early August and soon all have departed for the next months spent at sea, many in the Atlantic, others in the Mediterranean.

One of the most dramatic of all seabird cliffs in Great Britain is the Wick on Skomer: a great chasm, the result of much geological grinding and slipping millions of years ago, which has left a 200-foot (60-m) vertical basalt cliff opposite which are steep slabs below slopes in which Puffins nest. Others choose more level terrain beside the clifftop path, so that at times there are Puffins both to the right and left, waiting for an opportunity to take flight or simply standing and observing in turn the visitors watching and photographing them.

The Wick is vast and, throughout the breeding season, a very noisy avian city – largely from the screaming calls of Kittiwakes nesting on the lower sections of the cliff. Alas their numbers on Skomer, mostly at the Wick, declined from a peak of 2,557 pairs in 1984 to 1,441 in 2021. In huge groups along the central ledge are massed ranks of Guillemots, their trumpeting calls in contrast to the deep burring resonance from the more scattered but still numerous Razorbills. Early in the season, look out for the quite delightful 'butterfly' display flight of the Razorbills as they leave the ledges with wings beating in a slow accentuated motion. Both have mostly departed by the end of July, the Kittiwakes remain for a couple more weeks.

Suddenly there is relative silence, but small numbers of one seabird remain: the Fulmar, which in the updraughts along the cliff face shows astonishing prowess. It is the perfect glider, the 'Grey Glider of the North Atlantic'. At the inner end of the Wick is a single high ledge – you cannot miss it – within the safe recess of which Fulmars have nested for over 50 years; as they are long-lived birds perhaps one or two we see today are original colonists. Some will sit tight, incubating and later brooding, a small chick. Others come and go, swooping down or soaring just below the clifftop with only the occasional twist of the wings to maintain momentum, sometimes steering with their feet. When one lands there comes the greeting ceremony, a loud, often prolonged cackling by both adults in the pair.

Skomer has an impressive list of breeding landbirds and for many years was the last outpost of the Curlew in Pembrokeshire, reaching a peak of 24 pairs in 1974; 40 or so years later, there were just a couple of pairs by 2020. Oystercatchers, of which there are about 60 pairs, breed almost anywhere on Skomer. Canada Geese which first nested in the early 1990s, are now well established, Mallard nest annually, being joined on occasion by Shelduck, Teal and Shoveler.

Other notable breeding birds include Buzzard, Peregrine Falcon, Short-eared Owl, Chough and Raven, while do not be surprised if you hear a Pheasant, about 15 pairs nest in the areas of deep Bracken. Wood Pigeons nest on the ground, while Magpies which first nested in 1967 use the best Bramble bushes, Carrion Crows the rock outcrops and Jackdaws the cliffs. Wheatears, Skylarks and Meadow Pipits frequent the plateau, Sedge Warblers and Whitethroats the stream valleys, while Rock Pipits announce their presence from the cliff slopes.

Equally impressive is the list of vagrants recorded on Skomer, perhaps none more so than a Blackburnian Warbler from North America watched on 5 October 1961 on the sheltered cliff above the North Haven landing beach, the first for the Western Palaearctic and only recorded on two subsequent occasions both on Scottish islands (Fair Isle and St Kilda). Other trans-Atlantic passerines have included Dark-eyed Junco, Black-and-white Warbler, Bobolink and Rose-breasted Grosbeak, while American waders like Upland Sandpiper, Lesser Yellowlegs and Pectoral Sandpiper also grace the island list.

Vagrants from other parts of the globe have included a White-tailed Eagle in 1993, the first for Wales since 1910, a Frigatebird of one species or another, and

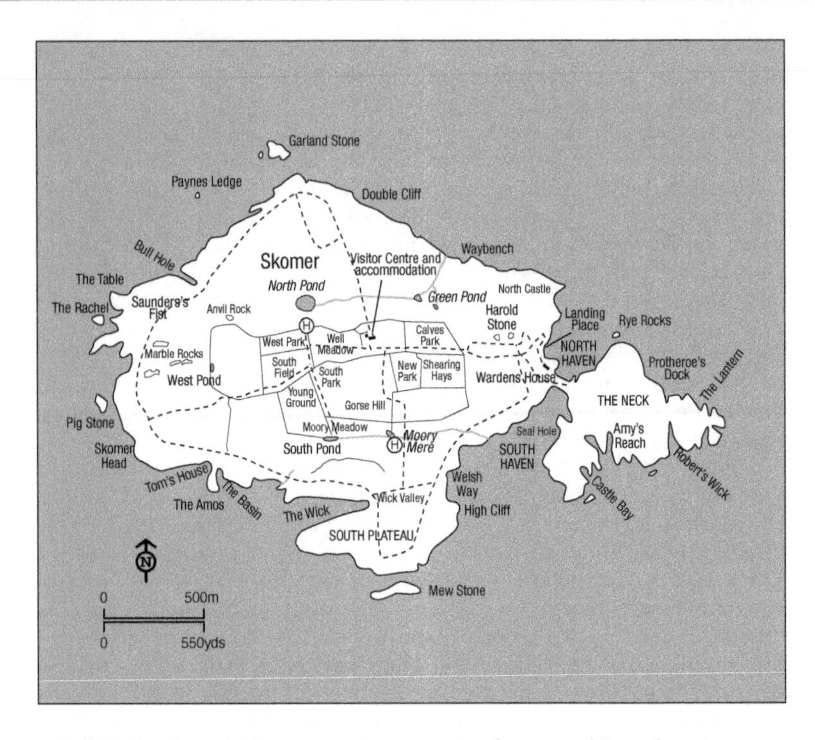

a Pallid Harrier which at one time was in the same binocular view as a Montagu's Harrier. Other rarities have included Black Stork, Isabelline Shrike, Dusky Thrush, Black-headed Wagtail, Citrine Wagtail, Red-flanked Bluetail, Western Black-eared Wheatear and in May 2014 the fourth Blyth's Reed Warbler for Wales which delighted observers with its full song.

Although such wonders enchant observers and make the headlines, landbirds from the mainland no great distance away which reach Skomer can prove exciting too and have included Kingfisher, Green and Great Spotted Woodpeckers, Mistle Thrush, Long-tailed Tit, Jay, House Sparrow and Bullfinch.

Timing

Skomer is open daily Mondays excepted other than on Bank Holidays from 1 April until late September by boat from Martin's Haven.

Access

From 2021, bookings can only be made on the WTWSW website (see above). The price of a day visit in 2021 to Skomer is not cheap, at £20 for the boat and a further £20 to land for an adult. The first boat departs at 10am with up to five crossings made, the last at 12 noon. Returns from the island usually commence at 3pm. The boat service operates weather permitting from 1 April, or from Easter if in March, until 30 September daily, save Mondays, bank holidays excepted. Further information is available from the Lockley Lodge Information Centre situated just downhill from the National Trust car park (fee payable) at Martin's Haven between 1 April and 31 August.

A steep climb of about 80 feet (25 m) takes visitors from the landing steps to the main track, from which a footpath system, mostly level, leads to the main seabird colonies of High Cliff and the Wick and beyond to the west and north coasts, while other footpaths cross the island plateau via the farm buildings which includes a small information room and toilets.

CALENDAR

Resident: Cormorant, Shag, Common Scoter passing from the small flock further east in St Bride's Bay, Buzzard, Kestrel, Peregrine, Pheasant, Oystercatcher, Herring and Great Black-backed Gulls, Woodpigeon, Little and Short-eared Owls, Skylark, Meadow and Rock Pipits, Wren, Pied Wagtail, Dunnock, Stonechat, Blackbird, Magpie, Chough, Jackdaw, Carrion Crow, Raven, Linnet, Reed Bunting.

December–February: No access.

March–May: Kittiwakes are already on the cliffs, while Guillemots and Razorbills have been making occasional visits to the breeding cliffs since December and now put in more frequent appearances prior to egg-laying in late April. Manx Shearwaters arrive in early March quickly followed by Puffins, though until egg-laying in April there may be days when Puffins are absent, or only return in numbers from the late afternoon onwards. The first passage migrants will be Ring Ouzels and Black Redstarts at the same time as Wheatears put in an appearance, some remaining to breed. As the weeks progress migrants will include Whimbrel, Common Sandpiper, Cuckoo, Swift, Sand Martin, Swallow, several pairs of which nest, House Martin, Redstart, most warblers, with Sedge Warbler and Whitethroat remaining to nest, Spotted and Pied Flycatchers.

June–July: The peak of the seabird breeding season, though by the end of July most Guillemots and Razorbills will have left the cliffs. If the water level is low, passage waders are likely at the North Pond.

August to November: Puffins depart during the first couple of weeks, Kittiwakes by mid-August, Fulmars in September, Manx Shearwaters depart from early September onwards. Wigeon and Teal become more frequent at the North Pond, wader passage continues. Water Rails may be heard, while Hen Harriers, Tawny and Barn Owls sometimes take up residence. The spectacular migration of Swallows, House and Sand Martins in the spring is now repeated by the return movements south. Other summer visitors will be departing to be replaced by Fieldfares and Redwings, Blue Tits and Great Tits, Chaffinches and Greenfinches. Throughout the whole period there is always the chance of rare migrants both from the east and the west to round off the island year.

104 SKOKHOLM

OS Landranger 157
OS Explorer OL36
Grid Ref: SM735050
Websites:
https://www.welshwildlife.org/skomer-skokholm/skomer/
http://skokholm.blogspot.com/

Habitat

Skokholm, 260 acres (105 ha) in area and almost 2 miles (3.2 km) south of Skomer, is a jewel on the southern horizon, being composed of Devonian Old Red Sandstone rather than the dark volcanic rocks of its larger neighbour. In prehistoric times when sea-levels were much lower, you could have walked to Skokholm and indeed, as with Skomer, there is evidence of early communities.

In 2020, the wardens Richard Brown and Giselle Eagle observed that earth excavated by Rabbits from a burrow contained a Stone Age tool, later identified as a Mesolithic bevelled pebble, perhaps used to prepare shellfish some 9,000 years ago, while fragments of pottery originated from a 3,700 year-old Bronze Age cremation urn.

The name Skokholm is of Norse origin, being Stokkr holmr 'the island in the sound'; across the centuries there were numerous spellings, with Skokholm used from 1875 onwards.The earliest written record of Rabbit-catching on the warren here dates from 1324 when the profit for Skokholm, Middleholm and Skomer was £14, pasturage from the same islands being little more than £2. Rabbit-catching only ceased in 1932.

One of the owners of Skokholm, though only for a short time, was the Marchioness of Pembroke, better known as Ann Boleyn; another was Sir John Perrot, described as 'physically strong, arrogant, avaricious and utterly unscrupulous who sought to enrich himself by all the methods at his disposal'. An inventory of his farm stock on Skokholm and Skomer following his death in 1592 (while held in the Tower of London on the charge of treason), reported 395 sheep, 132 lambs, six calves, 35 heifers, bulls and bullocks and 10 wild horses and colts.

Skokholm became part of the Dale Castle Estate, Dale the nearest village on the mainland, in 1713, with the first detailed description of the island dating from 1811.This reports the house as being built after a whimsical manner and when supplies were needed from the mainland signals of dense smoke were made by burning straw and peat. Most intriguing was the presence of deer which 'grew fat', the venison was judged superior to any produced on the continent of Pembrokeshire. Alas, there was no mention of wildlife.

Not until 1925 did a naturalist leave a written record after reaching Skokholm, this by the indomitable Bertram Lloyd who, together with Charles Oldham, walked 9 miles (15 km) from Little Haven to Martin's Haven to be rowed to the island to stay overnight. Lloyd made the first bird list, 29 species in total, reporting that Manx Shearwaters were present in vast hordes while Puffins nested all along the clifftops and far inland in thousands.

The following year, 1926, R.M. Lockley reached Skomer but inclement weather meant no crossing to Skokholm. He finally made it in 1927 and within a matter of weeks had negotiated a 21-year lease at £26 per annum, arriving to take up residence in the late autumn. Lockley immediately began to report on the island's birds, his first published note concerning a Lesser Whitethroat on 3 November.

In February 1928, Lockey had a tremendous stroke of good fortune when the wooden topsail schooner *Alice Williams*, having been abandoned by her crew, was wrecked on Skokholm. Lockley was able to salvage her cargo of coal, valuable timbers and artefacts, including the figurehead and wheel, both now secure in the island's Bird Observatory – the first in Great Britain when opened in 1933. Lockley's books about island life soon ensured that Skokholm and the other islands of the Pembrokeshire coast became probably the best known of all nature conservation sites in Wales, and among the best known in Britain.

Bird-ringing was carried out on Skokholm from 1933 to 1940 and again from 1946 to 1976, the results helping to elucidate many things regarding the movements of birds and the fortunes of the islands' seabird populations. In 2006, the island was purchased by the Wildlife Trust of South & West Wales following an appeal which resulted in over £600,000 being raised. The Bird Observatory reopened in 2012. The wardens live at the lighthouse, which had been purchased following a further appeal enabling major renovations and building work to be carried out, largely by volunteers, on the observatory buildings, lighthouse and landing jetty. The Bird Observatory has its own website (see above), with news on recent activities, monitoring and reports, while the Skokholm blog provides news and photographs daily. House Mice are the only resident land mammal in addition to Rabbits, though six species of bats have been recorded. Grey Seals are present throughout the year, although there are few suitable sites for pups. The Wild Goose Race and other tidal streams around Skokholm play host to Common Dolphin and Harbour Porpoise. The only reptile is the Slow Worm, while elvers manage to scramble up the stream from North Pond where it tumbles down the cliff.

The cliff slopes are largely dominated by Thrift and Sea Campion; the extensive heathlands of little more than 70 years ago have largely disappeared, the result of Rabbit grazing, and been replaced by Yorkshire Fog grass and Bracken. The few trees which grow to no great height are mostly around the observatory buildings while sea-cliff scrub occurs in a few more sheltered spots. The rich lichen flora includes the spectacular internationally important Golden Hair Lichen.

Species

Skokholm is surprisingly different to its sister island to the north. Spectacular Old Red Sandstone cliffs, carpeted in yellow and grey lichens, offer a very distinct backdrop to the seabird colony. There are no day-visitors, meaning that only up to 20 self-catering guests are present alongside the staff and researchers. This means that the path network, an essential feature on a fragile island shared with over 100,000 burrow-nesting seabirds, can be more subtle and take in much closer views of the cliff-nesting seabirds. The walk up from the jetty to the old farm, now home to guest accommodation, and the Bird Observatory buildings passes the few areas of denser scrub. These small stands of Grey Willow, Elder and Sycamore are thronged on fall days with migratory passerines. It is here that the three Heligoland traps and mist nets are situated, apparatus used for the systematic ringing of birds essential to our understanding of their life histories; all guests are welcome to join the staff and visiting ringers for a demonstration and to see a wide range of species up close.

Away from its more sheltered centre, the Skokholm landscape is dominated by maritime grassland and carpets of Sea Campion and Thrift; these areas, which during the summer months are important feeding grounds for breeding Choughs, Wheatears and Skylarks, are used by large numbers of thrushes, pipits and Starlings during spring and autumn. Three small seasonal ponds attract a remarkable variety of waders, with 37 species noted in the seven years from 2013 alone. Nightfall brings the return of Skokholm's most important inhabitants, with dark nights around the new moon or during poor weather the best times to see Manx Shearwaters coming and going from their burrows. About 100,000 pairs nest on the island; wherever there are burrows there will be Manx Shearwaters. One bird, a participant in an early project investigating bird navigation, was in June 1952 taken to Boston, USA, by a visitor returning home to his work with the Boston Symphony Orchestra. It was found in its burrow in the early hours of 16 June having travelled more than 3,200 miles. A letter that arrived on Skokholm later the same day reported on its release the other side of the Atlantic on 3 June.

Storm Petrels nest in field boundary walls, while a nocturnal walk with the wardens takes guests to an area where hundreds of these birds can be watched – using infrared viewing equipment so as to not disturb them. Guests are also welcome to join the staff in some of their monitoring work, activities such as emptying the moth trap and of course bird ringing. All of the migrants counted during the day are entered into a log which dates back to 1927, this being an island tradition which takes place each evening around the fire in Lockley's Cottage.

Skokholm is one of only two sites in Wales with a species list in excess of 300 birds, with a total of 301, the other being Bardsey (332). A mouth-watering tally over the years includes several first records for Britain, species such as Olive-backed Pipit, Swainson's Thrush and Western Bonelli's Warbler. Firsts for Wales have included American Golden Plover, Little Swift, White-throated Robin, Pied Wheatear, Savi's, Booted and Sardinian Warblers, Red-eyed Vireo, Rustic Bunting, Northern Oriole and Bobolink. Other scarce visitors have included Black Stork, Black Kite, Red-footed Falcon, Black-winged Stilt, Semi-palmated and Upland Sandpipers, Scops Owl, Pallid Swift, Roller, Richard's, Tawny and Red-throated Pipits, Isabelline and Desert Wheatears, Myrtle, Icterine, Melodious and Barred Warblers, Red-breasted Flycatcher, Red-backed and Woodchat Shrikes, Serin, Ortolan Bunting and Common Rosefinch.

Access

Self-catering accommodation for up to 20 visitors is available from mid-April to late-September. During the peak seabird monitoring season, from mid-May to the end of July, stays are for a week from Monday to Monday. Outside of

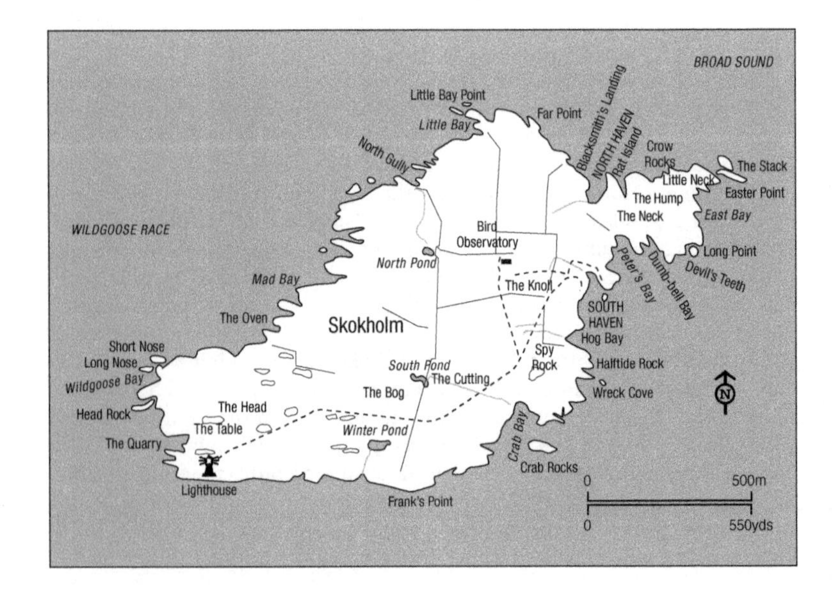

this period, boats run on Mondays and Fridays. There are no day trips to Skokholm. Enquiries and bookings to the Wildlife Trust of South & West Wales.

CALENDAR

Resident: Cormorant, Shag, Mallard, Common Scoter likely offshore throughout year (part of a small flock in St Bride's Bay), Buzzard, Peregrine, Oystercatcher, Herring, Lesser and Great Black-backed Gulls, Short-eared Owl, Skylark, Meadow and Rock Pipits, Wren, Pied Wagtail, Dunnock, Blackbird, Chough, Jackdaw, Carrion Crow, Raven, Reed Bunting.

December–February: No access. Red-throated and Great Northern Divers, Fulmars arrive early January, Gannets shortly afterwards offshore, Wigeon, Teal, Shoveler, Merlin, Hen Harrier, Guillemots and Razorbills make infrequent visits throughout period. A gull flock in Broad Sound to the north contains thousands of Kittiwakes and Black-headed Gulls. All resident species much reduced in numbers, and in hard weather may desert the islands altogether.

March–May: Guillemots and Razorbills which have made occasional visits during the winter are now more frequent especially during calm weather while Lesser Black-backed Gulls reoccupy their colonies. Manx Shearwaters arrive in early March quickly followed by the first Puffins offshore, the first landings mean spring has arrived on the island then follow on as previously then at end of March–May make a reference to the return of Storm Petrels. Passage migrants include Whimbrel, Common Sandpiper, Turtle Dove, Cuckoo, Swift, Hoopoe, Sand Martin, Swallow, House Martin, Black Redstart, Redstart, Whinchat, Wheatear, Ring Ouzel, most warblers, while Sedge Warbler and Whitethroat remain to breed, Spotted and Pied Flycatchers.

June–July: Grey Heron and Black-headed Gull visit from late July, Guillemots and Razorbills have all completed their breeding cycle by end of period and their colonies are once more deserted. Species that do not breed on the island continue to arrive throughout the period, with June seeing late overshoots and early failed breeders.

August–November: Wigeon, Teal, Water Rail, passage waders and terns. Puffins depart in early August. Martin and Swallow, Yellow and Pied Wagtail passage during early September, Robins arrive during August and take up their winter territories, main warbler passage finishes by mid-September, though Blackcaps may be seen until November. Fieldfare, Redwing, Mistle Thrush, Black Redstart, Goldcrest, Blue and Great Tits, Chaffinch, Goldfinch and Greenfinch during late autumn. Hen Harrier and Short-eared Owl are regular visitors, probably relating to those which roost on Skomer or on the mainland.

105 RAMSEY ISLAND

OS Landranger 157
OS Explorer OL35
Grid Ref: SM700235
Website: https://www.rspb.org.uk/reserves-and-events/reserves-a-z/ramsey-island/

Habitat

Ramsey Island is a National Nature Reserve, owned by the RSPB, just over half a mile (1 km) offshore from St Justinian's, St David's. Like the other islands the name is of Norse origin, the original Hrafrens-ey translates as 'Ravens Island' or 'Garlic Island' – both occur on Ramsey to this day. The Welsh name

Ynys Dewi is even older; dating from the 6th century, it translates as 'David's Island'. Ramsey has two hills rising from the flat plateau, Carnllundain at 450 feet (137m), and the slightly lower Carnysgubor, both affording wonderful views over the western cliffs and several spectacular bays, such as those of Aber Mawr and Aber Myharan, as well as inlets and sea caves. On the low eastern plateau is a series of fields, several small ponds and short valleys, and to the south an impressive expanse of maritime heath. A feature of Ramsey that makes it unique among Welsh islands are the satellite islets, the largest Ynys Bery 'Falcon's Island' to the south; while to the west a whole chain, the Bishops and Clerks described by George Owen, Pembrokeshire's Elizabethan chronicler, as preaching 'deadly doctrine to their winter audience'. Alas, his only reference to birds on these islets was to note 'these all yield store of gulls in the time of the year', presumably during the nesting season when eggs would certainly have been collected.

The first report on the birds of Ramsey was in a letter to the great Welsh botanist, geologist, antiquarian and philologist Edward Lhuyd, who died in 1709, from his 'ingenious Friend' the Reverend Nicholas Roberts of Llandewi Velfrey, Pembrokeshire. In it, he describes in detail four species of seabird nesting on the island: the Elgug which we know as the Guillemot, the Razorbill, the Puffin and the Harry-bird which we know as the Manx Shearwater. The St David's Cathedral Survey of 1717 described Puffins as breeding commonly on Ramsey but by 1894 Brown Rats had reached the island, with Puffins confined to a colony at the north end becoming extinct shortly afterwards.

Ramsey was purchased by the RSPB in 1992, by which time almost a thousand pairs of Manx Shearwater had re-established. Early in 2000 a rat eradication operation commenced, which by the end of the year was deemed successful. Since then the Manx Shearwater colony has continued to grow, in 2016 to almost 5,000 pairs while about 10 pairs of Storm Petrels are now well established too. Hopefully, if the Puffin colonies on Skokholm and Skomer continue to prosper, some might begin to recolonise here as well; there is certainly plenty of room.

Resident land mammals are Rabbits, Bank Voles and Common Shrews. Ramsey, with its deep-sea caves and numerous beaches, is the headquarters for Grey Seals in south-west Britain. Between 600 and 700 pups are born each year, mostly between August and November, while adults and immature seals are present in all months. Harbour Porpoises are resident in Ramsey Sound and the waters to the west of the island, where Common Dolphins are also regularly seen.

Species

The west cliffs hold most of the breeding seabirds, comprising Fulmars, Guillemots, Razorbills and also a small number of Kittiwakes, these having seriously declined during the recent past, while Manx Shearwater numbers continue to increase though an overnight stay is required in order to make their acquaintance. Peregrines and Ravens breed on the cliffs, with up to 10 pairs of Choughs in the sea caves while feeding flocks in late summer can number up to 65.

Ramsey is grazed by sheep and Red Deer and drenched in salt-spray during the winter, limiting the amount of shrub and tree cover. However, the sheltered Willows around the farmhouse and valley, as well as patches of Bramble and

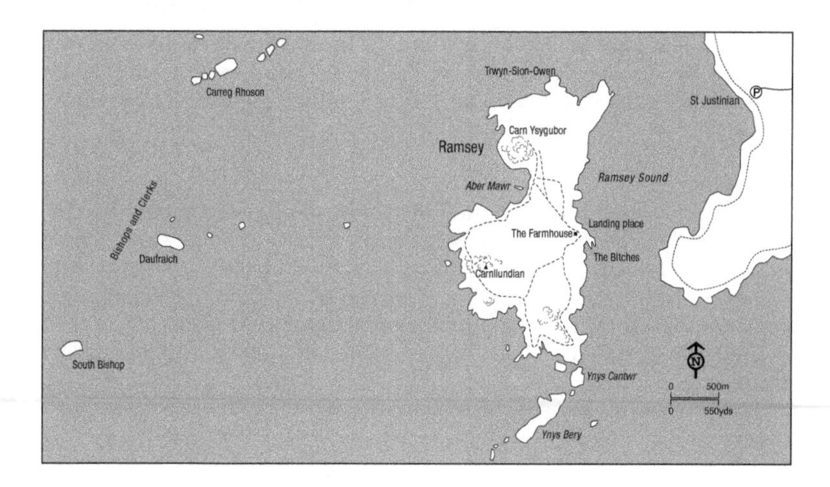

Gorse attract migrant birds and are well worth checking for breeding Whitethroat, Stonechat and Wheatear. Start with the area outside the warden's walled garden (staff won't mind you birdwatching here), this was the site of the Indigo Bunting in October 1996, with Blyth's Reed, Eastern Subalpine, Booted and Barred Warblers in recent years. Then head onto the south trail and up the valley. It is always worth checking the stone walls for Little Owls on the way.

Head south at the junction, onto the east coast, which is productive for breeding Linnet, and the bowl-shaped bay at Myharan, a hotspot for common migrants in late summer, with flycatchers, warblers and if you are lucky Wryneck. The south coast plateau is a great place to see Peregrine hunting across the bay. Climb north up and over Carnllundain; the boulder slopes are nesting sites for many Wheatear and the grassland above the beach at Aber

CALENDAR

Resident: Shag, Mallard, Buzzard, Kestrel, Peregrine, Oystercatcher, Herring and Great Black-backed Gulls, Woodpigeon, Little and Short-eared Owls, Skylark, Meadow and Rock Pipits, Wren, Pied Wagtail, Dunnock, Stonechat, Blackbird, Song Thrush, Magpie, Chough, Jackdaw, Carrion Crow, Raven, Starling, Linnet.

December–February: Red-throated and Great Northern Divers, Fulmars, Gannets, Merlin, Hen Harrier, Kittiwake, Guillemot and Razorbill.

March–May: Manx Shearwaters arrive early March, Storm Petrels late April. Guillemots and Razorbills make more frequent and longer visits to cliffs, though will not be fully resident until the eggs are laid in early May. Passage migrants include Whimbrel, Common Sandpiper, Cuckoo, Swift, Sand Martin, Swallow, House Martin, Black Redstart, Redstart, Whinchat, Wheatear, Ring Ouzel, most warblers, while Whitethroat remain to breed, Spotted and Pied Flycatchers.

June–July: Guillemots and Razorbills have completed their breeding cycle by end of period and their colonies are once more deserted.

August–November: Can be productive for seawatching after northwesterly winds, with Sooty and Balearic Shearwaters, skuas and terns. Black-headed Gulls, Kittiwakes, Sand Martin, Swallow, Pied Wagtail, Fieldfare, Redwing, Mistle Thrush, Black Redstart, Wryneck, Goldcrest, Blue and Great Tits, Chaffinch, Goldfinch. Kittiwakes depart the cliffs by mid-August, Fulmars by mid-September.

Mawr the best place to see Choughs feeding. The stone walls that crisscross the fields are worth scanning for migrants. Back at the farm, check the stone buildings, especially the yard outside the toilets – this was the unlikely location of Ramsey's second Myrtle Warbler in 2019, the first being in 1994.

Access

Thousand Islands Expeditions run daily crossings to Ramsey, weather permitting, from the lifeboat station at St Justinian's, 1 April or Easter (whichever is earlier) to 31 October. Landing tickets should be booked in advance with Thousand Islands at their office in Cross Square, St David's, Pembrokeshire SA62 6SL. Tel: 01437 721721, email: info@thousandislands.co.uk

Other companies offer shorter boat trips around Ramsey, with information available in St David's.

106 GRASSHOLM

OS Landranger 157
OS Explorer OL36
Grid Ref: SM598092

Habitat

R.M. Lockley made his first visit to Grassholm, one of the smallest Pembrokeshire islands, in July 1928 when he and Doris spent a night ashore during their honeymoon. In a vivid description of the natural history and especially the Gannet colony, he wrote 'I had seen nothing in nature so spectacular, a ballet of blue and silver'. Although landings are no longer permitted, a circumnavigation by boat will immediately confirm his words, while the island on a sunny day glows white when seen 7 miles (12 km) away from the west cliffs of Skomer.

The Mabinogion, the earliest prose stories of the literature of Britain, includes a tale of those who lived on Grassholm for 80 years. Confirmation of early inhabitants was made in 1890 when a small collection of pottery, flint flakes and burnt stones were collected and distinct traces of enclosures noted. In 1956, Lockley discovered the remains of a large stone-walled enclosure previously obscured, in the absence of grazing animals, by a thick mat of maritime grassland but now revealed by the advancing Gannet colony. Closer investigation subsequently revealed more earthworks and the remains of stone-built roundhouses, cairns and field boundaries – thus a community was active here in prehistoric times, probably reached on foot as sea-levels were then much lower. Even though a large part of the island is now obscured by Gannet nests, aerial photography has revealed more unrecorded earthworks, walls and buildings while a ground survey in 2012 confirmed there had been a much more intensive settlement than previously realised.

Lockley, who came to live on Skokholm in 1927, met fishermen whose grandfathers remembered Gannets nesting on Grassholm as far back as 1820 and 40 years on, in 1860, eggs from the small colony 'were greatly prized'. The first count of just 20 nests was made in 1883 by E. Lort-Phillips a big-game hunter and collector who camped on the island, finding it difficult to erect his tent such was the density of the Puffin colony – even so he was awakened in the night by a curious grunting sound under his ear, an aggrieved Puffin trying to emerge from its burrow. At least 200 pairs were nesting when members of the Cardiff Naturalists' Society under the leadership of the Cardiff trawler-owner J.J. Neale, arrived to camp at the end of May 1890. The following day, their solitude was shattered when a party of army and navy officers arrived and began dragging Puffins from their burrows and killing them with sticks while three 'in the costume and with the accent of gentlemen' were throwing Gannet eggs over the cliff. In Haverfordwest Court, on 9 August that same year, the seven defendants were each fined £1.16s plus costs, a total of £22.17s. This was the first such case under the Wild Birds Protection Acts of 1880 and 1881.

Vivian Hewitt, millionaire, pioneer aviator and avid collector of avian specimens and much else, was the first ornithologist to visit Grassholm after the First World War, having travelled by motor launch from Gower on 30 May 1922. Two years later Clemency Acland and Morrey Salmon reached the island after rowing for 9 miles in a breathless calm; they spent eleven and a half hours in the boat that day. Well worth it as their photographs revealed some 2,000 pairs of Gannet were nesting, the first time a colony had been counted in this manner. By 1933, the colony had increased to 4,750 pairs and the following year R.M. Lockley together with his wife Doris, Julian Huxley and a cameraman from London Films camped on the island to film the birds. The resulting film was *The Private Life of the Gannet* (the name chosen as Alexander Korda had recently had success with *The Private Life of the Henry VIII*) and it became the first wildlife film to receive an Academy Award.

In 1937, Grassholm was sold by the owners of Skomer to Martin Harman of Lundy who purchased it for his son John, later to be awarded a posthumous Victoria Cross at Kohima in 1944. The next owner was Malcolm Stewart, an ardent naturalist, yachtsman and visitor to Gannet colonies who sold the island to the Royal Society for the Protection of Birds in 1947 for £500. Two years later in 1949 the first count using aerial photography was made by John Barrett, then warden of the Dale Fort Field Centre, who reported 9,200 pairs. All subsequent surveys have used the same method to chart the continued growth of the colony.

Species

Although Gannets can be seen off most parts of the coastline of Wales, there is only one colony, that on lonely Grassholm, 7 miles (9.6 km) west of Skomer and 8 miles (12.8 km) from the nearest point on the Pembrokeshire mainland. In 2015 they numbered 36,011 pairs with an estimated 10,000 non-breeders also present. Gannet numbers have so increased that there is little space for other species save a thriving colony of Guillemots and Razorbills occupying ledges too narrow for their larger neighbours. Sadly there are no longer any Puffins, yet 150 years ago they nested everywhere on the island – perhaps as many as 100,000 pairs. They may have simply dug themselves out of existence, the collapsed burrows once clearly visible, now mostly covered by Gannet nests, offer a clue to their demise. A handful of pairs of Storm Petrels nest among the few remaining deep tussocks with Rock Pipits and a pair of Ravens the only other breeding birds. Oystercatchers, Turnstones and Purple Sandpipers can be seen around the coastline, and in late summer, Kittiwakes gather in large numbers. Grey Seals are always present, while Harbour Porpoise and Common Dolphins are frequently encountered during the boat voyage.

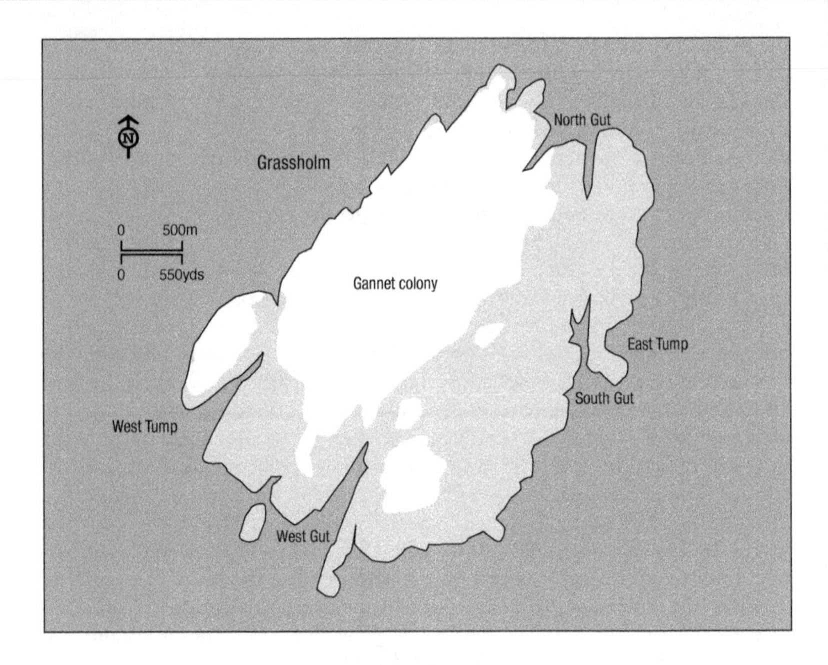

Small and isolated though Grassholm may be, there is an impressive list of casual, on occasions surprising, visitors including Grey Heron, Pintail, Quail, Woodcock, Short-eared Owl, Kingfisher, Hoopoe, Treecreeper, Robin, Goldcrest, Red-breasted Flycatcher, Red-backed Shrike and Snow Bunting. Of note was the sighting of a Zino's/Fea's/Desertas Petrel in July 2010.

Access

The RSPB has a strict no landing policy in place to prevent disturbance to the gannetry. Boat trips run daily, weather permitting, from the lifeboat station at St Justinians to circumnavigate the island to view the gannets from the water. Tickets must be booked in advance at the Thousand Island Expedition office in Cross Square, St David's, SA62 6SL or Tel: 01437 721721, email: info@thousandislands.co.uk

Dale Sailing also run trips to Grassholm, details at: www.pembrokeshire-islands.co.uk

CALENDAR

February–May: The Ravens' breeding cycle is well underway. Gannets return, and the island will no longer be silent until late October when all have departed. Guillemots and Razorbills commence their sporadic visits, culminating in egg-laying from late April. Migrating landbirds make brief visits.
June–July: Guillemots and Razorbills complete their breeding season and depart.

August–November: Young Gannets leave, though some adults will be present until late October. Passages of Manx Shearwaters offshore, later in this period large movements of Kittiwakes, Guillemots and Razorbills take place. Migrating landbirds make brief visits.

107 THE SMALLS AND THE CELTIC DEEP

OS Landranger 157
OS Explorer OL36
Grid Ref: SM466088

Habitat

Some 8 miles (13 km) due west of Grassholm – and in fact farther west than the most easterly coast of Co. Down, Northern Ireland – lie The Smalls: a cluster of rocks, the westerly point of a submarine reef extending 18 miles (29 km) from Wooltack Point on the Marloes Peninsula by way of Middleholm, Skomer and Grassholm. A Viking sword hilt of the 11th century was discovered on the Smalls reef in 1991 presumably from a shipwreck, of which there have been many across the centuries.

The highest point of the rocks is just 12 feet (4 m) above the surface at high tide and here a lighthouse commenced operation in 1776, being replaced by the present structure some 126 feet (38 m) high with a range of 18 miles in 1861 and manned by keepers until automation in 1987. Between Grassholm and the Smalls are two notorious reefs, the Barrels followed by the Hats. It was here that the *Christos Bitas* oil tanker disaster took place in October 1978 – fortunately after most of the Gannets had departed from the colony on nearby Grassholm.

Species

Not surprisingly, records of birds from this, the most remote part of Wales, are sparse but from March to October you can be certain of Gannets heading to and fro on their feeding forays from Grassholm. Manx Shearwaters from the larger islands are often to be seen, especially on summer evenings as they return to their colonies, while Kittiwakes often gather on the Smalls itself.

Alas only two lighthouse keeper's observations seem to exist, these for 1960 being published in the Skokholm Report for that year. It includes a list of 55 intrepid species of landbird, with some real surprises for this remote rock, like Grey Heron, Kestrel, Wren, Robin, Goldcrest and Snow Bunting. At dawn on 1 May no fewer than 42 Wheatears, 74 Willow Warblers and 166 Sedge Warblers sought sanctuary. That autumn there was an impressive passage of Grey Phalaropes; on seven days between 22 September and 5 October 1960, some 362 were sighted with no fewer than 227 feeding within 20 yards of the rock on 29 September.

Boat voyages by dedicated watchers (after all it is a journey of 25 miles from

CALENDAR

March–June Gannets and Manx Shearwaters return to their colonies on the larger islands. Small birds on migration gather at the lighthouse or are observed on passage.

June–October Gannets and Manx Shearwaters now with chicks on feeding forays and by September Manx Shearwaters, both adults and young, commencing migration to winter quarters. By late October the Gannet colony is deserted. Seabird passage includes all four skuas, Kittiwakes, Guillemots and Razorbills and hopefully rarer petrels. Turnstones on the rocks, phalaropes likely at sea.

Dale) in late summer to the south and west of the Smalls have proved very rewarding. In addition to seabirds including Wilson's Petrel, Storm Petrel, Great, Cory's and Balearic Shearwater, cetaceans have proved remarkable with sightings of Fin Whale, Minke Whale, Killer Whale, Risso's Dolphin, Common Dolphin, Bottlenose Dolphin and Harbour Porpoise.

108 CALDEY AND ST MARGARET'S ISLAND

OS Landranger 158
OS Explorer Outdoor Leisure 36
Grid Ref: SS140965

Habitat

Third-largest of the Pembrokeshire islands, Caldey and its tiny satellite St Margaret's Island are the least visited by birdwatchers. However, one of the earliest references to Pembrokeshire birds is that by John Ray and Francis Willughby who during their journey in 1662 noted that the islands had 'gulls, sea swallows and puits, their nests so thick that a man can scarcely walk but he must needs set his foot upon them'. The gulls presumably Herring Gulls the 'sea-swallows' terns, most likely Common Terns, while the 'puits' were Black-headed Gulls.

Species

The Black-headed Gulls and Common Terns have long gone but Herring Gulls thrive, the colonies total some 2,000 pairs, the second-largest colony in Great Britain. The Cormorant colony on St Margaret's – once the largest in Wales, 322 pairs in 1973 – now number some 120 pairs. On the cliffs of the same island are some 225 pairs of Kittiwakes, 1,700 Guillemots and 250 Razorbills, while a few pairs of Puffins are still present.

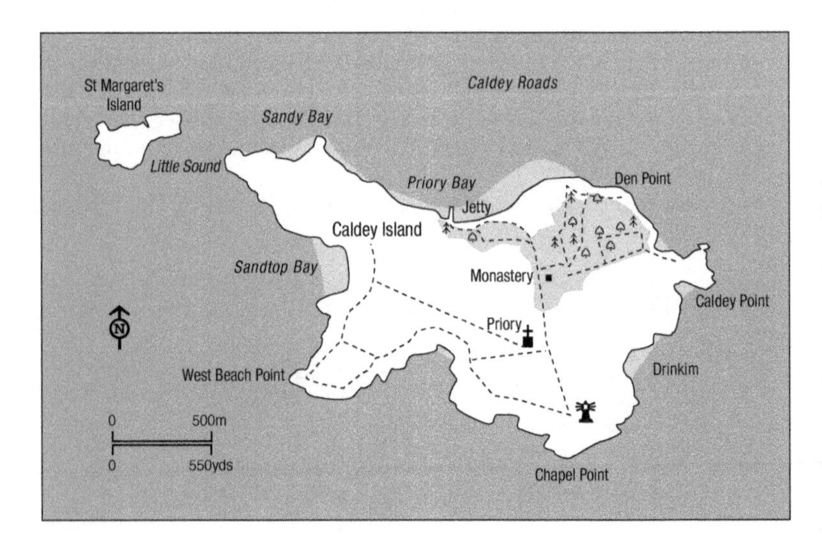

With its woodlands, farmland and gardens, Caldey attracts a wider range of breeding landbirds than the western islands, including breeding Blue Tit, Treecreeper, Robin, Whitethroat, Chiffchaff, Goldcrest, Goldfinch and House Sparrow. Scarce visitors have included Hoopoe, Alpine Swift, Yellow-browed Warbler, Black Redstart, Woodchat and Red-backed Shrikes. Usually 2 or 3 pairs of Chough breed here and the Herring Gull colony numbering around 1,700 pairs is the second largest colony in the UK. Small numbers of other seabirds are present but hard to see from the island.

Timing

April until September when visitor boats cease to make the crossing.

Access

Tenby is easily reached by way of the A477 St Clears to Pembroke road, leaving this at the Kilgetty roundabout to head south on the A478 to reach Tenby where there is a multistorey car park in the centre of the town. Boats leave Tenby South Beach every 15 minutes during the spring and summer (but not on Sundays) at a cost of £11 return. Once on Caldey a range of tracks and footpaths provide routes to most parts of the island. No landings are permitted on St Margaret's Island but this is best seen from one of the round Caldey boats which ply from Tenby.

CALENDAR

Resident: Cormorant, Shag, Buzzard, Peregrine, Kestrel, Pheasant, Oystercatcher, Turnstone, Great Black-backed Gull, Lesser Black-backed Gull, Herring Gull, Wood Pigeon, Raven, Carrion Crow, Jackdaw, Chough, Blue Tit, Treecreeper, Wren, Dunnock, Blackbird, Stonechat, Goldcrest, Rock Pipit, Pied Wagtail, House Sparrow, Goldfinch, Linnet, Chaffinch.

March–September: Guillemots, Razorbills and Puffins present until late July/early August. Gannets always offshore and quite often Manx Shearwaters on feeding forays from the immense colonies on Ramsey, Skokholm and Skomer, even more so during strong westerly winds. Always the chance of scarce landbird visitors; a scarcity of observers means much has always surely gone unrecorded.

109 ST DAVID'S HEAD AND VALLEYS

OS Landranger 157
OS Explorer OL35
Grid Ref: SM723279

Habitat

This west-facing headland riddled with small, vegetated valleys barely 3 miles (4.8 km) from St David's is ideal for observing spring and autumn migration. The largest and most often visited is Porth Clais, a valley which goes from the sea right the way into St David's, merging into Merry Vale, by the Cathedral. Other notable spots include the valleys of Nine Wells and Solva to the south-west and St David's Head itself.

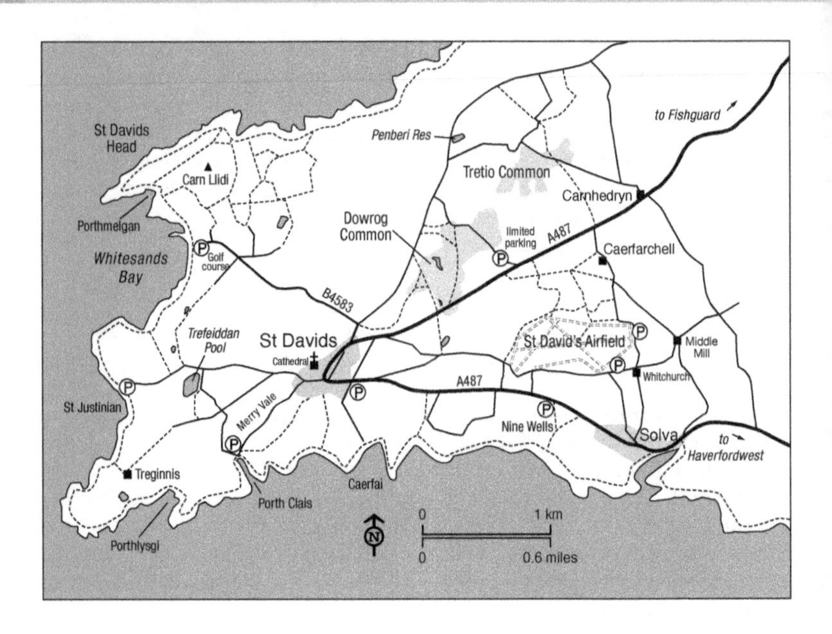

Species

Chough, Peregrine, Raven and Stonechat all breed and can be seen throughout the year. In spring and early summer the valley is alive with the songs of Grasshopper and Sedge Warbler, Whitethroat, Blackcap, Chiffchaff and Willow Warbler. Spring rarities have included Red-backed and Woodchat Shrikes, Golden Oriole, Subalpine Warbler and a Snowy Owl in 2018. Autumn in this area has attracted Yellow-billed Cuckoo (the fourth for Wales), Isabelline and Lesser Grey Shrike, Dusky Warbler and Red-eyed Vireo, along with County rarities such as Barred and Yellow-browed Warblers and Red-breasted Flycatcher.

Timing

Passage periods.

Access

For seawatching try the Head itself, access by the coastal path from Whitesands SM733272 (fee payable) where the B4583 terminates.

Passerines can often be found on the Head or in Porth Clais SM742240 and the valley known as Merry Vale, an excellent migrant hotspot extending from here into the middle of St David's by the Cathedral SM751254. Access to the valley is limited to the short footpaths by the car parks at each end of the valley.

CALENDAR

Resident: along the coast there are Cormorant, Shag, Gannet, Fulmar, Peregrine, Skylark, Meadow and Rock Pipits, Stonechat, Raven, Chough, Linnet and Yellowhammer. In the valleys Blackbird, Song and Mistle Thrush, Great, Blue, Marsh, Coal and Long-tailed Tits, Goldcrest and Chaffinch.

November–February: Fieldfare, Redwing, Siskin and Brambling in the valleys with Chiffchaff and Firecrest near-annual winter visitors.

March–June: Wheatears arrive first, then Sand Martin, Swallow, House Martin, Swift, Whinchat, Grasshopper, Sedge and Willow Warblers, Blackcap, Whitethroat, Lesser Whitethroat and Chiffchaff.

July–October: Tit flocks in the valleys, autumn passage migrants, mainly Sedge Warbler, Whitethroat, Lesser Whitethroat, Blackcap, Willow Warbler, Chiffchaff, Spotted Flycatcher.

110 ST DAVID'S COMMONS

OS Landranger 157
OS Explorer OL35

Habitat

The oceanic heaths of Anglesey, the Llŷn and Pembrokeshire have a character all of their own, occupying as they do low-lying land on windswept coastal peninsulas. The best example in Britain is the mosaic of heaths largely to the northeast of St David's, many of them owned by the National Trust, fragments – though some are large – set in an agricultural landscape of early potatoes and dairy herds. To extend the amount of open water available, the Wildlife Trust of South & West Wales has created a series of pools and pits on its Dowrog Common Nature Reserve. To the naturalist, the heaths contain riches and treasures for those able to search and with an eye for some of the diminutive plants and insects found here. Scenically, the commons are best from early summer onwards, first with masses of Heath Spotted-orchids in a variety of hues from white to lilac. As the summer proceeds, there are the pinks and purples of the three heathers and the saffron yellow of Western Gorse. For the specialist, such plants as Yellow Centaury, Pale Dog-Violet, Wavy St John's-wort, Three-lobed Water-crowfoot and Pillwort will cause great excitement, the latter a strange fern endangered throughout Europe and now listed as a Red Data Book species by the International Union for the Conservation of Nature. The Marsh Fritillary butterfly also occurs, its larvae feeding on the leaves of Devil's-bit Scabious and hibernating at the base of the plant in communal silk cocoons. Dragonflies and damselflies are numerous throughout the summer, from the largest such as Emperor, Hairy and Golden-ringed Dragonflies to the Small Red, Scarce Blue-tailed and Southern Damselflies, the latter a globally threatened species which has suffered a 30% decline in Great Britain since 1960.

Species

Sparrowhawks, Buzzards and Kestrels breed in the willow scrub and are normally not too difficult to see, neither are Red Kites which have recently colonized this area of West Wales. The wintering Hen Harriers which come to St David's Peninsula are one of the main attractions, with up to six going to roost on Dowrog. They are likely to be seen quartering any of the commons during the day, and some hang on until April before moving north. Other winter raptors include regular Merlins, Peregrines and Short-eared Owls, the latter having nested in the past. At one time Montagu's Harriers nested too but are now only rare visitors to the county. Stonechats, Grasshopper and Sedge Warblers, Whitethroats and Reed Buntings are among the small number of passerines that breed. There are always several pairs of Mallard, together with Moorhen and Coot.

Winter wildfowl numbers are small but sometimes include little parties of Whooper Swans. The Mallard flock increases during the autumn, to be joined by small numbers of Wigeon and Teal, occasionally Shoveler and Pintails. The absence of muddy margins to the pools means that wader numbers are small, those that have been noted include Jack Snipe, Dunlin, Greenshank and Green Sandpiper, while Snipe are present right through the winter. Bitterns occur in most winters at Dowrog, while more unusual visitors have included Little Bittern, Purple Heron, White Stork, Spotted Crake and Savi's Warbler.

Timing

Not critical. If, however, you wish to look for Hen Harriers going to roost, then visit in the late afternoon, though for the less hardy on a cold winter's day it is possible to watch from the car.

Access

Dowrog Common: An unclassified road leaves the A487 St David's to Fishguard road at SM782268 and crosses the common; there is a small car park close to the western boundary at SM772277. Follow the pony tracks across the common.

St David's Airfield: Several routes lead north off the A487 between Solva and St David's, alternatively turn off the same road where it leaves the northeast outskirts of St David's at SM763258, to eventually skirt the southern edge of the airfield. This site has also proved a good one to find passage Dotterel in both spring and summer. Other unusual birds have included Red-backed and Great Grey Shrikes.

Trefeiddan Pool: Easily viewed from the unclassified road which runs from St David's to the lifeboat station at St Justinian, the best point being on the side road at SM735253.

Tretio Common: An unclassified road leaves the A487 St David's to Fishguard road at Carnhedryn Uchaf and crosses the common. This area is not so well watched as the previous two sites and also lacks any open water, but nevertheless is well worth a visit.

CALENDAR

Resident: Mallard, Sparrowhawk, Buzzard, Red Kite, Kestrel, Pheasant, Moorhen, Coot, Woodpigeon, Skylark, Meadow Pipit, Pied Wagtail, Dunnock, Robin, Stonechat, Blackbird, Magpie, Carrion Crow, Raven, Linnet, Reed Bunting.

December–February: Whooper Swans, Wigeon, Teal, Shoveler, Hen Harrier, Merlin, Peregrine, Water Rail, Jack Snipe, Snipe and Short-eared Owl.

March–May: Cuckoo, Swallows and Sand Martins over the pools, Wheatear on passage, Grasshopper and Sedge Warblers, Whitethroat.

June–July: All residents and summer visitors present.

August–November: Chance of passage waders during August and early September, winter visitors including Hen Harriers from mid-October.

111 STRUMBLE HEAD

OS Landranger 157
OS Explorer OL35
Grid Ref: SM896413

Habitat

Strumble Head rises to about 100 feet (30 m), mainly as steep grass- and heather-covered slopes with rock outcrops. This is the western end of the Pencaer Peninsula, a 3-mile (4.8-km) run of coastline protecting Fishguard Bay to the east. A lighthouse has stood on the rocky islet of Ynys Meicel since 1908, but the keepers who once crossed the narrow footbridge have been absent for many years, for Strumble, as is the way with all lighthouses nowadays, is an automatic station controlled by computer from the other side of the country. In spring and early summer, the cliffs around Strumble Head are alive with colour, the flowers of Thrift, Spring Squill, Sea Campion and Scurvygrass, together with some surprises such as Primroses in sheltered spots and Cowslips which seem to thrive in the clifftop grassland.

It is, however, as a spot for observing migration, especially the passage of seabirds, that Strumble Head is noted – in fact it is the premier spot in Wales. Most movement takes place from late August until early October, with the best days being those where the wind is westerly or northwesterly (preferably a northwesterly gale which has been preceded by two or three days of strong southwesterly winds, which blow the seabirds up into the Irish Sea, then with the changing wind, allows them to exit, passing the Head). A Second World War building below a small car park gives watchers an ideal viewing platform and protection from the elements for seawatching, this being well described by Bill Oddie at the official opening in September 1988 as 'the Hilton of seawatching stations'.

Species

Save for Herring Gulls on Carreg Onnen and Fulmars towards Pen Brush, there are no breeding seabirds at Strumble Head. Choughs breed on this coastline and one does not usually have to go very far before seeing a pair, while the family groups from late June onwards can prove spectacular. There are usually Ravens about, Kestrels hover over the cliffsides, and the occasional Peregrine will sweep by. Stonechats and Linnets frequent the clifftop scrub, being joined in summer by Grasshopper Warblers and Whitethroats, and in the stream hollows even the occasional Sedge Warbler.

With the colonies on Ramsey, Skokholm and Skomer some 25 miles (40 km) away, Manx Shearwaters pass daily right through the summer months, the evening passage being as striking as that soon after dawn. Small numbers of Balearic Shearwater, which breeds on the Balearics and other islands in the Western Mediterranean, are seen from midsummer until late autumn. Sooty Shearwater are regular, while very small numbers of Great and Cory's Shearwaters are seen almost annually. Each October a handful of Leach's Petrels pass, usually fewer than ten a year but southwesterly gales in October may increase this to 50 or so, but Storm Petrels, although breeding on the Pembrokeshire islands, are comparatively scarce. To further whet the birdwatcher's appetite, Zino's/Fea's/Deserta's Petrel, Wilson's Petrel and Little Shearwater have been recorded.

Divers feed in the tiderace off the Head throughout the winter, Red-throated Diver being the most frequent. Of all wildfowl, Common Scoter is the most regular, passing in all months, with the lowest numbers in late winter and spring. Small numbers or individuals of several other species have been noted, including Wigeon, Tufted Duck, Velvet Scoter, Goldeneye and Red-breasted Merganser while Surf Scoters have been seen on a couple of occasions. Passage waders include Knot, Bar-tailed Godwit, Whimbrel, Curlew, Spotted Redshank, Redshank, Green Sandpiper and Grey Phalarope.

Great and Arctic Skuas pass regularly each autumn, with smaller numbers of Pomarine Skua, and a handful of Long-tailed Skua. Sabine's Gull are seen each autumn, again numbers reflect southwesterly gales, while Kittiwakes are always offshore. From mid-September onwards there are large numbers of Guillemots and somewhat smaller numbers of Razorbills, though still counted in thousands on good days, in flight southwards through St George's Channel, many passing close to Strumble Head, while small numbers hunt fish in the tideraces.

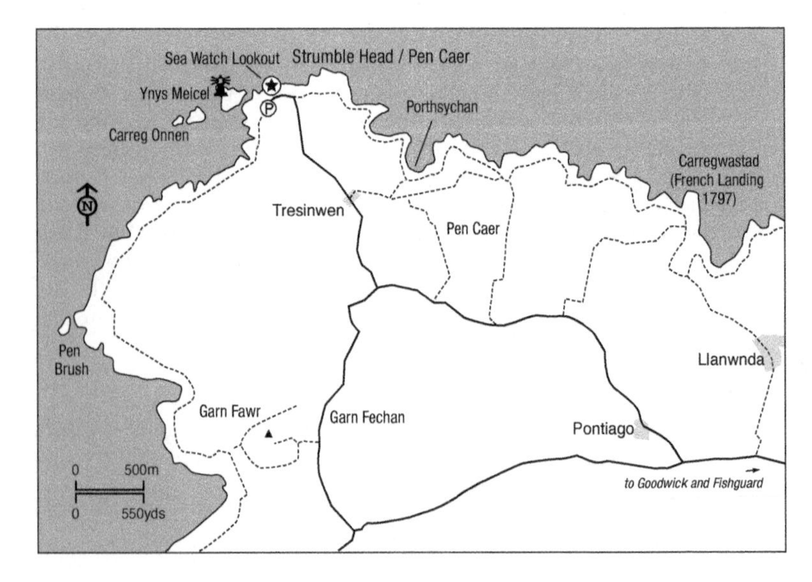

When seabird passage has died away, it is always worth searching the sheltered valleys to the east and west for small migrants. Among the highlights are the bands of tits and Goldcrests which work their way so actively along the cliffs in autumn. In the early part of September, large numbers of Sand Martins and Swallows fly in from the sea and then follow the coast to the south-west, to be followed from mid-October by at times immense passages of Skylarks, Meadow Pipits, thrushes, Starlings and finches, which are a special feature of late autumn watches at Strumble Head. Rare visitors are seen occasionally, among them Pallid Swift, Siberian Stonechat, Pallas's, Yellow-browed and Hume's Warblers, Rose-coloured Starling, Common Rosefinch and, perhaps the most remarkable of all, an Alpine Accentor on 30 October 1997, the first for Wales.

Not just birds either, Grey Seals are usually present close inshore with breeding beaches and caves no great distance away.

Timing

It is essential that one arrives at dawn to catch the main movements, both in numbers and in species, though it is always worth watching whatever the time of day. Much depends on the weather situation, the ideal being gale-force southwesterly winds which, on veering northwesterly and decreasing, have the effect of releasing large numbers of birds previously held back in the Irish Sea.

Access

An unclassified road from Fishguard Harbour terminates at Strumble Head, where there is car parking at SM894411. This allows watching from the car, but better still is the large former War Department building just below; this has been renovated as an observation centre and provides some degree of shelter, though beware: it is a popular spot, and it can be crowded on 'good' mornings during the early autumn.

CALENDAR

Resident: Fulmar, Gannet, Cormorant, Shag, Buzzard, Kestrel, Peregrine, Oystercatcher, Herring and Great Black-backed Gulls, Rock Pipit, Stonechat, Chough, Raven, Linnet.

December–February: Red-throated and Great Northern Divers, Wigeon, Scaup, Eider, Velvet Scoter, Goldeneye, Merlin, Black-headed and Common Gulls, Guillemot, Razorbill.

March–May: Possible returning Arctic and Great Skuas, small numbers of terns, mostly Sandwich Tern, from early April. On land, a range of summer visitors likely to be seen includes Whimbrel, Common Sandpiper, Cuckoo, Wheatear, Ring Ouzel, warblers and Spotted Flycatcher, while passing Sand Martins and Swallows with the occasional House Martin seen throughout April and May.

June–July: The few breeding species in evidence, good movements of Manx Shearwaters and Gannets offshore. The first terns begin to reappear by end of July.

August–November: This is the time to be at Strumble Head. An early morning start is highly recommended (along with a folding chair to sit on) and if you are able to spend a reasonable period there you should observe among others Sooty, Manx and Balearic Shearwaters, Leach's Petrel, Wigeon, Eider, Velvet Scoter, Ringed Plover, Knot, Dunlin, Black-tailed and Bar-tailed Godwits, Whimbrel, Curlew, Turnstone, Pomarine, Arctic, Long-tailed and Great Skuas, Little and Sabine's Gulls, Kittiwake, Sandwich, Common and Arctic Terns, Guillemot, Razorbill, Puffin, Swallow and Sand Martin, and from mid-October Skylark, Meadow Pipit, thrushes, Starling, finches.

112 FISHGUARD HARBOUR AND GOODWICK MOOR

OS Landranger 157
OS Explorer OL35
Grid Ref: SM948377

Habitat

The harbour area, where on the afternoon of 24 February 1797 an invasion force of French mercenaries surrendered – folklore recounts how they were routed by one Jemima Nicholls, a local shoemaker with a formidable reputation – is easily accessible from the town and has good parking. A ferry service operates to Rosslare in Co. Wexford, while a small number of inshore boats and pleasure craft are moored offshore. On the landward side of the harbour is an area of reeds known as Goodwick Moor, a Wildlife Trust for South & West Wales nature reserve.

Species

Up to 250 gulls may be present and these can include Mediterranean Gull, of which up to 12 individuals winter here (or on the Nevern Estuary, see Site 114) while over 35 different birds have been recorded on autumn passage. Little Gulls are almost as regular, along with the occasional Iceland, Glaucous and Yellow-legged Gulls. Late afternoon appears to be the best time as gull numbers build before they move off to roost on the water. A Ross's Gull, the first for Wales (there have been just three others since), was discovered during a Wildlife Trust field trip on 15 February 1981 and remained for just one more day. The tail-end of Hurricane Wilma in November 2005 brought at least 13 Laughing Gulls to Wales, of which six were found in Pembrokeshire, one of these was present at the harbour on 4 November.

The harbour may also play host to up to 25 Great Crested Grebes and Red-throated and Great Northern Divers and occasionally Black-throated Divers in winter. For the last 10 years or so, at least one pair of Black Guillemots has bred in the harbour walls, with individuals present in the area from February to October at what is the most southerly location this side of the Irish Sea. A small number of waders are also present on the beach, usually Turnstone, Ringed Plover, Redshank and Oystercatchers. The outer harbour wall is one of the best sites for wintering Purple Sandpiper in the county, with up to a dozen, but unfortunately for security reasons it is now off limits to the public.

Goodwick Moor Nature Reserve plays host in winter to small numbers of Water Rail, Snipe, the occasional Jack Snipe and even Bittern, while in summer listen and look for Cetti's Warbler, Reed and Sedge Warblers and Reed Buntings. The rarest bird recorded was a Pechora Pipit, which spent four days here in November 2007 – the only record for Wales of a bird that breeds eastwards from the Pechora region of north-east Russia to the Pacific, normally wintering in the Philippines and central Indonesia; there are just over 100 other British records, most from Shetland. Lower Town Harbour is also well worth a look and do not fail to check the River Gwaun from the bridge for Grey Wagtails and Dipper.

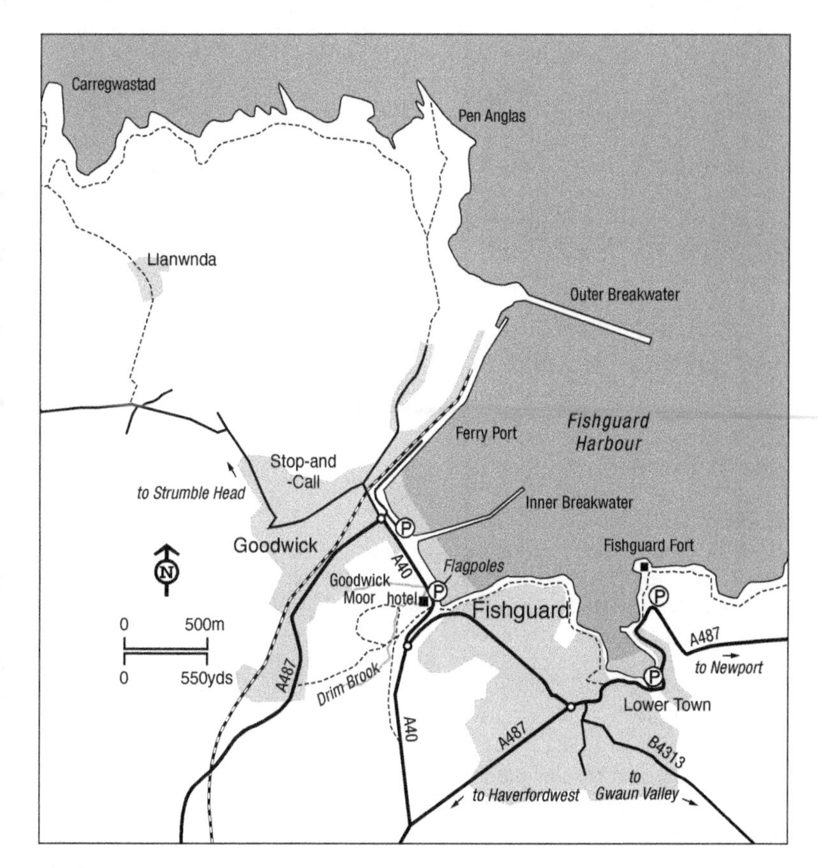

Timing

All year.

Access

There are car parks at Goodwick, by the entrance to the Ferry Port and by the 'Flagpoles'. A walk from either up the Inner Breakwater gives good close views of anything in either side of the harbour. An excellent viewpoint for the whole Harbour and especially to check for divers is Fishguard Fort SM962375. There is also a car park at Lower Town SM962371.

For Goodwick Moor, park beside the harbour at SM949376, from where it is a short walk up the hill past the Sea Front Hotel before a footpath takes you over the Drim Brook. There was a boardwalk trail here, through the Moor, but at the time of writing, it has been removed due to damage.

CALENDAR

Resident: Cormorant, Shag, Black-headed, Herring, Lesser and Great Black-backed Gulls, Oystercatcher and Peregrine, Fulmar, Gannet and Kittiwake offshore. On Goodwick Moor: Cetti's Warbler, Long-tailed and Marsh Tits, and Reed Bunting.

November–February: Within the harbour: Red-throated and Great Northern Divers, Great Crested Grebe, Mediterranean and

Common Gulls, Oystercatcher, Ringed Plover, Dunlin, Redshank, Curlew and Turnstone. Grey Heron, Water Rail and Snipe winter on the Moor, occasionally so do Bittern and Jack Snipe, Fieldfare, Redwing and Siskin may be seen by the entrance.
March–June: Mediterranean Gulls remain until late April. Waders drop in, some staying for a few days, including Ringed Plover, Dunlin, Redshank, Bar-tailed Godwit and Turnstone. Reed, Sedge and Willow Warblers, Chiffchaff on Goodwick Moor.
July–October: Build-up in the number of gulls, Mediterranean included, as well as passage waders.

113 DINAS ISLAND

OS Landranger 157
OS Explorer OL35
Grid Ref: SN015401

Habitat

Well, this is not quite an island despite the name, though you may ponder as to how long rising sea-levels will take before the waters surge through the valley between Pwllgwaelod and Cwm-yr-Eglwys. Indeed, the ruined church at Cwm-yr-Eglwys offers a stark reminder of the power of the sea, having suffered storm damage in 1850 and 1851 as did the churchyard, the level being reduced by some three feet before the Royal Charter Storm of October 1859 completed the destruction.

It was to Island Farm, the only habitation on the island proper, that R.M. Lockley came in 1942. Best known for his years on Skokholm and for writing a prodigious number of books, scientific papers and popular articles, he had left Skokholm in 1940 because of the wartime situation, first moving to Cwm-yr-Eglwys. Now a full-time farmer, there was only time for Lockley to author two books – *Inland Farm* and *Island Farm* – but still scope for natural history study, including the discovery of a major Grey Seal haul-out at what he called the Red Wilderness (see his book *Seals and the Curragh*).

Species

Deserving an immediate mention, though hardly a sighting to be repeated, is a male Moussier's Redstart on Dinas Island, 24 April 1988. One of the members of a party of five birdwatchers lagged a little behind his companions in the early afternoon to watch a male Stonechat when his attention was drawn to a small, brightly coloured bird perched on a rock nearby. Soon the rest of the party had returned and were able to watch it until just before 3pm, when it vanished not to be seen again. This the only British record of a Northwest Africa resident – the individual that reached Pembrokeshire is the only sighting away from the Mediterranean. Maybe there will not be another Moussier's Redstart anytime soon, but this serves as a reminder that rare birds are likely to appear anywhere, so who knows what else might appear on this spectacular headland. Meanwhile, one is guaranteed a small colony of Guillemots and Razorbills, Choughs and Ravens, Peregrine, Stonechat and Wheatear.

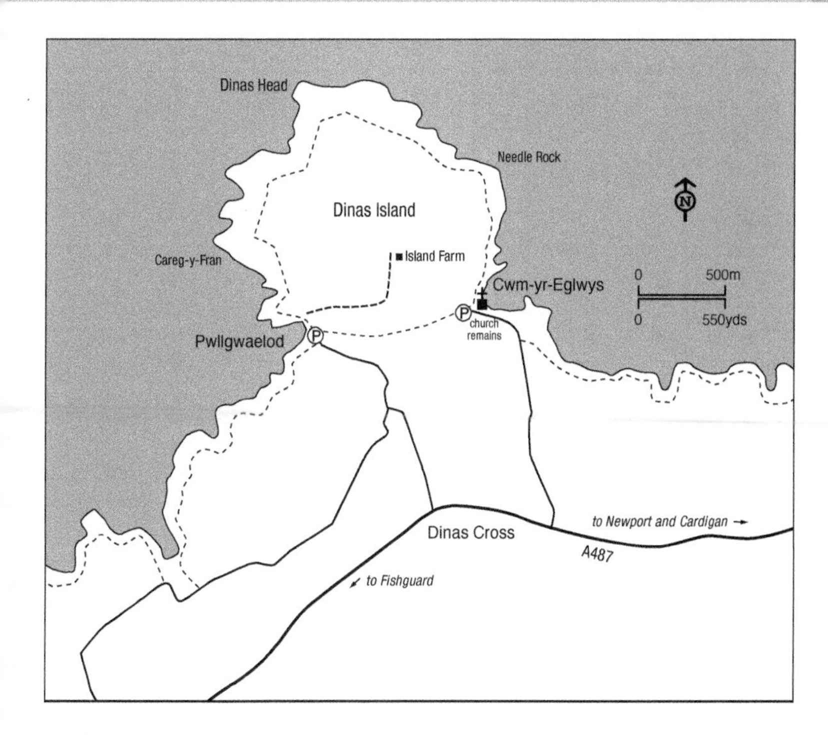

Timing

Between early March and late October, though who knows what a winter visit may reveal.

Access

From the A487 Cardigan to Fishguard road, one has a choice of two routes from the eastern side of Dinas Cross. A minor road descends the hill to reach the shore at Cwm-yr-Eglwys, another to Pwllgwaelod, with parking available at both sites. A section of the Pembrokeshire Coast Path provides a complete circuit of Dinas Island, though whichever direction you choose, there is a steep climb to begin with – at least it's downhill at the end.

CALENDAR

Resident: Cormorant, Shag, Red Kite, Buzzard, Kestrel, Peregrine, Oystercatcher, Herring Gull, Great Black-backed Gull, Meadow Pipit, Rock Pipit, Wren, Stonechat, Chough, Raven, Linnet.

December–February: Divers offshore, while Guillemots and Razorbills make occasional returns to the cliffs.

March–May: Wheatears the first of the summer visitors to arrive, while moving into April and early May major passages of Martins and Swallows, some lasting many hours. Guillemots and Razorbills commence egg-laying in late April.

June–July: Guillemots and Razorbills depart, Gannets feeding offshore.

August–November: Seabird passage well under way by early September, the sheltered valley looks ideal for small migrants and could well reward those willing to make regular visits, with Black Redstarts a possibility at the ruined church.

114 NEVERN ESTUARY

OS Landranger 157
OS Explorer OL35
Grid Ref: SN063395

Habitat

The River Nevern enters the sea at Newport, making a small sandy estuary which suffers from disturbance, especially during peak holiday times. Due to its small size it is easily watched, with a path running all along its southern shore, from the 'Iron Bridge' to the sea by the boat club. The close proximity to Fishguard Harbour allows gulls to move freely between the two sites, as do divers and grebes.

Species

Canada Geese often dominate the scene, with up to a thousand sometimes present, while Shelduck are present save when off on their late-summer moult migration. Wigeon and Teal overwinter, and sometimes there are Goldeneye, Goosander and Red-breasted Merganser. Little Egrets and Grey Herons fish in the shallows, while no fewer than 12 Spoonbills spent a day here in September 2005. From the 'Iron Bridge' look for Common, Wood and Green Sandpipers, the latter sometimes overwintering. The gull flock should as always be carefully scrutinised for Mediterranean Gulls which interchange with Fishguard Harbour (Site 112). For waders, a half or low tide is best and both spring and autumn

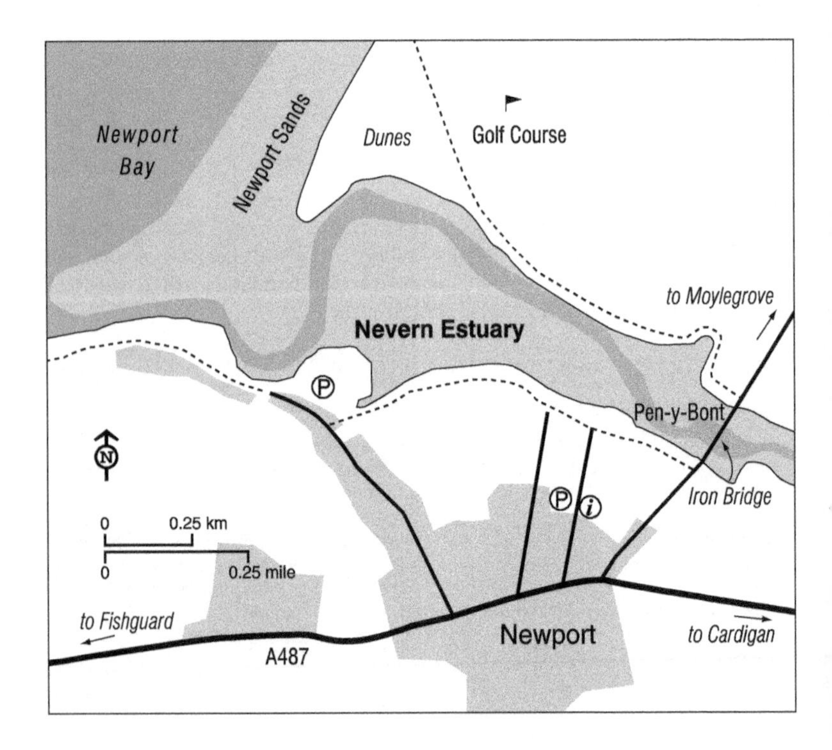

passage make this site well worth visiting. Kentish Plover have been seen in spring and in autumn Little Stint and Curlew Sandpiper, while Sanderling are likely along the open sandy shoreline. Mediterranean Gulls are likely in any month; Black-headed Gulls will be much in evidence throughout the autumn and winter, while the Yellow-legged Gull, a scarce winter visitor to the county, has been reported.

Timing

Winter and passage periods, preferably early in the morning or in the evening to avoid the disturbance from holidaymakers.

Access

The best vantage point is the 'Iron Bridge' at SN062395 or from the footpaths on each side of the estuary leading downstream from it.

CALENDAR

Resident: Mallard, Little Egret, Grey Heron, Black-headed, Herring, Great and Lesser Black-backed Gulls, Oystercatcher, Kingfisher, Meadow Pipit, Reed Bunting.
November–February: Canada Goose, Wigeon, Teal, Little Grebe, Common and Mediterranean Gulls, Ringed Plover, Dunlin, Redshank, Curlew, Fieldfare, Redwing, Rock Pipit, Siskin and Lesser Redpoll.
March–June: Passage waders including Ringed Plover, Dunlin, Curlew, Whimbrel, Black-tailed and Bar-tailed Godwits, Greenshank, Sanderling and Common Sandpiper. Reed, Grasshopper and Sedge Warblers in the reedbed by the 'Iron Bridge', Blackcap, Whitethroat, Willow Warbler, Chiffchaff among the scrub.
July–October: Roving tit flocks with good numbers of Long-tailed, Blue and Great Tits, Goldcrests and Treecreeper in the scrub by the side of the estuary. Numbers of gulls increase from August onwards, including Mediterranean and Common, passage waders mainly Ringed Plover, Dunlin, Redshank, with the possibility of Sanderling, Turnstone, Whimbrel, both Godwits, Common and Green Sandpipers.

115 PENGELLI FOREST

OS Landranger 145
OS Explorer OL35
Grid Ref: SN123396
Website: https://www.welshwildlife.org/nature-reserve/pengelli-forest/

Habitat

The largest block of ancient oak coppiced woodland remaining in West Wales, of which 162 acres (66ha) has been a Wildlife Trust nature reserve since 1979 to which, following an appeal in 2021 a further area was added so making 194 acres (78.5ha) though this new total will still only represent a third of its size in the 16th century. Then it was known as a 'greate woode', a source of much timber and of grazing for horses, cattle and sheep, a place where large numbers of Woodcock were taken in winter, a place where honey was collected from the nests of wild bees in summer. Over the intervening 350 years or so, the wood has shrunk as agricultural development and conifer forestation has nibbled at the perimeter, though fortunately the main block remains, possibly owing to the presence of several steep-sided ravines which make access difficult in parts.

Pengelli Forest may be divided broadly into two main habitats. Closest to the entrance is Pant Teg, mainly sessile Oak and Birch woodland on steep slopes. Proceed through this and you reach a large area where Birch, Alder and Ash, along with Oak, are dominant on poorly drained boulder-clay soil with a thick ground cover. Other trees of note here include Aspen, Goat Willow, Wild Cherry and, most interesting of all, three specimens of Midland Hawthorn, at this its only known station in West Wales. Woodland plants found in the forest include Golden-Saxifrage, Early Dog-violet, Wood Anemone, Common Cow-wheat, Moschatel and Wood Sage. Mammals include Badger, Grey Squirrel, Polecat and Dormouse, for which nest boxes are provided. This latter is particularly local in south-west Wales though undoubtedly overlooked. Woodland butterflies include Silver-washed and Small Pearl-bordered Fritillaries and Purple Hairstreak.

Species

All the typical woodland birds can be seen, Sparrowhawks, Buzzards, Red Kites and Tawny Owls being the birds of prey which nest in or close to the forest, and there have been occasional records of Goshawk. Great Spotted Woodpeckers breed, while Lesser Spotted Woodpeckers have also been seen in the past. In this even-aged woodland with few natural nest-holes, the provision of nest boxes was an early task undertaken by the Wildlife Trust management team. The timing was perfect, for it coincided with the westwards extension in range of Pied Flycatchers, which first nested here in 1981, with up to three pairs in recent years. Other nest box occupants include Redstarts, Great, Blue, Coal and Marsh Tits. The woodland warblers such as Garden Warbler, Blackcap, Chiffchaff and Willow Warbler are all present. Woodcock still come in winter as do Fieldfares and Redwings, while small numbers of Brambling occur with other finches.

Timing

Not material, unless it is the dawn chorus you wish to hear in May and the early part of June.

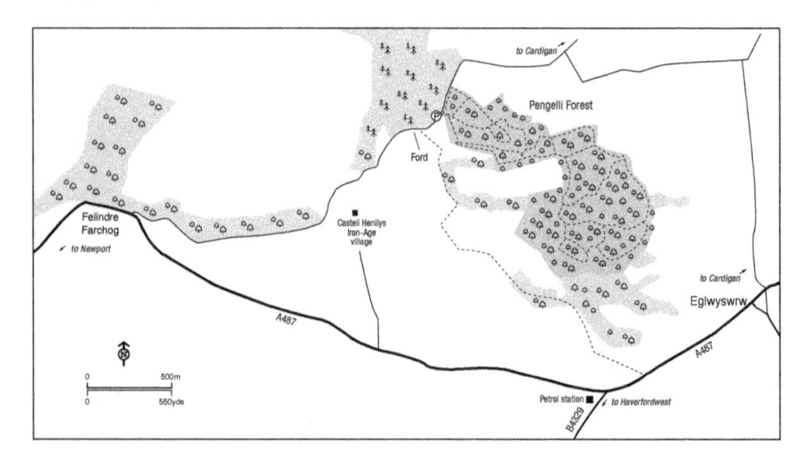

Access

Leave the A487 Cardigan to Fishguard road to SN104390, just east of Felindre Farchog, by the unclassified road which follows the bottom of the ridge, wends its way around the great Iron Age mound of Castel Henllys, crosses a stream by means of a ford, and eventually skirts the western boundary of the forest. Cars may be parked close to the access gate, from where a series of footpaths leads through a large section of woodland. A leaflet is available from the Wildlife Trust of South & West Wales.

CALENDAR

Resident: Sparrowhawk, Buzzard, Red Kite, Pheasant, Stock Dove, Woodpigeon, Tawny Owl, Great Spotted Woodpecker, Wren, Dunnock, Robin, Blackbird, Song and Mistle Thrushes, Goldcrest, tit family, Nuthatch, Treecreeper, Jay, Magpie, Jackdaw, Carrion Crow, Raven, Chaffinch, Greenfinch, Goldfinch, Bullfinch.

December–February: Woodcock, Grey Wagtail, Dipper, Redwing, Fieldfare, Brambling.

March–May: Summer visitors arrive, including Cuckoo, Redstart, Whitethroat, Garden Warbler, Blackcap, Chiffchaff, Willow Warbler, Spotted and Pied Flycatchers.

June–July: All residents and summer visitors present, many young birds about, though by end of period the forest will fall strangely silent, save for the buzz of numerous insects and the calls of young Buzzards overhead.

August–November: Summer visitors depart by early September, though the occasional Blackcap and Chiffchaff may remain until October and one or two may even overwinter. Woodcock arrive late October, shortly after the first Fieldfares and Redwings.

116 TEIFI MARSHES AND GORGE WOODLANDS

OS Landranger Map 145
OS Explorer Map 198 and OL35
Grid Ref: SN185455
Websites:
https://www.welshwildlife.org/nature-reserve/teifi-marshes/
https://www.first-nature.com/waleswildlife/sw-nnr-coedmor.php

Habitat

The Afon Teifi, one of the longest rivers wholly within Wales, 50 miles (80 km) in length, rises in the high wilderness above Strata Florida, flows through the great peat bog of Cors Caron, then meanders south and west to mingle with salt water at Llechryd or thereabouts before passing through the Teifi gorge to become truly estuarine. The wetlands, reedbeds and the western side of the wooded gorge, are a Wildlife Trust South & West Wales nature reserve which includes the spacious Welsh Wildlife Centre, while the woodlands of the Coedmore National Nature Reserve occupy the east bank. As with the lower estuary (Site 58), the Ceredigion–Pembrokeshire county boundary passes through this site.

The reedbeds look as if they have been here for centuries, but not so: barely a hundred years ago these were well-drained grazing pastures which included a rifle range. Now completely given over to wildlife, the reedbeds have spread and with that there has been a reduction in the numbers of wintering wildfowl. When mammoths roamed the land, glacial debris blocked the old course of the Teifi, the expanse now occupied by the marshes and reedbeds, while the melt-waters tore a new route through what is now a deep gorge. In 1188, Giraldus Cambrensis, on his travels through Wales, recorded that the Beavers resident on the Teifi were the only population in England and Wales. The Teifi is an excellent place for Otters, although patience, usually early in the morning on a dropping or low tide, is required.

The woodlands, largely Sessile Oaks on the west side of the gorge, each a haven for mosses, ferns and huge numbers of lichens, not forgetting bats and of course woodland birds. Largely hidden by the trees are the quarries where for centuries the durable Teifi Valley slate was extracted. Samuel Lewis writing in 1833 reported that the 'sylvan beauties of the scene ... rich groves, alternating with the naked rock, continue to excite the admiration of the traveller'. Take the footpath on the western side of the gorge and you will quickly appreciate his description.

Species

The Teifi Marshes are a stronghold for breeding Reed and Sedge Warblers, with a handful of Grasshopper Warblers and Lesser Whitethroats. Cetti's Warblers were numerous in the past, but numbers have declined, held back by cold winters, so that at the time of writing there are only a couple of singing males. Water Rails winter in good numbers and can often be heard screaming in the reeds and occasionally remain to breed. Bitterns occur in most winters, although seeing one is another story.

The woodlands of the Teifi Gorge host Chiffchaff, Willow and Garden Warblers, Blackcap, Greater Spotted Woodpecker, Buzzard, Sparrowhawk, Red Kite, Treecreeper, Nuthatch, Willow and Marsh Tits and occasionally Redstart and Pied Flycatcher.

The exposed mud in front of the Curlew Hide is a 'washing and loafing' area for the many gulls that feed, roost and breed on Cardigan Island. Careful observation has discovered the occasional Iceland, Glaucous, Mediterranean and even Ring-billed Gulls in the past. Of waders, there is almost always a Curlew in this area, and often Black-tailed Godwits, Redshank, Greenshank, Snipe and during passage periods both Green and Common Sandpipers. Few ducks visit the marshes these days, due to the encroachment of the reedbed, although during the winter months there may be as many as 100 Teal, 30 Mallard and a couple of Goldeneye and Goosander.

One of the sights at the Teifi Marshes in late autumn are the large flocks of Starlings, frequently in their thousands, which come to roost shortly before dusk. Such a feeding opportunity often attracts Sparrowhawks and Peregrines. Other species to look out for upstream of Cardigan are Kingfishers, Great Spotted Woodpeckers, Redstarts, Pied Flycatchers, and Reed Buntings. Rare and scarce visitors have included Little Bittern, Great White Egret, White Stork, Spotted Crake, Red-backed, Great Grey and Woodchat Shrikes, Firecrest and Bluethroat.

Timing

Low tide, mornings and evenings are best, when there is less disturbance from people walking or cycling along the old railway track through the marshes.

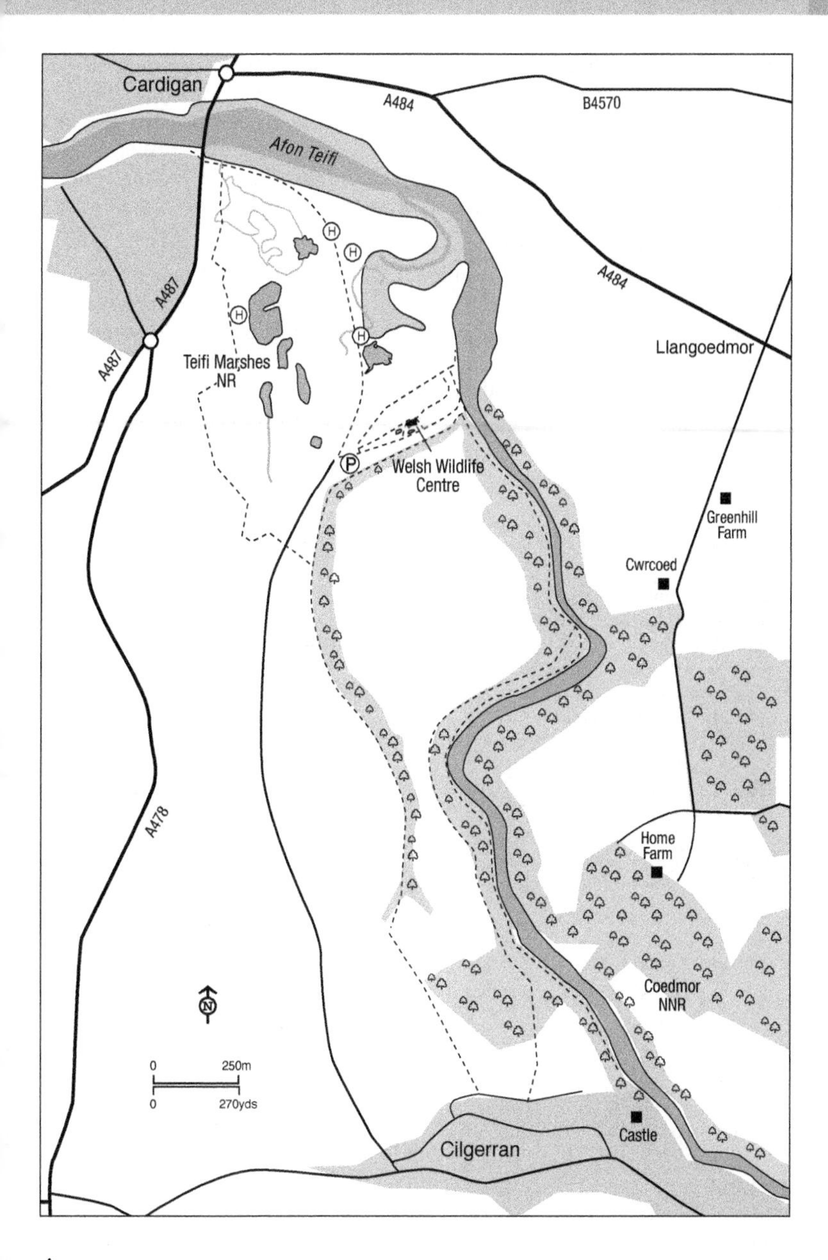

Access

Teifi Marshes – Welsh Wildlife Centre: Should be approached from Cilgerran, where a road is well signed to the Welsh Wildlife Centre at SN188452. The reserve is open throughout the year, with pay-and-display entrance car park and a visitor centre, shop and café. The latter on the top floor provides a panoramic view across the marshes and river. A series of footpaths radiate from the centre building, one going right through Cilgerran Gorge, the woodland trail eventually leading to the treetop hide. The longest, including several lengthy sections of boardwalk and a section of disused railway line, affords a complete circuit of the marshes. Several sections are suitable for disabled persons, including access to four hides positioned at strategic points. Alternatively, on foot, access from the back of what was Cardigan Mart (by Jewsons), along the disused railway line which goes all the way to Cilgerran.

Coedmore National Nature reserve:
East bank: take the A484 main road towards Carmarthen, at Llangoedmor, turn right onto an unnamed road, which travels in a semicircle to Llechryd. Approx. 2/3 mile (1 km) along this road there is a lay-by on the right and access into the reserve.

West Bank: There is a riverside footpath from Llechryd Bridge to Cilgerran. Either park by Llechryd Bridge or in Cilgerran, turn towards the river by the village shop; car park and toilets are to be found at the bottom of the hill in the middle of Cilgerran Gorge.

CALENDAR

Resident: Cormorant, Grey Heron, Little Egret, Mute Swan, Canada Goose, Shelduck, Mallard, Red Kite, Sparrowhawk, Buzzard, Peregrine, Moorhen, Oystercatcher, large gulls, Tawny Owl, Barn Owl, Kingfisher, Great Spotted Woodpecker, Grey Wagtail, Goldcrest, Marsh Tit, Nuthatch, Treecreeper, Raven, Chaffinch, Bullfinch, Reed Bunting.

December–February: Little Grebe, Wigeon, Teal, Goldeneye, Goosander, Snipe, Woodcock, Curlew, Redshank, Black-headed, Mediterranean and Common Gulls, Water Rail, Fieldfare, Redwing. Possibility of Bittern in the reedbeds.

March–May: Sand Martins usually the first summer visitor to arrive at the marshes, Sedge Warblers and Reed Warblers in full voice by end of April.

June–July: All breeding species present, young Buzzards and Kites calling overhead while hirundines feed over the largest pools.

August–December: The Starling roost comes into its own from October onwards, when visiting Fieldfares and Redwings appear.

117 LLYS-Y-FRÂN RESERVOIR

OS Landranger 145
OS Explorer OL35
Grid Ref: SN042246
Website: http://www.llys-y-fran.co.uk/

Habitat

There are few large expanses of open fresh water in the county of Pembrokeshire, Llys-y-frân Reservoir is the largest of these but due to its depth, and the high levels of disturbance, only small numbers of diving ducks winter here. The reservoir and surrounding park is managed by Welsh Water.

Species

Up to 20 Tufted Ducks, 10 Goosanders and similar numbers of Great Crested Grebes can be seen in winter. There have been occasional rarities, in the form of Smew and Great Northern Diver. A juvenile Pacific Diver spent a couple of

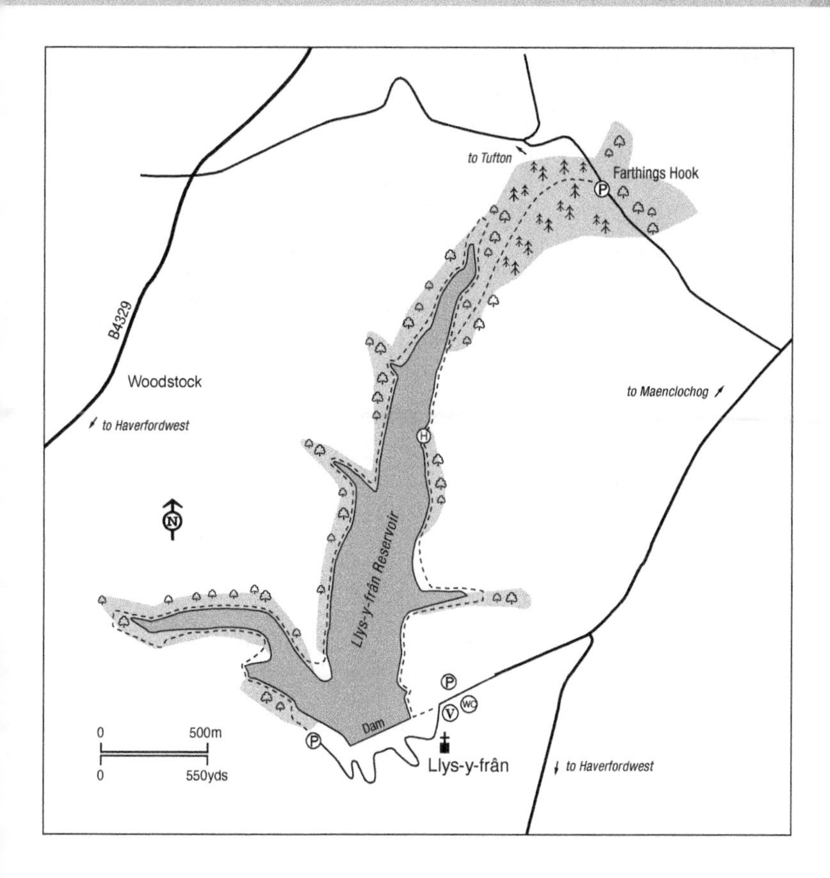

months here in late winter 2007 and 2008 – the second British record. There is, however, a large winter gull roost (January–March are the best months), mainly of larger gulls, including up to 12,000 Lesser Black-backed, the largest roost in Wales, among which keen-eyed observers might be able to pick out Mediterranean, Iceland, Kumlien's and Glaucous. For the last 10 years, up to 2021, an adult Ring-billed Gull has also been coming to roost at the reservoir during the winter months. Surrounding the reservoir is open countryside and areas of woodland, of which those at the top of the eastern arm are usually the most productive.

Timing

Winter.

Access

For the best views, park by the visitor centre (fee payable) and walk along the right-hand arm of the reservoir, which is less disturbed than the left.

CALENDAR

Resident: Canada Goose, Mallard, Grey Heron, Cormorant, Black-headed, Herring, Lesser and Great Black-backed Gulls on the water. In the woodland there are Buzzard, Sparrowhawk, Red Kite, Greater Spotted Woodpeckers, Great, Blue, Coal, Long-tailed and Marsh Tits, Nuthatch, Treecreeper, Goldcrest, Bullfinch, Chaffinch, Siskin and Lesser Redpoll.
November–February: Wigeon, Teal, Goldeneye, Tufted Duck, Goosander, Great Crested and Little Grebes, Common, Herring, Mediterranean and Black-headed Gulls, Reed Bunting.
March–October: In most years Great Crested Grebes remain throughout the summer, occasionally to breed. Arrival of summer visitors, including Swallow, House and Sand Martins, Common Sandpiper and in the surrounding woodland Willow and Garden Warblers, Blackcap and Chiffchaff.

118 ST BRIDE'S BAY

OS Landranger 157
OS Explorer OL35

Habitat

This large bay, a wintering area for many sea-duck and divers, has several key observation areas all easily reached by roads from Haverfordwest. Depending on the year and weather conditions, birds move up and down the bay, from Newgale in the north to Little Haven and Goultrop Roads in the south.

Species

Thirty years ago, as many as 1,000–2,500 Common Scoters wintered in the bay, more recently, numbers have only reached 500 or so, sometimes close inshore among which there may be Scaup, Velvet Scoter and Long-tailed Duck. Surf Scoters have been recorded on several occasions as well as a Black Scoter, off Newgale from 25 December 1991 and thus an unexpected Christmas present for the observer, remained until March 1992, the first of just two sightings in Wales. Up to 20 Red-throated Divers may also winter, usually in Goultrop Roads. Great Northern and Black-throated Divers, Great Crested and Slavonian Grebes may also be seen. Early autumn southwesterly gales often result in newly fledged Manx Shearwaters being blown ashore and stranded on the beaches, especially at Newgale – just a tiny proportion of those which have left the islands for their South Atlantic wintering grounds.

Behind the Duke of Edinburgh pub at Newgale, the small marshy valley is often flooded and so attracts many gulls and small numbers of wildfowl. Careful examinations of the pipit flocks have revealed up to 10 wintering Water Pipits. Rare visitors have included Green-winged Teal, Lesser Yellowlegs, Grey Phalarope and once along the nearby beach during stormy weather a Sabine's Gull.

Timing

Winter.

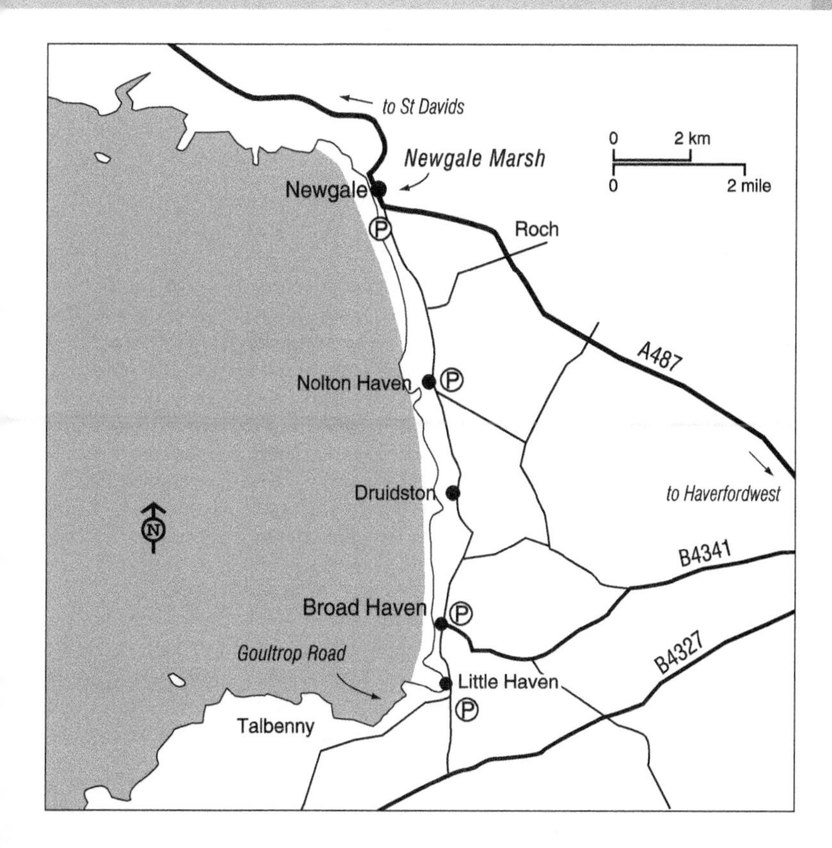

Access

At the northern end is Newgale (SM846223), 7 miles (11 km) from Haverfordwest on the A487. Minor roads leading from this reach to Nolton Haven (SM858185) and Druidston Haven (SM862171). At the southern side of the Bay and accessible off the B4341 are Broad Haven, SM861138, Little Haven, SM857129 and Goultrop Roads SM850123.

CALENDAR

Resident: Cormorant, Shag, Gannet, Fulmar, Peregrine, Herring, Lesser and Great Black-backed Gulls, Stonechat, Rock Pipit, Meadow Pipit, Raven, Chough and Linnet.

November–February: Common Scoter, Great Crested Grebe, Great Northern and Red-throated Divers in the bay, with possibility of Velvet Scoter, Red-breasted Merganser, Eider, Scaup, Black-throated Diver and Slavonian Grebe depending on the winter. Wigeon, Teal, Heron, Merlin, Curlew, Snipe, Redshank and Starling at Newgale.

March–June: Manx Shearwaters offshore, some wader passage, mainly at Newgale of Ringed Plover, Sanderling, Dunlin, Redshank, Curlew, Whimbrel, Bar-tailed Godwit.

July–October: Returning Common Scoters from July onwards, wader passage and storm-driven Manx Shearwaters.

119 MARLOES MERE

OS Landranger 157
OS Explorer OL36
Grid Ref: SM778082
Website: https://www.nationaltrust.org.uk/marloes-sands-and-mere

Habitat

In a hollow to the west of Marloes lies Marloes Mere, a freshwater marsh with open water, the only one close to the coast, which attracts wintering wildfowl and migrants alike. Agricultural changes on the surrounding land have reduced the mere's area but to compensate are two irrigation reservoirs, the northern one being the most interesting. There are two hides overlooking the mere, that in the north-east corner known as the Oriole Hide, having been funded by donations contributed by those who journeyed to Roch on the north side of St Bride's Bay to see the Baltimore Oriole which was in residence from 2 January to 24 April 1989. On the south side beside the track from the car park past Runwayskiln, the hide in memory of Mark Britton, a local birdwatcher who died far too young, provides splendid views.

Species

The pools and winter floodwater attract small numbers of duck, especially Wigeon and Teal. Pintail, Shoveler, Teal and Water Rail have bred in the past. In winter this is a roost site for Hen Harriers while Short-eared Owls, which nest on Skomer, are often present. On passage watch out for Marsh Harriers and

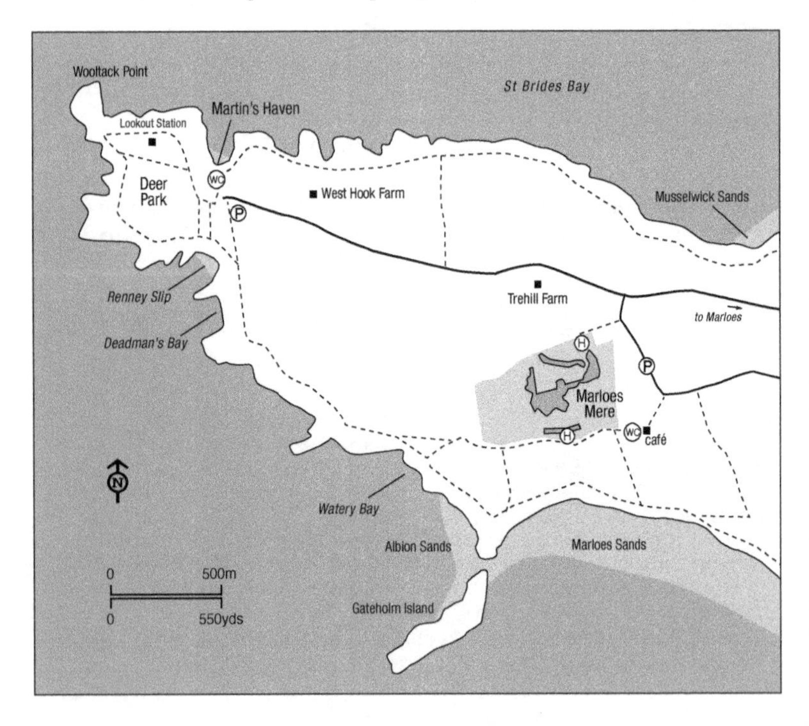

waders. Rare visitors have included Garganey, Spotted Crake, Quail, Crane, Bee-eater, Red-rumped Swallow and Blackpoll Warbler. The fields surrounding the Mere are also good for Chough, Stonechats and Linnet.

Timing

Winter and passage periods.

Access

A National Trust car park (seasonal charges apply, pay-and-display machine) provides an overview, though for closer observation there are hides in the north-east corner reached by a short footpath, and one that offers the best views and light conditions close to the footpath from the car park past Runwayskiln (former Youth Hostel, now a seasonal café).

CALENDAR

Resident: Canada Goose, Mallard, Little Grebe, Heron, Peregrine, Moorhen, Coot, Water Rail, Meadow Pipit, Stonechat, Linnet, Reed Bunting.
November–February: Wigeon, Teal, Pintail, Shoveler, Gadwall, Tufted Duck, Merlin, Hen Harrier, Short-eared Owl, Fieldfare, Redwing.
March–June: Ducks including Wigeon, Teal, Pintail, Shoveler, Gadwall and Tufted may remain all summer, occasionally to breed, Hen Harriers and Merlins may stay in the area until April, even May on occasions, passage waders sometimes visit such as Whimbrel, Greenshank and Green Sandpipers, as might Marsh Harriers. Sand Martin, Swallow, House Martin, Swift, Wheatear, Reed Grasshopper and Sedge Warblers, Whitethroats.
July–October: Winter ducks arrive from late August onwards, but mainly in October. Passage of Swallows, Sand and House Martins, Reed and Sedge Warblers, Whitethroats.

120 MARTIN'S HAVEN AND DEER PARK (WOOLTACK POINT)

OS Landranger 157
OS Explorer OL36
Grid Ref: SM760089
Website: https://www.nationaltrust.org.uk/marloes-sands/trails/martins-haven-the-deer-park-walk

Habitat

Martin's Haven is at the end of the road from Marloes. Just below the National Trust car park is the Lockley Lodge Visitor Centre (open spring and summer only) and a narrow valley leading to the beach – departure point between April and September for the boat to Skomer. En route to the beach there is a room devoted to information about the Skomer Marine Nature Reserve, as well as the all-important toilets.

A walk around the headland known as the Deer Park – note the high boundary wall – gives good views of Skomer and Middleholm and the area of fast currents in between, Jack Sound. The coves and sheltered beaches make excellent pupping beaches for Grey Seals, which can be easily observed from the clifftops, without disturbing the pups, from September to October.

Species

Choughs feed on the close cropped grasslands of the south and west cliff tops while tucked away at the south-west corner are several pairs of nesting Fulmars. Super views are obtained of Jack Sound with its fast currents, the nearest land the tiny island of Middleholm beyond which lies the mighty mass of Skomer. During the seabird nesting season when the tide is running north, as they merge into St Bride's Bay the turbulent waters attract Gannets, Guillemots, Kittiwakes and Razorbills as well as Harbour Porpoise.

The bushes in this valley are worth scrutiny during passage periods and have held Wryneck, Golden Oriole, Hoopoe, Red-backed and Woodchat Shrikes. Recent rarities have included Melodious and Greenish Warblers, and in 1992 a Spanish Sparrow, which the senior author probably saw but was otherwise engaged in transporting sand for a helicopter lift to Skokholm. A recent rarity here was an Isabelline Wheatear in 2013.

Timing

All year but preferably during passage periods.

Access

Park in the National Trust Car Park at Martin's Haven, fees apply.

CALENDAR

Resident: Great Black-backed and Herring Gulls, Peregrine, Kestrel, Chough, Raven, Rock Pipit, Meadow Pipit, Stonechat, Linnet, Reed Bunting.
March–June: Fulmar, Gannet (offshore), Swallow, House Martin, Swift, Wheatear, Reed Grasshopper and Sedge Warblers, Whitethroats and possible migrants. Guillemots and Razorbills and occasionally Puffins can be observed passing through Jack Sound or on their way to and from Skomer and St Bride's Bay. In the evening large groups, sometimes in their thousands, of Manx Shearwater can be seen offshore – a telescope will come in handy.
July–October: Passage of Swallows, Sand and House Martins, Reed and Sedge Warblers, Whitethroats, Wheatear. Occasional Wryneck, Red-backed Shrike.
November–February: Divers offshore. Possibility of wintering finches in the field edges of Martin's Haven car park.

121 THE GANN ESTUARY – PICKLERIDGE POOLS

OS Landranger 157
OS Explorer OL36
Grid Ref: SM808066

Habitat

The Dale peninsula has some of the best birdwatching sites in the county, in the form of the Gann, Dale Airfield and St Ann's Head. They are all in close proximity and depending on the time of year, tides and time available, each has its plus points. For waders, gulls and wildfowl it would be the Gann that takes pride of place, the lagoon a splendid remnant from when gravel was extracted to be used in the construction of the nearby airfield, now a top-notch site for passage migrants and wintering finches. Yellow-horned Poppy grows on the shingle bank at Dale; its unpleasant smelling sap is considered a cure for bruises, hence the alternative name of Bruisewort.

Species

The B4327 Haverfordwest to Dale road passes close to the shore at SM808067, where there is very limited roadside parking. Excellent views can be obtained across the flats, with a small wintering flock of Brent Geese, Wigeon, Teal, Curlew and Dunlin and at times Turnstones which feed on the beach. It is possible to walk north-east below the shingle ridge separating the lagoon from the estuary, and so obtain good views of both. Pickleridge Pools has a small wintering group of Little Grebe along with Goldeneye, Little Egret and waders. Beyond is an area of saltings and the sinuous, narrow, high-banked course of the Gann river, haunt of winter Greenshanks. There is a large winter gull roost in the bay and careful examination should reveal Mediterranean Gulls, up to 100

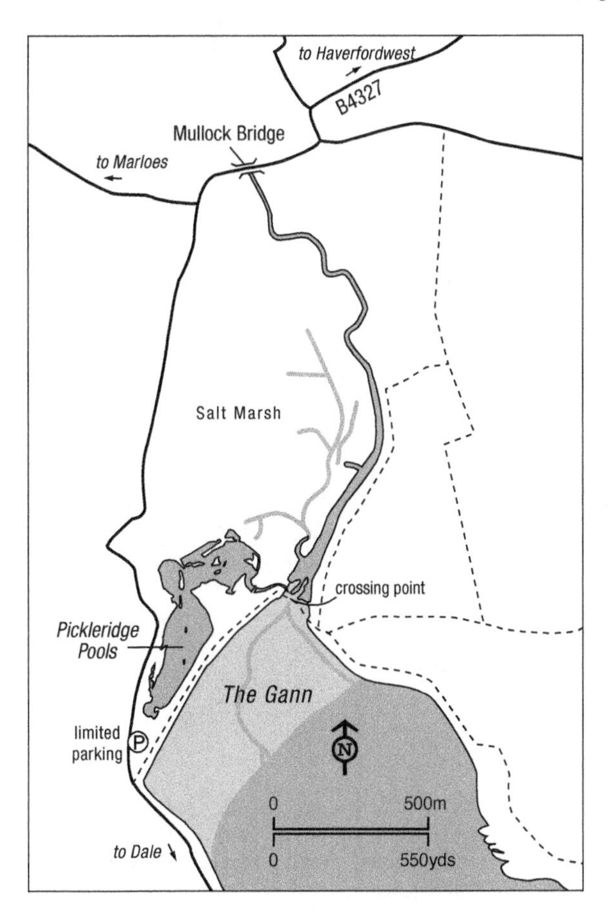

have been seen here in one day. Rare gulls have included Yellow-legged, Ring-billed, Little, Bonaparte's, Glaucous and Iceland. Sandwich Terns can often be seen fishing in the bay, anytime from April to September and several rare tern species have been seen here, including Caspian, White-winged Black, Forster's and American Royal. Grey Herons and nowadays Little Egrets are nearly always present while Great White Egrets and Spoonbills have both been recorded.

The Gann is probably best known as the county's top wader spot. Although numbers of individuals are generally small, this site has played host to a good variety of different species, including Avocet, Great White Egret, Kentish Plover, Temminck's Stint, Pectoral, Buff-breasted, Baird's and Semi-palmated Sandpipers, American Golden Plover and Long-billed Dowitcher.

Timing

Not critical, but the areas of intertidal mud and sand are much greater and observations need to be timed for the couple of hours or so either side of high-water, before the birds disperse.

Access

Very limited roadside parking near the shore from where one can walk along the bank observing both pools and foreshore, please do not venture onto the pools as this will disturb any birds present. Otherwise park in Dale itself, fee payable.

CALENDAR

Resident: Cormorant, Grey Heron, Little Egret, Mute Swan, Canada Goose, Shelduck, Mallard, Oystercatcher, Lesser Black-backed, Herring and Great Black-backed Gulls.

December–February: Red-throated and Great Northern Divers, Little and Great Crested Grebes, Brent Goose, Wigeon, Teal, Goldeneye, Red-breasted Merganser, Ringed and Grey Plovers, Lapwing, Dunlin, Snipe, Bar-tailed Godwit, Curlew, Redshank, Greenshank, Turnstone, Black-headed, Mediterranean and Common Gulls, Kingfisher.

March–May: Departure of winter visitors. Passage migrants on estuaries include Sanderling, Whimbrel, Sandwich Tern and Common Sandpiper.

June–July: Residents all present, but Shelducks virtually disappear by end of period when most undertake moult migration. First returning waders, with Lapwing and Curlew numbers rising rapidly during July.

August–November: Parties of Brent Geese start to arrive from late September. Wader movement throughout August and September, such as Little Stint, Curlew Sandpiper, Ruff and Black-tailed Godwit.

122 DALE AIRFIELD

OS Landranger 157
OS Explorer OL36
Grid Ref: SM798062

Habitat

This long-disused Second World War airfield has hosted several American waders in autumn, when large finch flocks take up residence. The clifftops

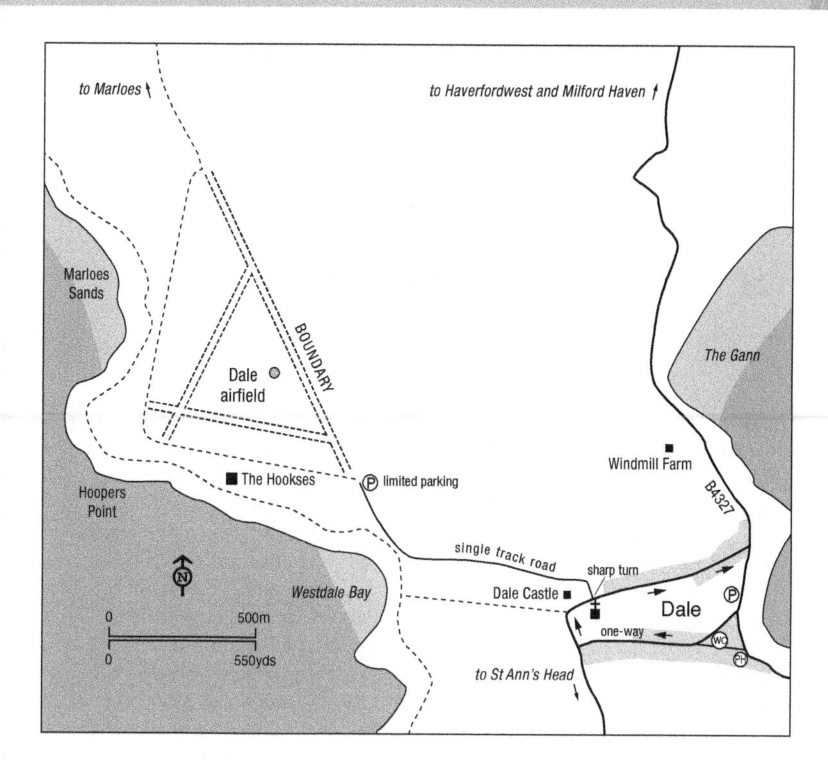

provide excellent feeding areas for Choughs and give panoramic views of Skokholm, Skomer and far out to sea to Grassholm, the Gannet colony glowing white even on a rainy day.

Species

From the entrance onto the airfield, take a large circular walk, checking for waders, pipits, larks and finches – the edges of the old runways are usually best. Check also any closely cropped turf, as it is on these areas that Buff-breasted Sandpipers have been seen. Other recent rarities have included Dotterel, Wood Sandpiper, Dotterel and Short-toed Lark.

Timing

Passage and winter.

Access

Although there is space for a couple of cars to park by the access to the airfield, the best advice is to park in the Dale car park and walk through the village to West Dale beach, then follow the coast path north, up onto the airfield.

CALENDAR

Resident: Meadow and Rock Pipits, Skylark, Linnet, Kestrel, Peregrine, Chough, Raven, Stonechat, Reed Bunting.

December–February: Snipe, Merlin, Hen Harrier, sometimes wintering Snow and Lapland Buntings.

May–July: Swallow, Whitethroat.
August–October: Occasional parties of waders, including Ringed Plover, Dunlin, Redshank, among which there may be Pectoral and Buff-breasted Sandpipers. Scarce passerines may be present, Red-backed Shrike, Wryneck, Snow and Lapland Bunting.

123 MILFORD HAVEN WATERWAY

OS Landranger 157
OS Explorer OL36

Habitat

From St Ann's Head to Haverfordwest by sea is 21 miles (33.6 km), from the broad waters of Milford Haven with its oil and gas terminals, to the reaches of the Daucleddau, and finally along the twisting course of the Western and Eastern Cleddau rivers between well-wooded banks to the upper reaches of the tidal limits. The numerous bays and creeks, together with the upper river, are fine areas for wildfowl and waders. For the most part the expanse of inter-tidal mud is limited, so that, unlike on the larger estuaries elsewhere in Wales, birds are generally not too distant even at low tide. There is an increasing use of the whole waterway by pleasure craft, though this has largely ceased by the time winter birds have arrived, while there are few summer residents other than Shelduck to be disturbed. These activities do, however, pose a potential threat. Much of the Western Cleddau is a Regional Wildfowl Refuge, designated in 1970, which means that no shooting can take place below the high-water mark. A footpath system around much of the upper part of the waterway provides access to some rarely visited areas of the shore.

As in many estuaries, large areas of saltings, in the case of Milford Haven just over 50%, have been colonised by Common Cord-grass, or *Spartina*, since it was first introduced in the 1940s. Marsh Pea, previously known from just one other site in Wales, was discovered in 1982 beside the Eastern Cleddau. Another interesting plant found in a few secluded estuary edges is Marshmallow, the roots of which were used in former days to make sweets of the same name. The woodlands on the steep, sometimes rocky slopes bordering parts of the estuaries are also of special interest to botanists. Wild Service-tree or Wild Checker *Sorbus torminalis*, found mainly on the Borders and in eastern England, is restricted to these estuarine woodlands; in autumn the few trees can easily be located by the bright red of the turning leaves, contrasting with the greens and yellows of the dominant Sessile Oak.

The limestone bluff at the junction of the Carew and Cresswell rivers and now part of the West Williamston Nature Reserve, provides a rather special habitat. Stone was quarried here for several hundred years, in the past most of it being transported by sea, using small vessels taken up narrow man-made creeks into the quarry workings for loading. Now the creeks are silted up and it's the haunt of passage Common Sandpiper and wintering Redshank. Where the soil cover is thin and scrub has been unable to take over, a rich flora has developed. Here one can find such gems as Hairy Rock-cress, Yellow-wort, Bee Orchid and Autumn Gentian.

Species

Breeding species are few and far between; there is simply not enough area of saltmarsh to provide space for birds such as Oystercatcher and Redshank. Small numbers of Shelduck attempt to breed, but with limited success, while the introduced flock of Canada Geese continue to thrive and increase. Cormorants fish along the whole length of the estuary and can be seen throughout the year, as can Grey Heron and the now familiar Little Egret – up to 50 of the latter may be present on the Cleddau but few remain into the summer and possibly breed.

Numbers of birds on the Cleddau increase in autumn to early winter, when up to 6,000 Wigeon, 2,000 Teal, 4,000 Dunlin, 800 Redshank, 500 Curlew and 400 Oystercatchers may be present. Small numbers of Great Crested Grebes, Brent Geese, Goldeneye, Pintail, Shoveler and Grey Plover also occur during the winter months. Numbers of most wintering species are declining due to warmer winters and hence a lessening need for birds to come to south-western Wales, while, on the other hand, the number of wintering Greenshank has increased over the last 30 years. Another species that has increased is Black-tailed Godwit, with up to 150 present on the estuary, mainly in the area of the Pembroke River. Green and Common Sandpipers regularly overwinter on the Cleddau, though you will have to search hard in the creeks and narrow pills.

Other waders that appear from time to time include Knot, Little Stint, Curlew Sandpiper and Ruff, while Avocets are a little more unusual. As on other estuaries in Wales, gulls are now very much a feature and those who carefully study the flocks will be regularly rewarded with sightings of Mediterranean, Little, Iceland and Glaucous Gulls, along with the rarer Ring-billed and Yellow-legged Gulls.

Timing

Timing is not so critical on the upper reaches as on many other estuaries. The river is narrow, and even at low tide only a limited area of foreshore close to the bank is available for feeding birds, so that observations can be made at reasonable ranges. On the lower reaches, and especially at Angle Bay and in the Pembroke river, the areas of intertidal mud and sand are much greater, and so observations need to be timed for the couple of hours or so either side of high-water before the birds disperse.

Access

Sandy Haven. Access at Sandy Haven (SM855074) or near Sandy Haven Chapel (SM857086). This is a regular haunt of wintering Greenshank.

Llanstadwell. The tiny creek and ridge at SM958050 is easily viewed from the car; a good spot for checking gulls.

Westfield Pill. Formerly a tidal creek, now dammed when the nearby Neyland Marina was built. The site is a reserve of the Wildlife Trust of South & West Wales and is best accessed through the marina at SM967056 or by parking on the west side of the high-level bridge on the A477 at SM966061 and following the path down the slope to the pill. The dam area is often a haunt of Kingfisher, while Firecrests and wintering Chiffchaffs are often recorded in the woodland surrounding the site. Ravens nest underneath the bridge. Rare species recorded here in the past include Night Heron, Mandarin, Temminck's Stint and Hoopoe.

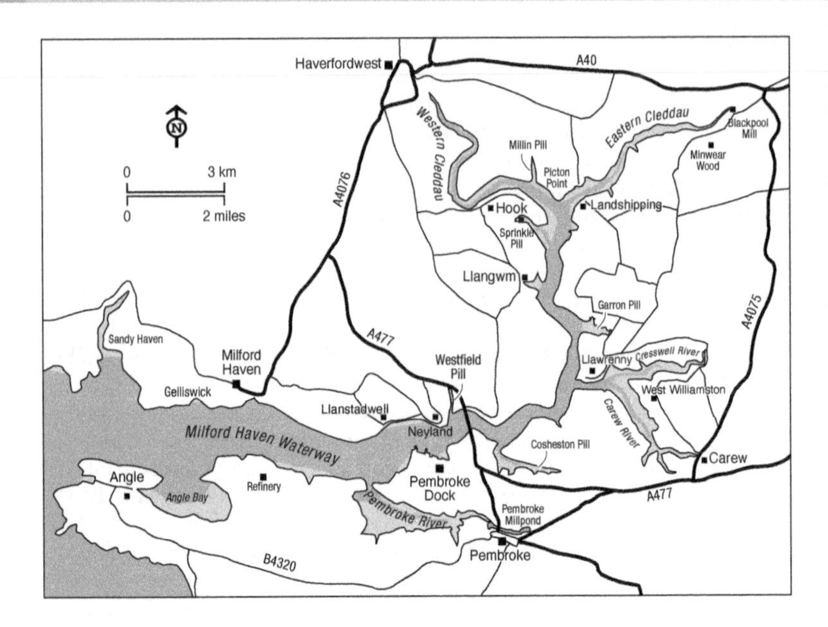

Western Cleddau. A footpath follows much of the perimeter of the upper reaches of Milford Haven. Currently access is best at Lower Hook Quay (SM990120), from where one can walk west along the shore to Little Milford (SM966119), where a road comes close to the river, or south-east to Sprinkle Pill. The extreme upper reaches are easily viewed from the Fortunes Frolic footpath which follows close to the eastern bank, leaving Haverfordwest near the old gasworks at SM957154.

Eastern Cleddau. The road runs to the shore at Picton Ferry (SN009122) and directly opposite on the south bank at Landshipping (SN011117). This latter site gives a fine view of the estuary and opportunity to see the main species, careful scanning with a telescope should reveal Grey Plover and Pintail.

West Williamston. WTSWW Reserve. There is a small car park at SN033058, from where one can walk to the estuary and then north-west to the point where the Carew and Cresswell rivers meet.

Pembroke River. On the north bank there is easy access to the shore at East Pennar (SM961021). This area often contains large numbers of waders and wildfowl during the winter.

CALENDAR

Resident: Cormorant, Grey Heron, Little Egret, Mute Swan, Canada Goose, Mallard, Oystercatcher, Lesser Black-backed, Herring and Great Black-backed Gulls.

December–February: Red-throated and Great Northern Divers, Little and Great Crested Grebes, Wigeon, Teal, Pintail, Goldeneye, Red-breasted Merganser, Ringed and Grey Plovers, Lapwing, Dunlin, Snipe, Bar-tailed Godwit, Black-tailed Godwit, Curlew, Redshank, Greenshank, Turnstone, Black-headed and Common Gulls, Kingfisher, Shelduck.

March–May: Departure of winter visitors. Passage migrants on estuaries include Sanderling, Whimbrel and Common Sandpiper.
June–July: Residents all present, but Shelducks virtually disappear by end of period when most undertake moult migration. First returning waders, with Lapwing and Curlew numbers rising rapidly during July.

August–November: Occasional parties of Brent Geese from late October, Mallard numbers increase during September, soon followed by Wigeon, Teal and other winter ducks including Pintail. Wader movement throughout August and September, both passage and winter visitors including those which occur in only small numbers in this far south-west corner of Wales, such as Little Stint, Curlew Sandpiper and Ruff.

124 BLACKPOOL MILL AND MINWEAR WOOD

OS Landranger 157
OS Explorer OL36
Grid Ref: SN060145 / SN054138
Website:
https://naturalresources.wales/days-out/places-to-visit/south-west-wales/minwear-forest/?lang=en

Habitat

Flowing from the north-east, the Eastern Cleddau becomes estuarine just upstream from Blackpool Mill, a handsome four-storeyed building dating from 1813, one of the finest examples of a grist mill in Great Britain, which remained in use grinding corn until 1950. Generations of Dippers have nested below the mill – a short while spent on the nearby spectacular arched bridge built by Baron de Rutzen, a Polish nobleman, provides an opportunity to observe the birds flying to and fro.

Minwear Forest formed part of medieval Narberth Forest, described by George Owen in his *Description of Pembrokeshire* as one of the best standing woods in the county, large quantities of timber being converted into charcoal to fuel the iron foundry near Blackpool Mill. Richard Fenton, writing at the end of the 18th century, noted 'the whole forest from an over anxiousness of gain, was at once laid low'. Today the woods, mostly coniferous but with welcome pockets of broadleaves, are easily accessible.

The eastern arm of the Cleddau Estuary has few access points. One of the best is Minwear View (SN055137). The woods here are one of the few sites in the county where Wood Warblers can be heard. Blackpool Mill (SN060145) gives a good view of the Eastern Cleddau, good for Kingfisher, Grey Wagtail and Dipper, while the woods provide an excellent selection of species.

Species

Blackpool Mill: Kingfisher, Dipper and Grey Wagtail all frequent this stretch of the Eastern Cleddau, with the bridge itself being a good vantage point or upstream at the water treatment works. Marsh Tits, Nuthatch, Treecreeper and Goldcrest are present in the woodland surrounding the 'Leap Walk'. In spring and summer, Chiffchaff, Willow and Garden Warblers may be heard singing from this area.

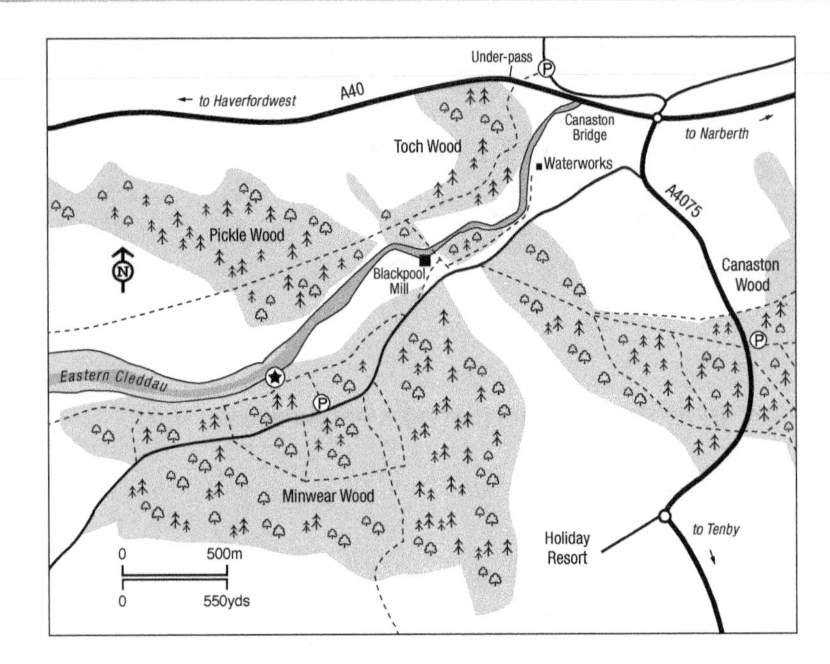

Minwear Wood: Follow the path from the car park, through the trees down to the estuary. This is a good area for wintering Common Sandpiper. The woods themselves can be productive and are one of the few sites in the county that have breeding Wood Warblers (try the area down from the car park or the section between it and Blackpool Mill). Tree Pipits can be found in the area around the enclosures and alongside the Tenby road.

Timing

Not so critical on the upper reaches as on many other estuaries, but generally early morning is best.

Access

Small car parks along the road at Blackpool Mill and at Minwear Wood.

CALENDAR

Resident: Cormorant, Grey Heron, Mute Swan, Canada Goose, Shelduck, Mallard, Oystercatcher, Kingfisher, Dipper, Marsh Tit, Treecreeper, Nuthatch, Great Spotted Woodpecker, Tawny Owl, Sparrowhawk, Goshawk, Buzzard and Grey Wagtail.

December–February: Little Grebes, Wigeon, Teal, Goldeneye, Goosander, Lapwing, Dunlin, Snipe, Curlew, Redshank, Green Sandpiper, Common Sandpiper, Black-headed Gull.

March–May: Passage migrants on estuary include Whimbrel and Common Sandpiper. In the woods returning warblers – Willow, Wood, Garden, Chiffchaff, Whitethroat – and Tree Pipits.

June–July: Residents all present.

August–November: Mallard numbers increase during September, soon followed by Wigeon and Teal. Fieldfare and Redwing from October onwards.

125 PEMBROKE MILL PONDS

OS Landranger 157
OS Explorer OL36
Grid Ref: SM983016

Habitat

William Marshal, 1st Earl of Pembroke, who died in 1219, had the castle that dominates the Lower Mill Pond rebuilt, while it was here the future Henry VII was born in 1457; his statue stands on the nearby bridge. At the time of writing, a statue of William Marshal on horseback is nearing completion. In the Middle Ages there was a tidal mill here, perhaps similar to that which can be seen not many miles away on the Carew River (see Site 123), while the harbour hosted shipping including vessels employed in a prosperous wine trade with France.

To ensure a rail link between Pembroke and Pembroke Dock, a high embankment was constructed towards the eastern end of what is now the Middle Mill Pond, creating the now secluded Upper Mill Pond with shallow open water and reedbeds. A Wildlife Trust of South & West Wales nature reserve since 1979, this is one not to be missed.

Species

At one time the herd of Mute Swans which frequented the Middle Pond numbered up to 100, but now no more than 30 with several pairs breeding. Mallard, as might be expected, are always present while winter visitors include Teal and Tufted Duck. Little Grebe, a bird that has quietly increased in Pembrokeshire since the 1990s, nest with their numbers being supplemented by winter visitors. Cormorants, Coot, Moorhen, Grey Heron are always present. The Upper Mill Pond is much frequented by Black-tailed Godwits when the tide on the nearby estuary covers their feeding area, up to 150 have been reported. Other waders reported include Common and Green Sandpipers and Lesser Yellowlegs. A striking feature during the summer months are the flocks of Swallows, House and Sand Martins and Swifts feeding over the Middle Pond.

Timing

Early morning best for the Lower and Middle Mill Ponds; if visiting the Upper Mill Pond check the tide tables for the nearby Milford Haven Waterway and arrive from half-tide onwards to watch the Black-tailed Godwits arriving.

Access

Limited parking beside the bridge between the Middle and Lower Mill Ponds, otherwise use the town car park. The excellent footpaths beside the Middle and Lower Ponds are suitable for wheelchairs. The open water of the Upper Mill Pond is only visible from Little Gates beside the railway embankment. A boardwalk and footpaths from the small car park and lay-by beside the A4075 on the eastern edge of the town provide a route alongside the reedbeds and through the woodlands of the nature reserve.

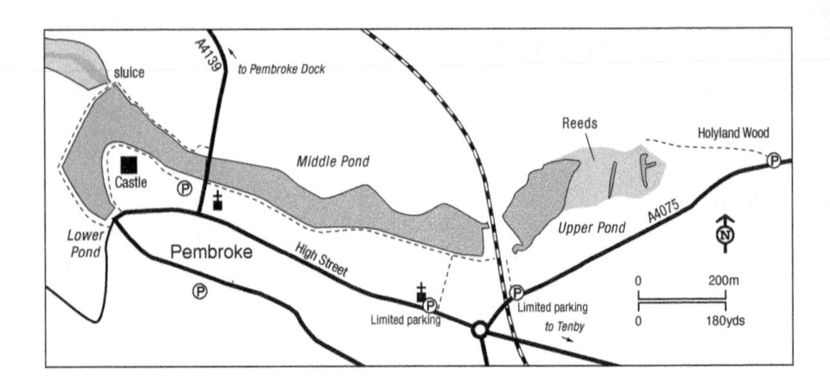

CALENDAR

Resident: Mute Swan, Mallard, Grey Heron, Little Grebe, Coot, Moorhen, Herring Gull, Great Spotted Woodpecker, Grey and Pied Wagtails, Wren, Blue, Coal, Great, Marsh and Long-tailed Tits, Treecreeper, Nuthatch, Reed Bunting.

December–February: Teal, Tufted Duck, Black-tailed Godwit, while the number of Little Grebes rises.

March–May: Summer visitors like Sedge and Reed Warbler, Whitethroat, Blackcap, Chiffchaff and Willow Warbler arrive.

June–July: Swifts, Swallows, House and Sand Martins over the Middle Mill Pond, Black-tailed Godwits begin to arrive.

August–November: The best season for passage waders, carefully check the banks of the Lower Mill Pond. From mid-October onwards Fieldfares and Redwings arrive and suddenly Starlings become more numerous.

126 ANGLE BAY

OS Landranger 158
OS Explorer OL36
Grid Ref: SM872030

Habitat

Angle Bay is a large bay on the south side of the Milford Haven waterway, opposite Dale and the Gann Estuary. At the western end of the bay is Angle Village, while the more productive eastern end is close to the Valero Oil Refinery at Rhoscrowther.

Species

Angle Bay attracts good numbers of wintering and passage waders and wildfowl, including Wigeon and Brent Geese, with up to 2,500 of the former and 50 of the latter. In winter, there are often Slavonian Grebes in the bay among the 20 or so Great Crested, and divers of all three species are quite usual. Large numbers of gulls are always present alongside which there are likely to be several Mediterranean Gulls, with numbers building from late July to up to 30 in the autumn and winter. Little Egrets are commonly seen fishing in the

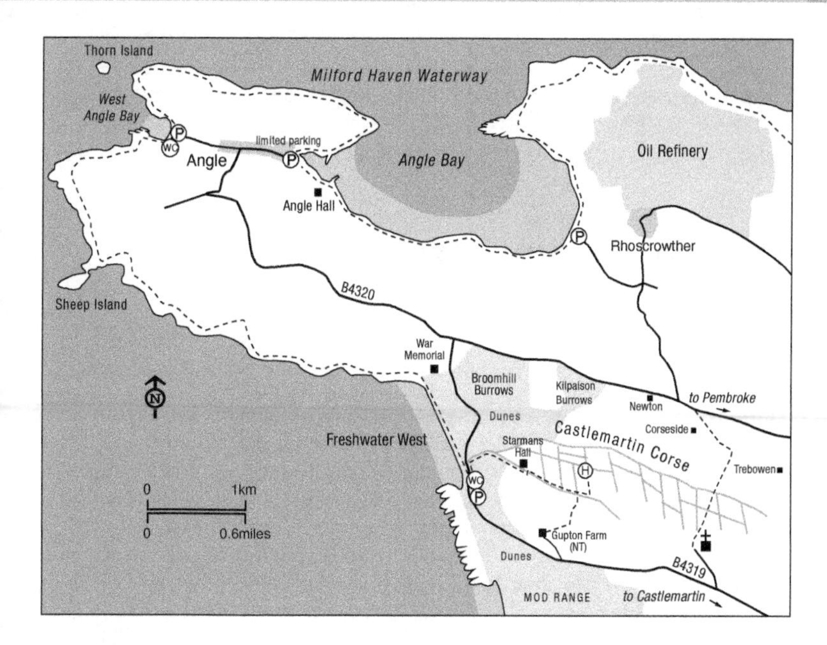

creeks, while a Squacco Heron was present for a month in autumn 2010. During passage times there may be good numbers of both Bar-tailed and Black-tailed Godwits, Curlew, Whimbrel, Dunlin and Ringed Plover, among which there are usually Turnstone, Sanderling and in autumn occasionally Little Stint and Curlew Sandpiper.

Timing

Half tide is probably best, with birds closer to shore. At high tide waders tend to roost at either end of the bay.

Access

Access is possible from Angle village by walking east along the south side of the bay from near Angle Hall (SM868027); or from Kilpaison, for which you leave the B4320 Pembroke to Angle road at SM908004 on the unclassified road for Rhoscrowther, and after nearly 1 mile (1.6 km) turn left at SM904016 for Angle Bay.

CALENDAR

Resident: Cormorant, Grey Heron, Mute Swan, Canada Goose, Shelduck, Mallard, Oystercatcher, Lesser Black-backed, Herring and Great Black-backed Gulls.

December–February: Red-throated and Great Northern Divers, Little and Great Crested Grebes, Wigeon, Teal, Brent Goose, Goldeneye, Red-breasted Merganser, Ringed and Grey Plovers, Lapwing, Sanderling, Dunlin, Snipe, Black and Bar-tailed Godwit, Curlew, Redshank, Greenshank, Turnstone, Black-headed, Mediterranean and Common Gulls.

March–May: Departure of winter visitors. Passage migrants on estuaries include Sanderling, Whimbrel and Common Sandpiper.

June–July: Residents all present, but Shelducks virtually disappear by end of period, when most undertake moult migration. First returning waders, with Lapwing and Curlew numbers rising rapidly

during July. Mediterranean Gulls start to return in late July.
August–November: Parties of Brent Geese arrive from late October. Mediterranean Gulls among the gull flocks. Wader movement throughout August and September, such as Little Stint, Curlew Sandpiper, Ruff and Black-tailed Godwit all possible among the Curlew and Dunlin.

127 CASTLEMARTIN CORSE

OS Landranger 158
OS Explorer OL36
Grid Ref: SR885993

Habitat

Castlemartin Corse is the wetland stretching inland from the beach and dunes of Freshwater West with areas of reedbed, fen meadow and further upstream extensive grazing wet meadow. Described in 1794 by Charles Hassall, an agricultural pioneer and colourful figure, as 'a tract of several hundred acres, a perfect bog of little value' he quickly set out to drain it. The engineer who oversaw the work was awarded a gold medal in 1800 for being 'the most successful reclaimer of waste land'. We will never know what ornithological gems were lost as a result, though less than a century ago Montagu's Harriers regularly nested, sometimes successfully when fortunate enough to escape the attention of egg-collectors. Water levels are not currently managed, and the reedbed is being invaded by scrub. It can be very dry or flooded, depending on the weather. Many an ornithologist would welcome some active management, pools of permanent water, a scrape or two – then what delights would enchant us.

Species

Under wet winter conditions the Corse can be excellent for wildfowl, and the reedbed supports all the usual breeding species. The site can be good for raptors at any time of year, with Buzzard, Red Kite, Marsh Harrier resident and Hen Harrier, Merlin and Short-eared Owls present during the winter months. The National Trust's Gupton Farm (on the south side) and Kilpaison and Broomhill Burrows (on the north) support farmland birds, feeding Chough and wintering Lapwing and Golden Plover. Great White Egret has visited the reedbed and a Crane has fed in stubble fields on Gupton Farm.

Timing:

All year.

Access

Park at Freshwater West and follow the directions inland from the back of the National Trust car park (map at toilet block). The main track goes inland parallel to the wetland. There is a hide about 500m inland which is useful for raptor viewing. There is a circular walking route with an overview of the wetland from the National Trust's Gupton Farm campsite (on the south side of the Corse, access from the Castlemartin road), where there is a car park and information point.

CALENDAR

Resident: Water Rail, Moorhen, Kestrel, Sparrowhawk, Buzzard, Raven, Skylark, Meadow Pipit, Linnet, Cetti's Warbler, Reed Bunting, Stonechat. Marsh Harriers have been seen in all months, Red Kite with increasing frequency.
December–February: Hen Harrier, Merlin, Short-eared Owl, Bittern. Wigeon, Teal, Shoveler, Greylag, 1000+ Golden Plover, 1000+ Lapwing. Up to 400 Snipe and a few Jack Snipe winter in the reedbed and fen meadow. Black-tailed Godwit visit the wet meadows upstream.
March–May: Regular migrants include Whimbrel, Garganey, Wheatear. Sand Martins breed nearby. Hirundines and Swift feed over the reedbed. Breeding birds include Reed Warbler, Sedge Warbler, Grasshopper Warbler. Sand Martins breed in the low cliffs at the south end of the main beach.
June–July: Breeding warblers.
August–November: Summer migrants depart, Swallows often until the end of October. Small numbers of returning ducks towards the end of October.

128 CASTLEMARTIN–STACKPOLE COAST

OS Landranger 158
OS Explorer OL36
Grid Ref: SR966930

Habitat

A line of carboniferous limestone cliffs extends westwards from the vicinity of Stackpole Quay for some 11 miles (17.6 km) to just beyond Linney Head, the most southwesterly point in Pembrokeshire. The coastal scenery for the whole distance is superb, with numerous bays, narrow inlets, sea caves, stacks, blow-holes and natural arches. At only a few points on the Castlemartin Peninsula, mainly secluded bays to which access is prohibited, is the cliff line broken. The two exceptions are the sandy beaches, much loved by holidaymakers, of Barafundle and Broad Haven. The latter lies at the entrance to the drowned inlets of the Bosherston Lakes.

The majority of the Stackpole and Castlemartin peninsula is a MOD live-firing Tank Range, with limited access, while to the west lies the popular surfing beach and sand dunes of Freshwater West, famed for the shell house scene in a Harry Potter movie and a marshy inland area, Castlemartin Corse (Site 127). To the east there is the Stackpole Estate, encompassing Bosherston Lily Ponds (Site 129) and St Govan's Head National Nature Reserve, with parts managed by the National Trust and others by Natural Resources Wales.

Species

Most seabirds, and certainly those easily visible, are at Stack Rocks, often referred to as the Elegug Stacks, after the local name for the Guillemot. The stacks, of which there is one large and three small, lie just off the mainland and must constitute one of the finest places in the country to see cliff-nesting birds at close quarters. They hold the greatest number of Guillemots and Razorbills on the mainland coast of Wales. If you are unable to visit one of the island colonies, then this is the place to come; the cliff edge is within a few yards of the car park. Numbers have been increasing for some years: almost 15,000

Guillemot are present, together with about a tenth of that number of Razorbills, many of which nest in crevices and holes, both on the stacks and on the coast cliffs opposite. Kittiwakes by contrast have been declining at the south Pembrokeshire colonies and fewer than 10 pairs now nest at the stacks. The liberal whitewash of droppings is often the only clue to the presence of several pairs of Shags, as incubating birds sit tight in gloomy recesses. Fulmars nest here, as do Herring Gulls, Lesser Black-backed Gulls, and on the top of the stacks several pairs of Great Black-backed Gulls. Just east of the stacks is an especially dramatic seabird colony, not in terms of numbers but because of its geological splendour. This is the Cauldron, an aptly named giant blowhole and natural arch through which Guillemots and Razorbills fly to reach their nest-sites: an awesome spectacle as the sea surges through the arch in times of storm, though the seabirds, often nesting within just a few feet of the high-water mark, apparently seem unconcerned.

Eastwards from Mewsford Point (SR942939) to New Quay (SR975931) there are a number of colonies, most not visible from the cliff top, the largest, with some 700 Guillemots being at Mewsford Arches from above which the birds can be seen flying to and fro. Further to the east at Stackpole Head SR996942 some 1,700 Guillemots occupy ledges mostly unseen from the cliff top though the numbers of birds flying to and fro an immediate indication of a thriving colony.

Peregrines nest at several places, while there is usually a pair or two of Kestrels. In winter, Hen Harriers are frequent visitors to the range and the sand dunes in the far west. Several pairs of Ravens nest, while in the autumn and early winter a small and noisy flock may be encountered. The large number of sheep which graze on the range during the winter non-firing period are a major factor in encouraging the ten pairs of Choughs. The number of nest sites and good feeding areas ensure this is indeed one of the most important areas in Wales for the species. The disturbed soils, the result of tank activity, provide excellent feeding areas, while the close-cropped turf and the sheep dung are ideal for ants, beetles and other invertebrates on which the Choughs feed. In frosty weather they often resort to the sand dunes, and even to the seaweed debris at the top of the storm beaches.

Timing

Not critical, but an early morning visit to such places as Stack Rocks is always most pleasant: the clifftop should be your own, and the seabird city on the stacks will seem especially boisterous. Do not neglect, if visiting the Castlemartin area, including Stack Rocks and St Govan's Head, to check the range firing times; see Stack Rocks below.

Access

St Govan's Head. Turn off the B4319 for Bosherston and continue through the village and so to the coast. When visiting the chapel, tucked into a steep cleft in the cliff. Access is possible eastwards to St Govan's Head, a fine vantage point for observing seabirds, mostly Manx Shearwaters and Gannets from the nearby islands. It could also prove to be of interest during spring and autumn for other passing seabirds, but as yet has received little attention.

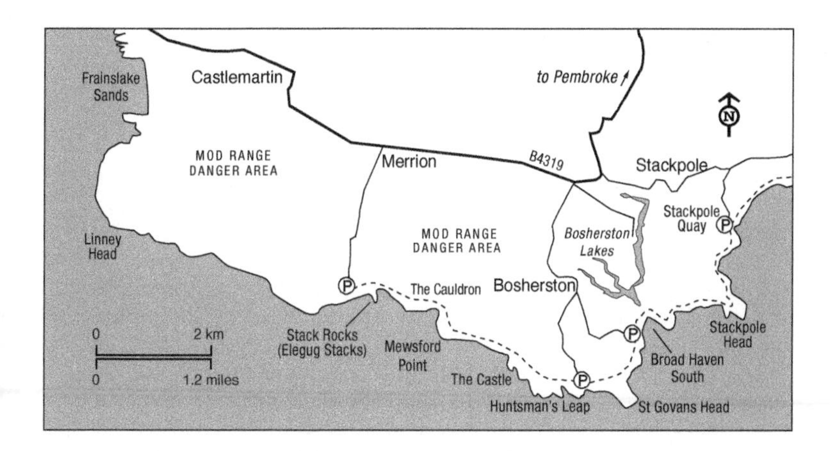

Stack Rocks. Turn south off the B4319 just west of Merrion Barracks, but please note that the road across the range is for much of the year open only at weekends, on bank holidays and on some evenings. It is closed when firing takes place, and then the gate is shut and red flags fly. Excellent general views of the range, and of the tank activity, may be obtained from the viewing point at SR927977, close to aptly named Cold Comfort on the hill to the west of the hamlet of Warren.

Broadhaven South. On the southern outskirts of Bosherston, a road bears east to a large car park directly above the beach (seasonal charges apply, pay-and-display machine).

Stackpole Head. There are two turnings off the B4319 from Pembroke to Stackpole village, and then to the National Trust car park at Stackpole Quay (seasonal charges apply, pay-and-display machine) (SR992957), from where the coastal footpath leads to the Head past the delightful Barafundle beach.

CALENDAR

Resident: Cormorant, Shag, Kestrel, Peregrine, Oystercatcher, Rock Pipit, Stonechat, Chough, Jackdaw, Raven.
December–February: Red-throated and Great Northern Divers, Fulmars return to the cliffs in early January, while Guillemots and Razorbills return sporadically to the cliffs throughout whole period, though on many occasions if you are not there shortly after dawn they will have returned to sea.
March–May: Manx Shearwater and Gannet appear offshore. First summer migrants arrive Black Redstart, Wheatear by end of March.
June–July: Breeding seabirds all present. Guillemots and Razorbills have normally all left the colonies by the end of July.
August–November: Fulmars depart the cliff in early September. Late summer migrants include Wheatear. By end of period Fulmars are coming back to the cliffs, while even Guillemots and Razorbills may make brief early morning visits to the cliffs.

129 BOSHERSTON LAKES

OS Landranger 158
OS Explorer OL36
Grid Ref: SR968948
Website: https://www.nationaltrust.org.uk/stackpole

Habitat

Bosherston Lakes or Lily Ponds, with parts managed by the National Trust and others by Natural Resources Wales, has a main central pool and three connecting arms. These shallow lakes were created in the 18th and 19th centuries by the construction of dams at strategic points across what were formerly marshy valleys. The lakes are one of the finest lowland freshwater systems in southern Britain and are noted for their rich aquatic flora, most striking being the large expanse of White Waterlilies, a spectacle which draws large numbers of people throughout the summer months. Otters are resident here, and with patience and a little luck one can be blessed with a sighting.

Close by the lakes stood Stackpole Court, a fine house of the Cawdors until it was demolished in 1967; fortunately, the fine range of outbuildings was left untouched for here is one of all too few colonies of Greater Horseshoe Bat – the rich feeding areas over the lakes, the nearby range and a variety of winter hibernation sites in caves and artificial caverns prove ideal.

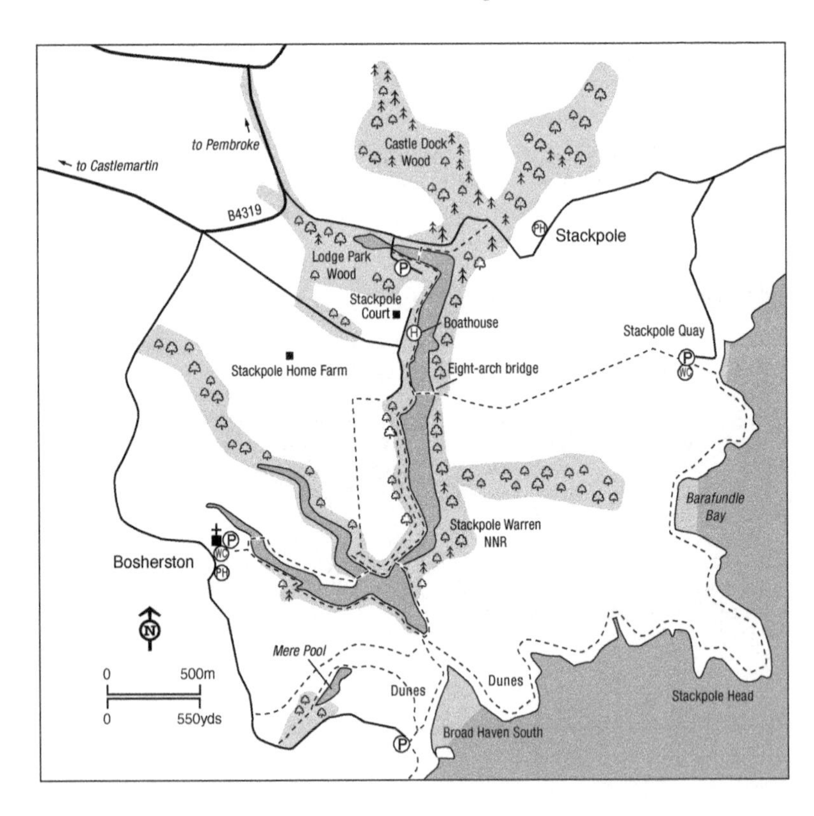

Species

Bosherston Lakes have few breeding wildfowl but provide instead a wintering site of local importance for a range of species. Coots predominate and up to 100 may be present in midwinter, with smaller numbers of Tufted Duck, Goldeneye and Goosander. In the past, large numbers of Pochard were recorded here during the winter months but their numbers have dropped dramatically in the county, as elsewhere in Wales, so much so that they are almost a county rarity. Among these wintering wildfowl there have been very occasional reports of Scaup, Long-tailed Duck, Ring-necked Duck and even Ferruginous Duck. Bitterns now occur most winters, especially during periods of hard weather, Water Rails are regularly heard from the reedbeds, though generally only in winter. Firecrests winter along the lakes, roosting in the Holm Oaks. Otters are resident but rarely seen by visitors, preferring the early mornings when there are fewer people about.

Timing

Early morning is generally more productive as there is less disturbance.

Access

Bosherston Ponds. Park in the car park close to Bosherston Church (seasonal charges apply, pay-and-display machine), and then follow the footpaths to and around the ponds, crossing by means of a footbridge to reach the coastal footpaths at Broad Haven beach. Alternatively, there is a small car park just over the bridge, near Lodge Park, at SR976964 (again charged). From the bridge, walk up the hill to Lodge Park, with excellent views over the north end of the eastern arm, then follow the footpaths around the ponds.

Broadhaven South. On the southern outskirts of Bosherston, a road bears east to a large car park directly above the beach (seasonal charges apply, pay and display machine).

CALENDAR

Resident: Mute Swan, Mallard, Little Grebe, Grey Heron, Moorhen, Coot, Kingfisher, Nuthatch, Treecreeper, Great Spotted Woodpecker, Marsh Tit.

December–February: Great Crested Grebes, Bittern, Wigeon, Gadwall, Tufted Duck, Goldeneye, Goosander, wintering Chiffchaff and Firecrest.

March–May: Summer migrants arrive, Swallow, Sand Martin, Swift, Sedge Warbler, chance of Garganey on the lakes. Chiffchaff, Blackcap, Willow and Garden Warblers in the woods.

June–July: In the woods, Willow Warbler, Chiffchaff, Blackcap, Garden Warbler and Spotted Flycatcher.

August–November: Summer migrants depart, Swallows often until the end of October. Small number of returning ducks towards the end of October.

RADNORSHIRE (MAESYFED)

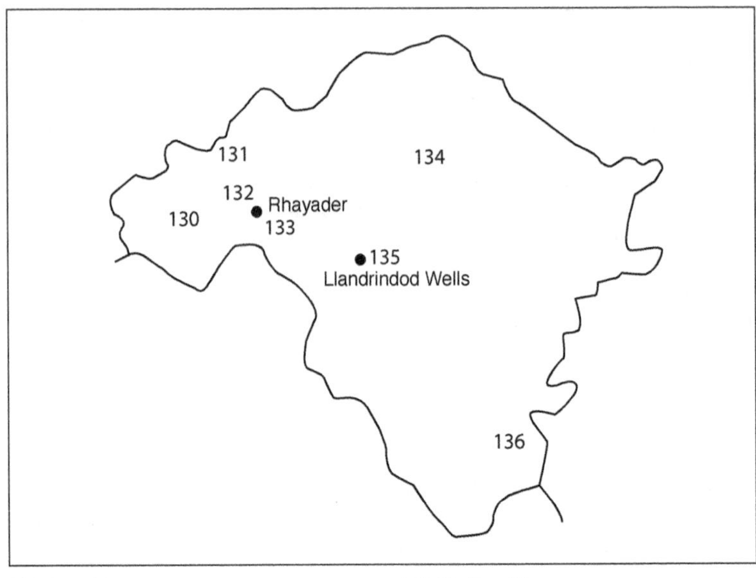

130 Elenydd
131 Gilfach Farm
132 Gigrin Farm
133 Wye-Elan Woods
134 Maelienydd
135 Llandrindod Wells Lake
136 Glasbury

Radnorshire is a very rural county, being described by Shakespeare in *King Lear* as 'With shadowy forests and with champions rich'd, With plenteous rivers and wide-skirted meads'. Over 55% of the county is over 1,000 feet (300 m) above sea-level, with about a hundred hilltops above 1,500 feet (450 m) and the highest point Rhos Fawr at 2,165 feet (660 m). A century or more ago much of the high ground was heather moor; now, following decades of change, much is largely improved grassland, while the fact that some heather remains is largely due to management for Red Grouse. The common lands are significant, Radnorshire having more than any other county in Wales, and where there has been less habitat change these support important bird communities.

The county avifauna reports: 'A quiet and fine early morning in late May or early June in an ungrazed Radnorshire Sessile Oak woodland with a good Bilberry or heather ground layer and a healthy, well-developed understorey is probably the county's best birdwatching experience.' Such woodlands are gems, they support the greatest variety and density of breeding birds not just in the county but probably anywhere in the British Isles. As elsewhere, many of the mainly oak woodlands have been replaced by vast stands of conifers, though it is not all bad news for here Long-eared Owls, Nightjars, Crossbills and Siskins have prospered.

Some 54 miles of the River Wye's 155-mile journey to the sea is either within the county or forming part of the boundaries with Breconshire and Herefordshire, while other major rivers include the Ithon and Teme, the latter on another section of the eastern county boundary. The Afon Elan on its journey has been impounded by no fewer than five mighty reservoirs before it reaches the Wye, while the Claerwen Reservoir and its feeder river sits astride the border with Ceredigion. Natural lakes are small and not noted for their birds – indeed, the most watched open-water site in the county is in the heart of Llandrindod Wells itself.

130 ELENYDD

OS Landranger Map 147
OS Explorer Map 200
Grid Ref: SN824704
Websites:
https://www.elanvalley.org.uk/visit/visitor-centre
https://www.first-nature.com/waleswildlife/e-nnr-claerwen.php

Habitat

These lonely uplands on the Breconshire–Cardiganshire–Radnorshire border, an area of about 137 square miles (220 sq km), give rise to numerous streams which chatter down steep valleys to feed the Claerwen and Elan rivers, tributaries of the Wye, which the Irfon in the south will eventually join, while those to the west gradually merge to form the Afon Teifi on a journey to the sea in Cardigan Bay. The group of five reservoirs: Craig Goch, Penygarreg, Garreg-ddu, Caban-coch and Claerwen, the last two shared with Breconshire, supply water to Birmingham and the West Midlands, and are a renowned tourist attraction in central Wales. All can be viewed easily from public roads. Some coniferous plantations surround the lower reservoirs, but as one proceeds up the valleys these are left behind and the uplands, which rise to 1,800 feet (550 m) at Bryn Eithiniog, are a great wilderness of largely heath and blanket mire. To the south between the Claerwen and Caban-coch reservoirs is Gors Goch – here more bog-mosses. Some 1,950 acres (789 ha) of this upland wilderness north of the Claerwen is a National Nature Reserve managed jointly by the Elan Valley Trust and Natural Resources Wales, the Claerwen having the largest area of blanket bog in the central Wales uplands. Grazing has been reduced to protect some of the plants and to provide more cover for ground-nesting birds like Snipe and Curlew.

Species

The density of Ravens on these wild uplands is perhaps higher than recorded anywhere else in the world, with a breeding pair every 3.03 square miles (4.9 sq km) and as many again non-breeding birds. Their deep kronking calls are the voice of this remote region, which is also notable for several other breeding birds, including small numbers of Teal, Hen Harrier and Merlin. The numbers of breeding waders have declined dramatically over the last 50 years: Golden Plover from 30 pairs in 1976 to just eight in 2010, Lapwing from 26 pairs in

1970 to just a single pair at best by 2010, Dunlin from 13 pairs in 1976 to five in 2009, Curlew from 18 pairs in 1976 to less than five by 2004 and Redshank from five pairs to none. The number of breeding Snipe, however, appear to remain stable at around a dozen pairs, as do the number of breeding Red Grouse at about eight to 15 pairs. Ring Ouzels have declined but other upland breeding birds still do well and include Red Kite, Tree Pipit, Whinchat, Stonechat, Wheatear, Linnet and Reed Bunting. The woodlands close to the reservoirs may proffer Buzzard, Goshawk, Mistle Thrush, Redstart, Blackcap, Chiffchaff, Wood Warbler, Pied and Spotted Flycatchers, Siskin, Lesser Redpoll and Crossbill.

Diving ducks occur in small numbers, mostly Tufted Duck and Goldeneye, scarce visitors having included Ferruginous Duck, Common Scoter and Smew. Goosander first nested in the county in 1972, when nine young were seen in the Elan Valley, numbers build up out of the breeding season to about 70. Mandarin first recorded in the county in 1985 nested that same summer and soon established a tiny population including nest sites at both the Caban-coch and Garreg-ddu Reservoirs. Common Terns very occasionally arrive at the reservoirs while on late summer passage and Little Tern has been noted. If the water levels are low, it is worth looking for migrant waders, and there are usually Common Sandpipers as they breed along the shore. Ospreys now seem to be regular, if uncommon, spring and autumn passage migrants along the Wye, with birds occasionally being seen at the reservoirs, where some individuals have stayed for short periods; surely it will not be too long before they nest?

Timing

Not critical, though early morning is by far the best and should enable you to avoid the large number of visitors to the reservoir sides. Even at the peak of the holiday season, however, if you make a short walk away from the roadside you seem to have the hills to yourself.

Access

Well signed from Rhayader, with an unclassified road terminating at a car park below Claerwen Reservoir. In the Elan Valley, the road follows the shoreline of the reservoirs save for the upper reaches of Penygarreg and the whole of Craig Goch. From the head of Craig Goch reservoir you have a choice: return the way you have come; turn east over Penrhiw-wen to Rhayader; or continue west on up the valley and eventually over the hills to Cwmystwyth in Ceredigion. If you wish to explore the tops, take the ancient trackway on foot southwest from SN897716 for 4 miles (6.4 km) to the head of Claerwen Reservoir. As an alternative to the reservoirs, do not overlook the superb hanging oak woodlands at Crwch (SN923640), a circuit of Carn Gafalt (SN940646), indeed a climb to the summit. The Elan Valley Visitor Centre at SN927646 is an essential calling point before heading to the reservoirs and the high ground; it includes two electric car charging points while the Elan Valley Trail of some 5 miles round Garreg-ddu Reservoir is suitable for wheelchair users.

CALENDAR

Resident: Mandarin, Mallard, Goosander, Red Kite, Hen Harrier, Buzzard, Kestrel, Merlin, Peregrine, Red Grouse, Coot, Snipe, Skylark, Meadow Pipit, Grey and Pied Wagtails, Dipper, Stonechat, Mistle Thrush, Raven and Reed Bunting.

December–February: Tufted Duck, Goldeneye, very occasionally Pochard.

March–May: Upland waders arrive by end of March, including Golden Plover, Lapwing, Dunlin, Snipe, Curlew. Common Sandpipers return to reservoirs during April. Small parties of Dotterel are possible on the high ground as they pass on northward migration. Other summer visitors include Cuckoo, Tree Pipit, Wheatear, Whinchat, Ring Ouzel.

June–July: All residents and summer visitors present, though the upland waders have largely departed by late July.

August–November: Chance of passage waders, even terns, at the reservoirs during August and early part of September, following which winter ducks begin to appear. Snow Buntings are very rare visitors to Radnorshire, just 31 on nine occasions between 1995 and 2012, but how often do observers head for the high ground in midwinter? Hardy souls could well be rewarded for their efforts.

131 GILFACH FARM

OS Landranger Map 147
OS Explorer Map 200
Grid Ref: SN962718
Website: https://www.rwtwales.org/nature-reserves/gilfach

Habitat

Gilfach Farm has been described as caught in a time warp. This upland farm, which has not seen any changes since the 17th century, is a jewel in the landscape. In area, Gilfach extends over 418 acres (169 ha) and ranges in elevation from 788 feet (240 m) to 1,509 feet (413 m) in a small valley to the northeast of Rhayader. The patchwork of pastures, where the Marteg meets the River Wye, has fortunately missed the changes that virtually everywhere else has undergone, and was purchased in 1988 by the Radnorshire Wildlife Trust. In addition to the land, the ancient longhouse, the heart of the farm, was purchased and restored. Through this same valley, though now long disused, passed the railway line from near Newtown to Talyllyn Junction near Brecon, a single-track line of 56 miles (90 km). At Gilfach, in addition to the open line there is a tunnel of some 1,115 feet (340 m), which now provides a home for at least five species of bat. Otters and Salmon occur along the River Marteg where that most handsome of dragonflies, the Golden-ringed, can be encountered during the summer months, while over a quarter of the total number of lichens in Wales can be found growing here. If you want the full story, a superb 1995 book, *Portrait of Wildlife on a Hill Farm* by A. McBride and T. Pearce, describes Gilfach and is worth buying just for the illustrations by Darren Rees.

Species

Buzzards and Red Kites can usually be seen soaring over the valley, as can Ravens, while Merlins, Peregrines and Woodcock arrive here mostly during the winter months. Tree Pipits, Wheatears, Ring Ouzels and Whinchats are four fine summer visitors which choose these mid-Wales uplands, coming to share the home of resident Linnets and Yellowhammers, the latter now Red-listed in Wales, as numbers have seriously declined during the past twenty-five years. Common Sandpipers, Dippers and Grey Wagtails frequent the river, while Pied Wagtails are never too far away. In the woodland, you should have no difficulty in due season to find Redstarts, Wood Warblers and Pied Flycatchers. Both Tawny and Barn Owls are present in the valley, as are Sparrowhawks, Stock Doves, all three woodpeckers, Mistle Thrushes, Willow Tits, Nuthatches, Treecreepers and Jays.

Timing

Not too critical, though clearly the summer months for the greatest range of species.

Access

Leave the A470 about 2.5 miles (4 km) north of Rhayader, where there is a picnic site beside the Marteg bridge and turn east along the valley road until you re-cross the river to Gilfach itself. The unfenced land and farm trail is open at all times for general access and to walk the self-guiding nature trails, including a short easy-access trail for the less mobile from the Old Farmyard. The visitor centre is open all year round, although the best times to visit are April – November.

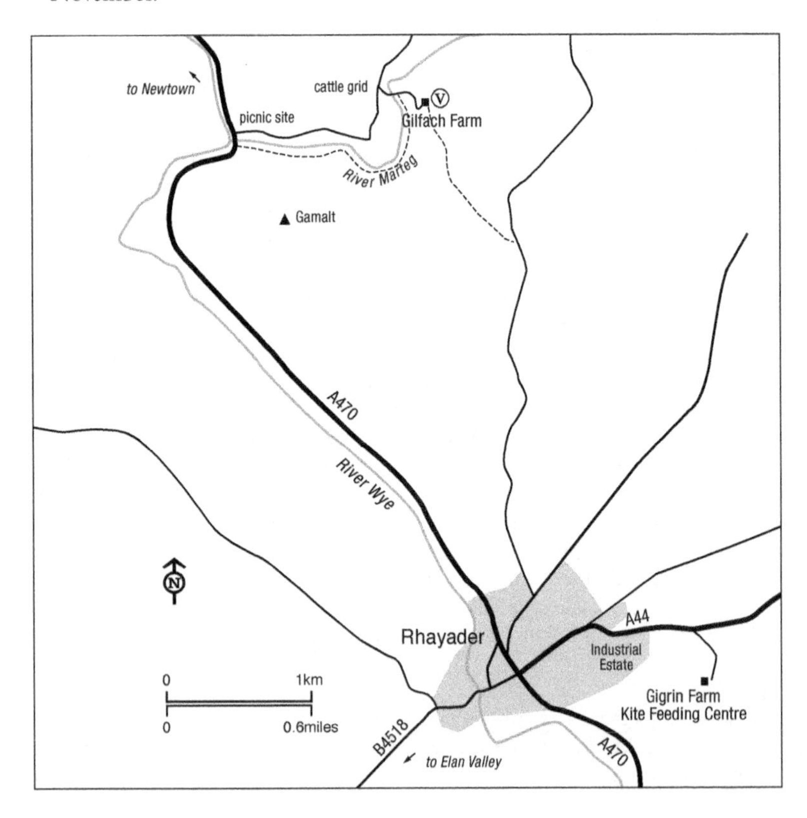

CALENDAR

Resident: Red Kite, Buzzard, Stock Dove, all three Woodpeckers, Grey and Pied Wagtails, Dipper, Wren, Dunnock, Robin, Blackbird, Song Thrush, Mistle Thrush, Goldcrest, Long-tailed, Willow, Coal, Blue and Great Tits, Nuthatch, Treecreeper, Jackdaw, Raven, Greenfinch, Bullfinch, Reed Bunting.
December–February: Merlin, Peregrine, Woodcock, Redwing, Fieldfare, Siskin.
March–May: Winter visitors departing. Arrival of summer visitors including Common Sandpiper, Cuckoo, Tree Pipit, Redstart, Whinchat, Wheatear, Spotted and Pied Flycatchers.
June–July: Breeding season in full swing, though by the end of period most young birds will have fledged.
August–November: Departure of summer visitors, early October and the first Redwings and Fieldfares should be arriving, possibility of Woodcock which usually arrive in early November, especially if the weather is cold.

132 GIGRIN FARM

OS Landranger Map 147
OS Explorer Map 200
Grid Ref: SN980678
Website: https://www.redkitefeeding.co.uk/

Habitat

The Gigrin is a family-run upland sheep farm of approximately 160 acres (65 ha) immediately southeast of Rhayader, where the land rises from 700 feet (213 m) to 1,200 feet (366 m). The Powell family, who have farmed here since 1954, had already developed a nature trail and begun to open parts of the property to the many visitors interested in wildlife who come to the Elan Valley. In 1992/93 Gigrin Farm became the Official Red Kite Feeding Station, following a request from the RSPB. It was hoped that a public Red Kite feeding station would reduce the number of young birds dying over the winter, while also drawing people away from nesting sites where undue disturbance had a negative impact on nesting. Numbers increased from a handful in the early years to over 400 in recent years. Now there are four large hides, and an interpretative centre, offering visitors superb opportunities to see Red Kites at close quarters. Kite feeding takes place every day of the year at 2pm GMT (3 pm BST).

Species

At the end of the 19th century, the outlook for the Red Kite in Great Britain was bleak. It had already disappeared from England and Scotland, while the handful that remained in the remote valleys of mid-Wales faced at best an uncertain future, at worst extinction. As one chronicler of Red Kite fortunes was subsequently to write: 'The number of young successfully reared was certainly heartbreakingly small, and nest after nest was plundered so that the population sank to its very lowest point; at worst the number of individual birds may have fallen to as few as nine or ten, but is more likely to have been between 12 and 20.' Fortunately, salvation came just in time when naturalists began to provide protection, and shortly afterwards a Kite Protection Fund was established with the support of the British Ornithologist's Club and then the RSPB. Despite considerable efforts to protect the small number of nests, the population never exceeded 10 breeding pairs until 1948 when 11 were noted, though in most years only a handful of young fledged and nest robberies continued. However, numbers slowly grew and, in 1970, 26 pairs fledged 17 young; in 1980, 43 pairs fledged 27 young; and in 1989, 69 pairs fledged 47 young – growth that continued through the 1990s, so that by the last year of the century numbers had reached over 200 pairs. This represents a triumph of resolution, determination and considerable personal sacrifice, though not, it must be said, without at times major personal differences between some of those involved in directing the protection measures.

It was only in 1971 that responsibility for protecting and monitoring this most precious of birds was taken in hand jointly by the Countryside Council for Wales (now Natural Resources Wales) and the RSPB. Almost a quarter of a century was to pass, and then, with the number of pairs in Wales for the first time exceeding 100, both organisations decided in 1993 to cease their direct involvement and support. This move caused great disappointment among the watchers who, to ensure the monitoring and protection continued, founded in 1994 the Welsh Kite Trust, an independent charity whose five trustees had been at the heart of Red Kite conservation in Wales for over four decades. Numbers of Red Kites have continued to prosper, so that the population in Wales now exceeds 500 pairs and the species is nesting in all counties. The successful work of the Red Kite Trust and all those across many earlier decades could not be more complete.

The number of Red Kites to be seen at Gigrin varies with the season, the highest counts occur during the winter months when the feeding station is in full swing and over 400 birds can be present. The feeding station also attracts large numbers of corvids, with sometimes up to 100 Ravens together with Carrion Crows and Jackdaws providing an equally remarkable spectacle. To support such gatherings, about 60 pounds of meat is put out each day, though this can easily double during periods of hard weather. Although the main feeding is during the winter, it continues on a small scale during the summer, when up to 20 Red Kites can be seen daily. The interpretative centre has a remote camera which provides close-ups of the feeding area for those who wish to watch in comfort, while a touchscreen computer adds further insight into the world of the Red Kite and the other raptors of mid-Wales.

Although Red Kites and Ravens are clearly the stars of the show, other birds of upland Wales such as Buzzards and Kestrels can usually be seen; indeed, over 90 species have been recorded at Gigrin. At the time of writing, ambitious plans include the creation of a wetland area with several lagoons which will provide a further dimension to this exciting location.

Timing

All year, though October to March for the peak of activity at the feeding station.

Access

Follow the signs from to Gigrin Farm which is just east of the A470 on the southern edge of Rhayader at SN980678 and open daily from 10am to dusk. Admission charge. Further information on the website or from Chris Powell, Gigfrin Farm, Rhayader, LD6 5BL (tel: 01597 810243).

CALENDAR

Resident: Red Kite, Sparrowhawk, Buzzard, Kestrel, Tawny Owl, Meadow Pipit, Grey and Pied Wagtails, Robin, Blackbird, Mistle Thrush, Magpie, Jackdaw, Carrion Crow, Raven, Chaffinch.
December–February: Merlin, Peregrine, Fieldfare, Redwing, Brambling, Linnet.
March–May: Summer visitors including Cuckoo, Tree Pipit, Whinchat arrive.
June–July: Breeding season coming to an end, with big movements of Swifts, Swallows and martins over the hills.
August–November: Summer visitors quietly slip away. Kite feeding commences in earnest in October and numbers rapidly build up. Autumn passage of Skylarks, pipits, thrushes, Starlings and finches from mid-October onwards.

133 WYE-ELAN WOODS

OS Landranger Map 148
OS Explorer Map 200
Grid Ref: SN968657

Habitat

The woodlands here are on the steep valley sides close to where the Elan joins the Wye south of Rhayader. The river forms the boundary with Breconshire; some of the reserve, including Carn Gafallt, is to the west and therefore in the neighbouring county. At present, the holding comprises the woodlands of Dyffryn, Cwm yr Esgob, Glanllyn and the heather-dominated plateau of Carn Gafallt, a total of 1,037 acres (420 ha).

There is a rich flora, while the woodland butterflies include Purple Hairstreak, Pearl-bordered, Small Pearl-bordered and Silver-washed Fritillaries. Badger, Fox, Polecat in the woodlands, though the latter is often only recorded as a road casualty, while Otters frequent the river.

Species

All woodland birds of western Britain are to be found here, with many populations, particularly those of Woodcock, Redstart, Wood Warbler and Pied Flycatcher, at a high level. Other species include Blackcap, Garden Warbler, Willow Warbler, Spotted Flycatcher and Jay. Ravens breed and regularly announce their presence as they are seen flying along the valley sides or across the uplands. Siskins breed in the valley, and there are usually Redpolls about as well. It is well worth searching for Hawfinches, while another species to look for is Tree Sparrow, a small flock has wintered. Other winter visitors include often numerous Fieldfares, Redwings and Bramblings, and of rarer visitors at such times both Waxwing and Great Grey Shrike have been noted.

Leave the woods behind, and on the approaches to the plateau the birds encountered include Skylark, Tree Pipit, Whinchat, Wheatear, Ring Ouzel and Mistle Thrush. At least eight species of raptor occur here, including Red Kite – this one of its last strongholds a century or more ago – Hen Harrier, Goshawk, Sparrowhawk, Buzzard, Kestrel, Merlin and Peregrine. The nearby feeding station (see Site 132 Gigrin Farm) means that good numbers of Red Kites and Ravens can be seen overhead in the valley during the winter months. The river, too, is a special feature, with Goosander, Common Sandpiper, Kingfisher (most

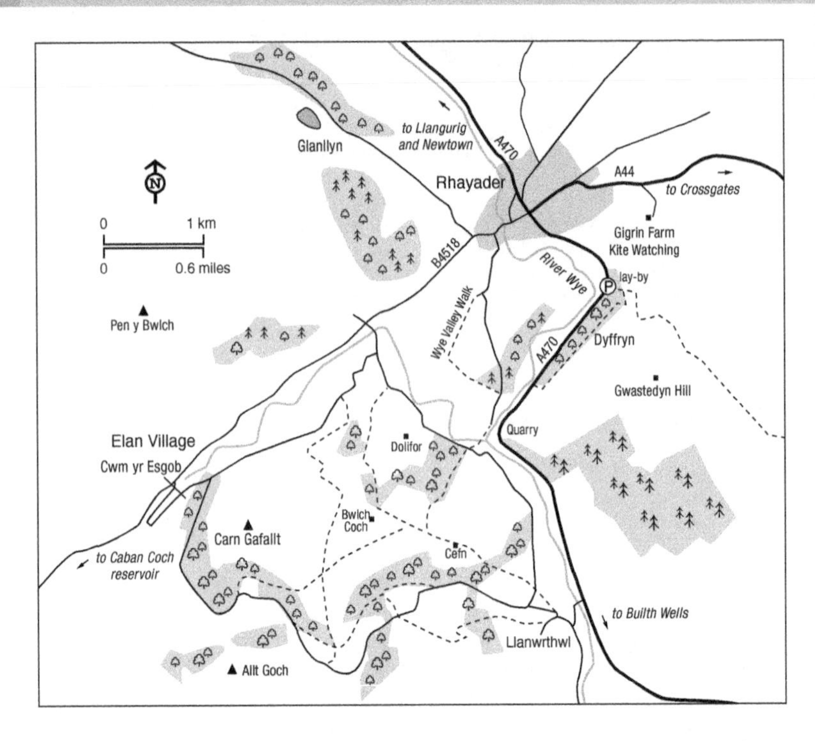

of Radnorshire's Kingfishers are to be found along the Wye), Grey Wagtail and Dipper. More unusual visitors to the valley or the surrounding uplands have included Marsh Harrier, Osprey, Hobby, Bar-tailed Godwit, Whimbrel, Little Stint, Wryneck, Yellow Wagtail and Firecrest.

CALENDAR

Resident: Grey Heron, Goosander, Red Kite, Goshawk, Sparrowhawk, Buzzard, Kestrel, Peregrine, Red Grouse, Pheasant, Woodcock, Stock Dove, Woodpigeon, Tawny Owl, Kingfisher, Green and Great Spotted Woodpeckers, Skylark, Meadow Pipit, Grey and Pied Wagtails, Dipper, Wren, Dunnock, Robin, Blackbird, Song and Mistle Thrushes, Goldcrest, Long-tailed, Marsh, Willow, Coal, Blue and Great Tits, Nuthatch, Treecreeper, Jay, Raven, Chaffinch, Greenfinch, Goldfinch, Siskin, Redpoll, Yellowhammer.

December–February: Fieldfare, Redwing, Brambling.

March–May: Chiffchaff the first summer visitor to arrive, usually before end of March, though not so frequent in this part of Wales as one might expect; other arrivals as April proceeds are Common Sandpiper, Cuckoo, Tree Pipit, Redstart, Whinchat, Garden, Wood and Willow Warblers, Spotted and Pied Flycatchers. Male Goosanders depart during May.

June–July: Goosander broods on river. All residents and summer visitors about, though many not very evident by end of July.

August–November: Gradual and largely unnoticed departure of summer migrants. Fieldfare and Redwing together with a few Bramblings from mid-October. Resident tits, often joined by Goldcrests and Treecreepers, forage in bands through the woods, so some areas temporarily devoid of birds while in another spot they seem to be everywhere – just one of the attractions of the winter woodlands.

Timing
Not critical, but a dawn visit in May or early June must be the aim.

Access
Access to Dyffryn Wood is possible at all times from a lay-by at SN979672, on the A470 just south of Rhayader. The road closely follows the river through the gorge and provides possibilities for observation at several points. Carn Gafallt is also open at all times; footpaths climb from the minor road between Llanwrthwl and Elan Village.

134 MAELIENYDD

OS Landranger Map 148
OS Explorer Map 200 and 254
Grid Ref: SO130709

Habitat
Extending northeast from Ithon Valley is the ridge of Maelienydd, rising to 1,368 feet (417 m) at its highest point but much of the area being no more than

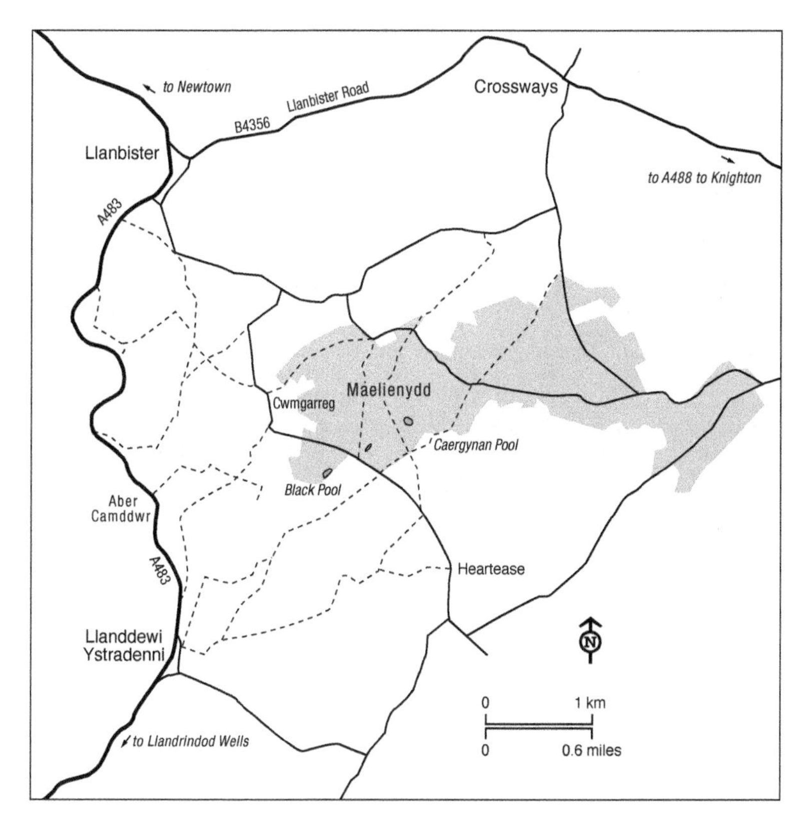

950 feet (290 m) in altitude. This is one of the few reasonably sized areas of upland rough grazing remaining in Radnorshire, a bonus feature being the presence of several small pools and adjoining marshy areas. Habitats such as this have been rapidly disappearing as improvements for agriculture have taken place, and in addition to Maelienydd the only significant ones that remain are those at Penybont Common and Rhosgoch, the latter a National Nature Reserve. Although major routes are not too far away, this is a lonely upland, though Cistercian monks were not deterred from building their monastery at Abbey Camddwr where the last native Prince of Wales 'Llewellyn the Last' rests, actually without his head which was taken away by the English.

Species

A handful of Teal nest and there are usually a few at the pools, where they are joined by Mallard. Red Kites hunt over these uplands, though Buzzards, and during the winter Hen Harriers, are more likely to be seen. Ravens are almost a constant companion as you explore the area. Good numbers of Golden Plover occur in winter, but sadly there seem to be no recent breeding records. Large flocks of Lapwing gather each autumn, when up to 1,000 birds have been recorded. Waders noted at the pools include Little Stint, Curlew Sandpiper, Dunlin, Curlew, Black-tailed and Bar-tailed Godwits, Spotted Redshank, Redshank and Greenshank. Among the passerines, look out for Skylark, Meadow Pipit, Grey Wagtail, Whinchat, Stonechat, Wheatear and Reed Bunting.

Timing

Not critical, but early morning is best.

CALENDAR

Resident: Mallard, Teal, Red Kite, Buzzard, Raven, Skylark, Meadow Pipit, Grey Wagtail, Reed Bunting.

December–February: Hen Harrier, Merlin, Peregrine, Golden Plover, Snipe.

March–May: Cuckoo, Tree Pipit, Whinchat, Wheatear, Whitethroat are among the summer visitors, while Swallows are not infrequent flighting over the high ground.

June–July: All residents and summer visitors present. In hot weather, large numbers of Swifts may hunt for insects high over the uplands.

August–November: Summer visitors depart, Wheatears may remain until late September, even early October in some years. Passage waders throughout August and September, Ringed Plover, Dunlin, Curlew and Green Sandpiper being the most frequently recorded. From mid-October winter visitors include Fieldfare and Redwing.

135 LLANDRINDOD WELLS LAKE

OS Landranger Map 147
OS Explorer Map 200
Grid Ref: SO063605

Habitat

As the name suggests, the story of Llandrindod Wells is linked with the health-giving waters, known from at least Roman times when there was an important settlement at Castell Collen, a short distance to the north-west. The town developed during the mid-18th century as a spa, with the lake, dug in a marshy hollow, becoming a much-loved part of the community. Today the Lake is known as a Carp fishery and for its Great Crested Newts, Frogs and Toads – the latter moving here each spring in prodigious numbers – while Otters are regular visitors. On sunny days Common Blue Damselflies can be seen swarming over the open water while Southern Hawkers compete for territory around the lake edge.

The wildlife importance of the lake was recognised by its designation in 2010 as a local nature reserve, but alas this seems pretty meaningless as Powys County Council who own the site have allowed boating, save for one small corner, and more fishing including the use of floating ground bait. Meanwhile, Brown Rats abound, feasting on the grain used to feed the ducks. Small wonder that, from a wildfowl point of view, Llandrindod Wells Lake is but a shadow of what it once was.

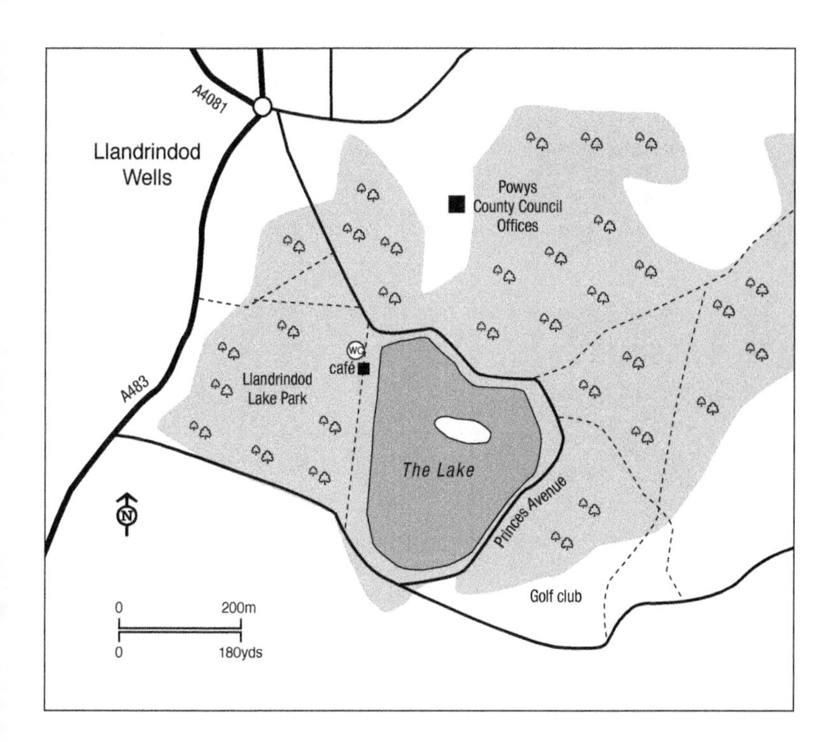

Species

The lake is good for wintering ducks including up to 50 Goosanders which start to arrive in November, some of which take advantage of the offerings of bread from passers-by. Mute Swans are present, while ducks have included Wigeon, Mallard, Tufted Duck and occasionally Gadwall, Pintail, Pochard and Goldeneye. Cormorants and Little Grebes both occur, while one or two pairs of Great Crested Grebe nest each year, together with several pairs of Coots and Moorhens. The only record of a Slavonian Grebe for the county is one that remained at the lake for four days in February 1996. Common Sandpipers occur on passage, and a Purple Sandpiper one day in March 1987 was the first for the county. Ospreys make occasional visits, one being observed in June 2005 to catch a large Carp – much to the delight of those watching from the lakeside café. Wintering Black-headed Gulls are very evident and a few pairs have nested, while Kittiwakes have made just two brief visits. Sightings of terns are equally remarkable: Black, Sandwich, Common and Arctic Terns have all been recorded. The lakeside woods include Great Spotted Woodpecker, Song Thrush, Redstart, Blackcap, Chiffchaff, Wood Warbler, Nuthatch, Treecreeper, Pied and Spotted Flycatchers, while just beyond Tree Pipit, Garden Warbler, Whitethroat and Linnet are likely to be found.

Access

There is ample parking around the lakeside.

CALENDAR

Resident: Canada Goose, Mallard, Great Crested and Little Grebe, Heron, Coot, Moorhen, with Great Spotted Woodpecker, Great and Blue Tits, Treecreeper and Nuthatch in the woods.

November–February: Goosander, Goldeneye, Tufted Duck, Pochard, Cormorant, Black-headed, Common and Herring Gulls.

March–June: Black-headed and Herring Gulls, occasional passage waders, Blackcap, Garden and Willow Warblers, Chiffchaff, Spotted and Pied Flycatchers in the woodland surrounding the lake.

July–October: Black-headed Gulls and occasional waders, tit flocks among the trees.

136 GLASBURY

OS Landranger Map 161
OS Explorer Map OL13
Grid Ref: SO180393

Habitat

The Wye forms the boundary between Radnorshire and Breconshire, and shortly after Erwood the river at last squeezes from its gorge to commence a meandering course across the water-meadows to the west, and then to the east, of Glasbury. The remnants of long-disused channels can still clearly be seen, while there are several pools which retain water in all but the driest summers. Nearby at Pwll Patti (SO166393) is a roadside hide overlooking an oxbow lake surrounded by marsh, which was the old course of the River Wye.

The area floods in winter and attracts good numbers of wintering wildfowl and swans.

Species

This is probably the best spot in the county for wildfowl and waders, which share both banks of the river and so two counties. Up to 10 Little Grebes winter here and are occasionally joined by Great Crested Grebes. Up to 40 pairs of Grey Herons nest in Radnorshire; they are often to be seen feeding on this section of river or flying past. Several pairs of Mute Swans breed, while in late summer a flock of up to 100 may congregate. Whooper Swans also occur, despite a preference for upland lakes and reservoirs. Canada Geese only breed in Radnorshire in very small numbers, but during the winter, a flock moves to feed on the pastures around Glasbury where it can number as many as 250 birds. Normally the Wigeon flock is up to 100 strong, but cold weather further east results in a rapid rise in numbers. Teal frequent the meanders, but the most numerous duck is Mallard, with flocks of up to 200 having been reported. Goldeneyes are regular throughout the winter, while hard weather brings Pochard and Tufted Duck and occasionally Scaup and Long-tailed Duck. Common Scoter and Garganey have been recorded on several occasions. Goosanders are regularly seen, with family parties totalling up to 100 females and ducklings having been recorded along the course of the Wye during the summer, and in winter at Glasbury they can number 30 or so.

Common Sandpipers breed and may occasionally overwinter. Little Ringed Plover and Oystercatchers have become established as breeding birds on the Wye hereabouts, the former since 1977, the latter since 1990. Ringed Plovers

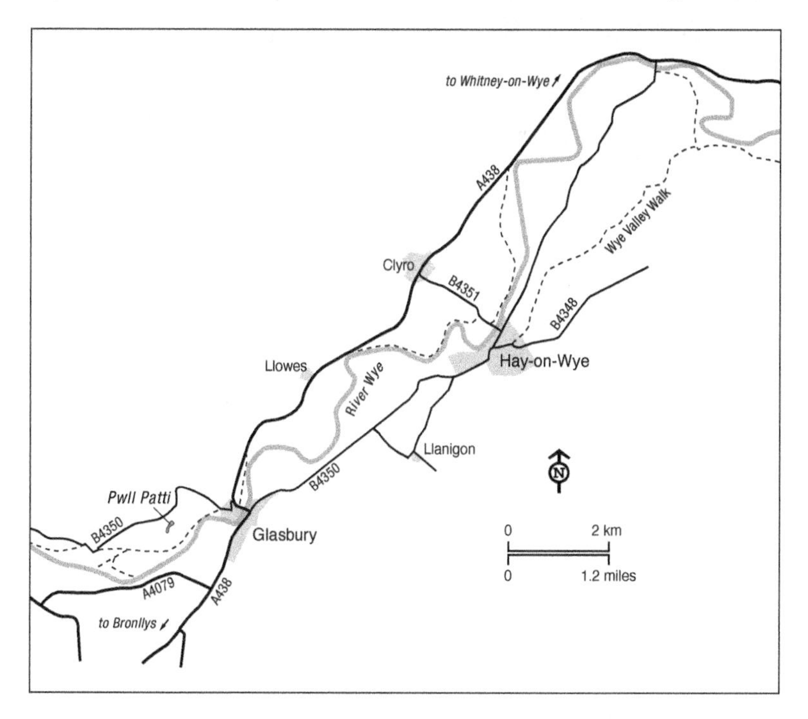

are occasionally seen, while the meadows are an attractive feeding area for several hundred Golden Plovers and Lapwings during the winter, together with up to 50 Curlews. Other waders have included Jack Snipe, Bar-tailed Godwit, Redshank, Greenshank, Green and Wood Sandpipers. In recent years Little Egrets have become regular here and both Cattle and Great White Egrets have been reported.

Timing

Not critical, though as elsewhere early morning visits are much to be preferred.

Access

Start at the bridge over the Wye at Glasbury, the A438, A4079 and B4350 roads all give good general views. On foot, you can walk north from the bridge at Hay-on-Wye for nearly 2 miles (3.2 km) following a section of the Offa's Dyke Path. To the southwest of the town, the Wye Valley Long Distance Footpath follows the river for nearly 2 miles (3.2 km) before returning to the B4350 at SO201421. The Radnorshire Wildlife Trust reserve at Pwll Patti (SO166393) has a birdwatching hide, park in the lay-by.

CALENDAR

Resident: Grey Heron, Canada Goose, Mute Swan, Mallard, Goosander, Moorhen, Coot, Kingfisher, Grey and Pied Wagtails, Dipper, Reed Bunting.

December–February: Little Grebe, Cormorant, Whooper Swan, Wigeon, Teal, Pintail, Shoveler, Pochard, Tufted Duck, Goldeneye, Golden Plover, Lapwing, Curlew.

March–May: Sand Martins arrive in late March and Common Sandpiper arrive in mid-April, by which time water levels are dropping and wintering species have departed.

June–July: The few breeding species are present, Mute Swan and Canada Goose numbers rise, chance of passage waders during late July.

August–November: Passage waders likely until well into September, including Ringed Plover, Dunlin, Redshank, Greenshank, Green and Wood Sandpipers. As autumn proceeds the winter ducks arrive, followed by Whooper Swan during November.

WEST GLAMORGAN AND GOWER (GORLLEWIN MORGANNWG AND GŴYR)

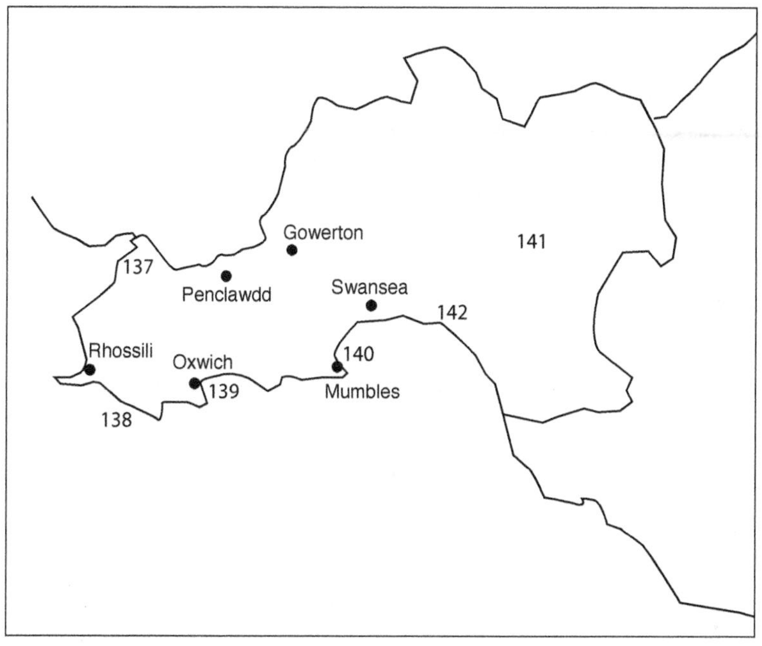

137 Burry Inlet South Shore and Whiteford Point
138 South Gower Coast
139 Oxwich
140 Swansea Bay to Mumbles
141 Cwm Clydach
142 Crymlyn Burrows & Crymlyn Bog

West Glamorgan and Gower include a mixture of urban and industrial areas, including Swansea and Neath–Port Talbot, the stunning coast of the Gower peninsula – a land set apart from the rest of Wales – and the Burry Inlet, while at no great distance inland the wooded valleys beyond Neath, with their numerous waterfalls and hanging oak woodlands, blend into the uplands of Breconshire. In many areas this has been replaced with large conifer plantations, introducing another habitat.

Gower's coastline of cliffs, bays and beaches attracts the crowds, but secluded areas are also home to wildlife. Choughs and Dartford Warblers have recently colonised the area and there are small colonies of seabirds dotted along the cliffs. Nearby, the south side of the Burry Inlet is nationally important for wintering wildfowl and waders. Swansea Bay is ornithologically well known for gulls, although a shadow of its former self. It was here that two students found the UK's first Ring-billed Gull in 1973. Subsequent observations have turned up nearly every gull on the British list, not to mention large numbers of Mediterranean Gulls.

137 BURRY INLET SOUTH SHORE AND WHITEFORD POINT

OS Landranger Map 159
OS Explorer Map 164
Grid Ref: SS447965
Websites:
https://www.swansea.gov.uk/whitefordburrowsnaturereserve
https://www.nationaltrust.org.uk/whiteford-and-north-gower
www.first-nature.com › sw-nnr-whiteford
http://www.ggat.org.uk/cadw/historic_landscape/gower/english/Gower_003.htm

Habitat

The southern coast of Burry Inlet, in contrast to the northern shore (see Site 43), is a lonely region of extensive saltmarshes, in places over 1 mile (1.6 km) wide, one of the largest areas of moderately grazed saltmarsh in Britain. Except between the villages of Penclawdd and Crofty, the road is well back from the shore, following the northern escarpment of Gower. Cockles have been dug on the Burry Inlet since Iron Age people descended from their hill camps, of which there is plenty of evidence in Gower, through the woodlands and so to the flats. Based at Penclawdd, the cockle industry was one of the most important in Wales, in 1884 said to be worth £15,000 per annum, and provided employment for a large number of women. Each day, with their donkeys, they would follow the receding tides on Llanrhidian Sands as they raked for cockles to be sent to markets all over South Wales. Few now make their living here from the industry.

To the west of the sands, beyond Great Pill, lies the massive dune system of Whiteford Burrows, extending 2 miles (3.2 km) northwards from the hamlet of Cwm Ivy and Landimore Marsh, so that its furthest point is almost halfway across Burry Inlet. The lighthouse is one of the last remaining examples of a cast-iron Victorian lighthouse. The whole of the burrows are owned by the National Trust, the first of its Enterprise Neptune purchases, part of a campaign to protect key sections of the British coastline. A large part is now a National Nature Reserve designated in 1965 and managed by Natural Resources Wales, comprising woodland and limestone grasslands, sand dunes and sandy shores bounded on the west by Carmarthen Bay, on the east by the Burry Inlet. There are so many good things about Whiteford – its remoteness, the birds, plants including Dune Gentian, Moonwort, Yellow Bird's-nest, Fen Orchid, Petalwort, Bird's-foot Trefoil, Lady's Bedstraw and Dune Pansies, and insects, common and rare, all enthral the visitor.

An exciting opportunity presented itself during the stormy winter of 2013–14, when the seawall protecting Cwm Ivy was breached and salt water poured in. Rather than rebuilding defences that had kept the land drained for hundreds of years, the National Trust decided to let nature take its course and Cwm Ivy is now a pristine saltmarsh overlooked by two bird hides.

Species

The birds that gather on the saltings and further out on the great flats are part of the same flocks observed on the Carmarthenshire side of the estuary. Careful counts include: 1,000 Brent Geese, 700 Shelducks, 5,000 Wigeon, 1,800 Teal, 3,500 Pintail, 7,000 Oystercatchers, 10,000 Knots, 1,000 Dunlins, 1,000 Black-tailed Godwits, 1,000 Curlews, 1,000 Redshanks, 500 Grey Plover and 200 Turnstones. Good numbers of Snipe winter on the saltings, a few Spotted Redshank overwinter, while Whimbrel occur on passage, mostly in the spring. Little Egrets colonised the area during the 1980s and up to 80 can be seen on the estuary and in its creeks and nest nearby. More recently the Great White Egret has also colonised the area, with between five and ten seen throughout the year, particularly around Llanrhidian; hopefully this species will breed in the not too distant future.

Occurring throughout the year, Eider is of special interest, as few occur in South Wales. First noted at Whiteford in 1919, but old fishermen said they had been here for as long as they could remember, and old fishermen have long memories. Numbers vary, and in recent years have numbered about 100, fewer than in the past when the total was about 200. The south shore of Burry Inlet is an important area for nesting Shelduck, with perhaps up to 25 pairs present. Red-breasted Mergansers are seen in small numbers in almost all months. Common Scoter are present in winter just west of Whiteford Point, part of the huge flock which spreads right across Carmarthen Bay, with small numbers entering Burry Inlet between July and April. Great Crested Grebes may be seen in all months, especially in the Llanrhidian area, with numbers rising to nearly 50 in the autumn. Much smaller numbers, usually fewer than five, Slavonian Grebes winter off Whiteford. Black-headed Gulls occur throughout the year, the handful present in early summer soon being joined by hundreds more so that by late July and into August up to 7,000 can be present; even though many will move on, normally well over 1,000 winter here. Kittiwakes occasionally move into the estuary and are more frequently seen off Whiteford Point, while

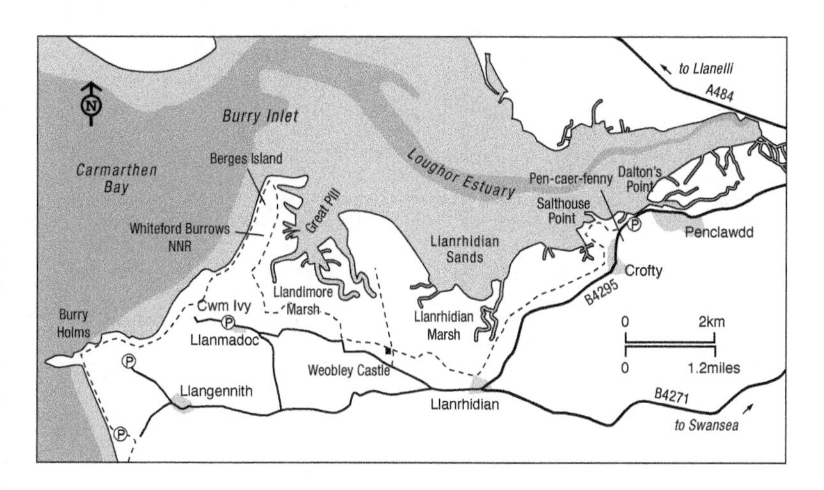

small numbers of Sandwich and Common Terns occur on passage in most years.

In winter, there is a good chance of watching raptors such as Hen Harrier, Merlin, Peregrine and Short-eared Owl over the saltings. Recent rarities include Black Brent Goose, Lesser Yellowlegs, Gull-billed Tern, Shorelark, Red-rumped Swallow and Isabelline Wheatear. A Collared Pratincole at Dalton's Point, Penclawdd on 27 May 1973 was the first for Wales, while another first was a Desert Wheatear which spent two days at the same location in November 1989.

Timing

Best just after high-water, as the tide recedes to reveal the first feeding areas for waders and before divers, grebes and sea-ducks move too far offshore. Choose a period of spring tides if you can when using the hide at Berges Island, erected in memory of Sir William Wilkinson one-time Chairman of the Nature Conservancy Council and President of the British Trust for Ornithology and inspired by the sight of three Black Terns flying past. As it was completed seven Spotted Redshanks flew past.

Access

The B4295 at Penclawdd runs close to the shore and provides a good viewing point for this section of the upper estuary, as does Salthouse Point (SS523958), an old causeway on the shore near Crofty. A minor road leads to the point from the crossroads in Pen-caer-fenny. Beyond Crofty, a minor road bears off the B4295 and follows the edge of the saltings for about 2.5 miles (4 km) until it rejoins the main road in Llanrhidian. West of here is some 4 miles (6.4 km) of saltmarshes intersected by many tidal channels, and with no prime vantage points and no good access other than by the track leading north from Weobley Castle (SS477926). Better to pass quickly on to Whiteford, parking in Cwm Ivy

CALENDAR

Resident: Cormorant, Grey Heron, Little Egret, Great White Egret, Shelduck, Mallard, Eider, Red-breasted Merganser, Common Scoter, Buzzard, Oystercatcher, Ringed Plover, Lapwing, Snipe, Curlew, Redshank, Turnstone, Great Black-backed, Lesser Black-backed and Herring Gulls, Skylark, Meadow Pipit, Stonechat, Raven, Linnet, Yellowhammer, Reed Bunting.

December–February: Red-throated and Great Northern Divers, Great Crested, Slavonian and Black-necked Grebes, Brent Goose, Wigeon, Teal, Pintail, Shoveler, Scaup, Long-tailed Duck, Goldeneye, Hen Harrier, Merlin, Peregrine, Golden and Grey Plovers, Knot, Sanderling, Jack Snipe, Black-tailed and Bar-tailed Godwits, Spotted Redshank, Common Gull.

March–May: Winter visitors depart, while passage waders are seen well into May when the last Whimbrels fly north. Wheatears usually from mid-March and from the end of the month Swallows and Sand Martins passing on most days.

June–July: Seabirds move into Carmarthen Bay and at times are seen close offshore, especially Manx Shearwaters and Gannets. Waders start to return, Green Sandpiper often one of the first, though you will need to look for this species in the creeks rather than on the open shore.

August–November: Wader passage at its peak early in period, with overwintering species also arriving. A large increase in gulls with many Common Gulls roosting on the estuary. Winter finches, along with Snow Bunting, possible from November.

where there is an information kiosk at SS437940. From here, a footpath leads north for over 2 miles (3.2 km), giving fine views of the reserve and the Burry Inlet. There is a bird hide on Berges Island, details of which are available at the kiosk. Alternatively, take the much shorter route to the Cwm Ivy wetland with its two hides.

138 SOUTH GOWER COAST

OS Landranger Map 159
OS Explorer Map 164
Grid Ref: SS417872
Websites:
Wildlife Trust of South & West Wales welshwildlife.org
https://www.nationaltrust.org.uk/rhosili-and-south-gower-coast
https://naturalresources.wales

Habitat

In the far south-west of Gower, beyond the great sweep of Rhossili Down, 632 feet (192 m), with the village of Rhossili – home of Seaman Evans who died on the return from the South Pole with Captain Scott – nestling at its foot, lies Worms Head. This narrow promontory is connected to the mainland by an extensive rocky causeway, revealed as the tide recedes, which allows access for the more adventurous and agile for several hours either side of low-water. The Head is some 37 acres (14.8 ha) in extent and comprises four small hills, rising like a dragon's serrated back (the name 'Worm' in Norse means dragon or serpent). Much of the headland is covered in Red Fescue, this providing ample food for the sheep which graze the table-top plateau of the Inner Worm. Maritime plants include Spring Squill, Sea-beet, Golden and Rock Samphire, Tree Mallow, Rock Sea-lavender, Sea Stork's-bill and Buck's-horn Plantain.

The mainland coast runs east some 4 miles (6,4 km) to Port Eynon with access, including some steep climbs and descents, by way of the Wales Coast Path. Much of the coastline is owned by the National Trust, and is the most dramatic cliff coast in Glamorgan. One section is a National Nature Reserve, while another at Deborah's Hole is owned by the Wildlife Trust of South & West Wales. A short distance to the east in 1823, geologists from Oxford University discovered in the Paviland Cave part of a human skeleton covered in red ochre which quickly became known as the 'Red Lady of Paviland'. Later it became apparent that the remains were those of a young man who died around 33,000–34,000 years ago. The cave was his home, his hunting ground a vast plain stretching to the south, for sea levels were far lower in those days.

Species

The seabird colonies are the most important feature and are what attract most birdwatchers to this spectacular spot. Some 140 pairs of Kittiwakes nested at Worms Head in 1986; since then there has been a steady decline, with 114 pairs in 1991, 119 pairs in 2001, 20 pairs in 2010, 11 pairs in 2018. There are also 170 pairs of Guillemots and a small number of Razorbills. Puffins still cling on, with just seven birds in 2017. Passage waders frequent the rocky causeway, and whatever the season, even at the height of the summer, you will find Oystercatchers and Turnstones there.

Small numbers of Herring, Lesser and Great Black-backed Gulls nest on the coast along with several pairs of Fulmar, Shag, Meadow and Rock Pipits, Whitethroat, Stonechat and Jackdaw. Choughs recolonised the area in the 1990s, after an absence of nearly 100 years, and now five pairs nest, while the post-breeding flock can be as many as 21. Another colonist at that time was Dartford Warbler, first recorded in Wales on South Gower in 1969 and becoming established in several counties from 1992 onwards, with 20 pairs on Gower in 2015 but only five pairs in 2017 and then two subsequently due to severe winter weather.

Common Scoters, outliers from the immense flocks in Carmarthen Bay, often move close in under the Worm, and sometimes 2,000 congregate in Rhossili Bay. Eiders also wander here, though the main flock is at Whiteford 8 miles (12.8 km) to the north. During the winter months Common Scoter, Eider, Red-throated Divers and numerous auks can be seen offshore. Great Northern and Black-throated Divers may also turn up.

The scrub and wooded valleys all along the coast are a magnet for passage migrants. Recent observations have found Yellow-browed Warblers on an annual basis and other rarities including Pallas's Warbler, Booted/Sykes Warbler, Common Rosefinch, Ortolan and Little Buntings.

Port Eynon Point, the most southerly point on the Gower coast and another Wildlife Trust of South & West Wales nature reserve, is the best site for seawatching. Careful observations have recorded up to 10,000 Manx Shearwaters passing in a day in summer, while an autumn day can witness up to 30 Storm Petrels and smaller numbers of Sooty and Balearic Shearwaters, terns, Kittiwakes and skuas, mainly Arctic and Great, although Pomarine and Long-tailed have been seen. A feature easily overlooked here is Culver Hole – 'culver' derives from the Old English word 'culfre' meaning dove or pigeon. Described as the oddest of all dovecotes, a massive cleft in the cliffs is sealed by a 60ft-high (20 m) stone wall with several large openings, and inside are some 30 tiers of nesting boxes for pigeons. These provided a most welcome food source in earlier centuries and in local stories were a hiding place for smuggled goods.

Timing

Visits to Worms Head are entirely dependent on the tide if you wish to cross from the mainland. A board telling visitors of safe crossing times is provided daily by the Coastguard observation building on the cliff above the causeway. Cross as the tide begins to recede from the rocky causeway and you should then have several hours ashore. For those wishing to view from a distance, the cliffs opposite the Head offer a fine vantage point.

Access

The B4247 ends in Rhossili village, where there is an ample car park. From here, a wide cliff path leads southwest to Rhossili Point, where the old Coastguard lookout is now a small information kiosk. Nearby, a path descends to the rocky causeway, and once on the Head itself there is a footpath to the westernmost point, though visitors are asked, in the interests of the seabird colonies, not to proceed to the outer section between mid-March and July. Worms Head and the whole 6 miles (9.6 km) of cliff stretching eastwards to

Port Eynon (SS468844) is a nature reserve managed by the Wildlife Trust of South & West Wales, the National Trust and Natural Resources Wales. There is a coastal footpath, and this must be one of the best sections of cliff coast in the whole of south-west Britain, with some superb flowers including Yellow Whitlow-grass which flowers between March and May on the cliffs and old walls of Gower, its only location in Great Britain, while not forgetting the Choughs.

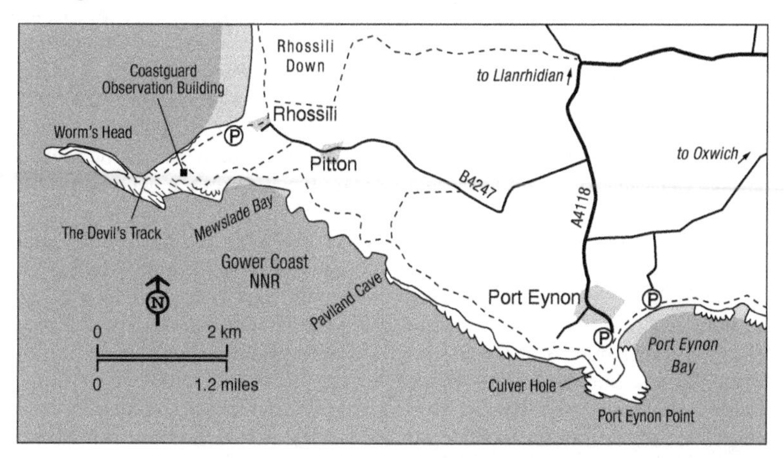

CALENDAR

Resident: Cormorant, Shag, Eider, Common Scoter, Kestrel, Peregrine, Oystercatcher, Turnstone, Lesser Black-backed, Herring and Great Black-backed Gulls, Rock Pipit, Stonechat, Chough, Jackdaw, Raven.

December–February: Red-throated and Great Northern Divers, Purple Sandpiper. Fulmars and Kittiwakes return to their colonies and both Guillemots and Razorbills make occasional visits to the cliffs.

March–May: By early May all seabirds have commenced nesting. Passage migrants include Whimbrel, Common Sandpiper, Wheatear, Black Redstart and Ring Ouzel.

June–July: Manx Shearwater, Gannet and occasional terns offshore. By end of period Guillemots and Razorbills have left the cliffs, having completed their breeding cycle.

August–November: Kittiwakes leave the cliffs in mid-August, Fulmars during early September. Seabird passage possible offshore until late September. Autumn passage may produce Whitethroat, Lesser Whitethroat, Sedge and Reed Warbler, Yellow-browed Warbler, Redstart, Whinchat, Firecrest.

139 OXWICH

OS Landranger Map 159
OS Explorer Map 164
Grid Ref: SS505873
Website:
https://naturalresources.wales/days-out/places-to-visit/south-west-wales/oxwich-national-nature-reserve/?lang=en

Habitat

From Pennard Pill westwards for 2.5 miles (4 km) is the broad, shallow sweep of Oxwich Bay, the second-largest bay on Gower, backed by dunes which in turn trap an extensive area of saltmarsh, freshwater marsh and lagoons, all of which drain to the sea by way of Nicholaston Pill. The dunes show all stages of succession, from the embryonic dunes of the shoreline, the erosion systems and blowouts of the seaward faces, through to the oldest dunes where trees such as birch and oak have become established. Wynford Vaughan Thomas described the great park round Penrice Castle, with the deep woodlands which encircle it and the marshlands at its foot, as a landscape composition hard to beat in the whole of South Wales. There are numerous fine plants to be found here, including such rarities as the Welsh Gentian and a maritime variety of Round-leaved Wintergreen. Among the unusual insects is a handsome Strandline Beetle *Nebria complanata*; one wonders how these fare when the beach debris, beneath which they shelter, is tied up or burnt.

Oxwich National Nature Reserve occupies 714 acres (289 ha); for the birdwatcher its fens and marshes are of great interest. They were originally reclaimed in the 16th century, and for over 300 years provided rich summer grazing. Fishponds dug in the early 19th century remain, the largest, of some 18 acres (7.2 ha), aptly named Serpentine Broad. The reed swamp extends for some 100 acres (40 ha), one of the largest areas of this habitat in south-west Britain. Fifteen species of dragonfly and damselfly have been recorded, and plants such as Mare's Tail, Common Bladderwort, Flowering Rush, Bogbean and Marsh Marigold thrive. Otters are resident, and patient watching from the hide can be most revealing.

Species

Although Grey Herons no longer breed in Penrice Woods, close to Oxwich, they are always to be seen flying over, or wading in the shallows at the pool margins. Little Egrets first nested on Gower in 2005 and Bitterns are regular winter visitors, while Little Bittern and Purple Heron have been recorded and possibly bred in the past. Cormorants come to fish from their colonies further west in Gower, and are often seen perched in the trees. Another scarce reedbed species, Bearded Tit, was first recorded here in 1965, and bred in small numbers up to 1980, but unfortunately there have not been any recent sightings. Although rarely proved, it is almost certain that Water Rails nest here and they can certainly be heard throughout most months. Moorhens and, especially, Coots are much more obvious and are seen on the open water. Although Kingfishers do not breed, they are regularly seen, especially during the late summer and autumn, when they move to coastal localities for the winter months. Marsh Harriers regularly pass this way on migration. The populations of Sedge and Reed Warblers have stimulated much study in recent years, an estimate putting the population of the latter as high as 10 pairs per acre (25 pairs per ha), making a total population of around 1,000 pairs. There are perhaps as many as 10 pairs of Cetti's Warbler too, their song well described as 'the most astonishing loud song among small European birds'.

Small numbers of Mallard breed, while Teal, Shoveler, Pochard and Tufted Duck may have done so in the past. Shelduck broods are occasionally seen, while Garganeys are scarce but regular passage migrants in early spring. Wader records tend to be few, a lack of muddy margins being the main reason. The

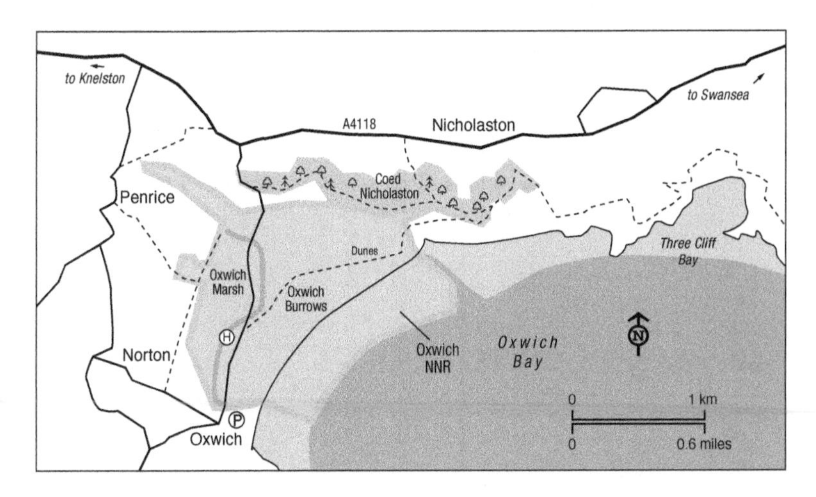

drumming displays of Snipe in spring are sadly a thing of the past, though many come to winter here as do Jack Snipe, while the passage of too many feet on the beach has meant the end of breeding Ringed Plovers. Yellow-browed Warblers have been recorded in October, while a Savi's Warbler in May 1987 was the first for Glamorgan.

Timing

Early mornings are by far the best, especially during the summer months, when Oxwich village, car park and beach can be packed with visitors.

Access

The A4118 is the main road west from Swansea through south Gower. Leave this at SS501883 and head for Oxwich village; the road at one point cuts through the freshwater marsh, and for much of its length follows the boundary between marsh and sand dunes. There is a large car park beside Oxwich

CALENDAR

Resident: Little Grebe, Cormorant, Grey Heron, Shelduck, Teal, Mallard, Sparrowhawk, Buzzard, Water Rail, Moorhen, Coot, Cetti's Warbler, Raven, Reed Bunting, Green and Great Spotted Woodpeckers, Nuthatch and Treecreeper frequent the woodlands.

December–February: Bittern, Gadwall, Shoveler, Tufted Duck, Jack Snipe, Woodcock in the damp woodlands.

March–May: Ringed Plover, summer migrants arrive, Swallows and Sand Martins over the marsh from early April, Sedge and Reed Warblers in full residence and song by end of April.

June–July: All breeding species present. First autumn passage commences in late July, with Green and Common Sandpipers and Kingfisher, a chance of terns in Oxwich Bay.

August–November: Wader and tern passage during August and early part of September. Summer migrants leave, with last warblers present until mid-September, when large gatherings of Swallows and Sand Martins before their departure for winter quarters. Wintering ducks arrive and a chance of Bittern in November.

Reserve centre, and a footpath system from here through the dunes and along the edge of the saltmarsh. An excellent all-weather track leads to the observation hide which provides good views over the open water and the freshwater marshes. The walk through the woods which hug the steep western side of the bay brings one to Oxwich church, once the site of the largest rookery in Gower, beyond which you can continue to the point. Not far away, Crawley and Nicholaston Woods are also well worth visiting and contain marked footpaths.

140 SWANSEA BAY TO MUMBLES

OS Landranger Map 159 and 170
OS Explorer Map 164
Grid Ref: SS618903

Habitat

From Mumbles Head to Port Talbot, the shoreline of Swansea Bay extends for over 11 miles (17.7 km), the whole length backed by the conurbation of Swansea and its suburbs – residential and holiday in the west, the docks and industry to the east. At first glance this is perhaps not an area to excite the ornithologist, but careful observations by dedicated local observers have revealed this as a site not to be missed. The tidal flats, especially in the west, extend up to 1 mile (1.6 km) from the shore, while the docklands should not be neglected as a birdwatching haunt. At the very heart as far as birdwatchers are concerned is Blackpill (SS620907), one of the great birdwatching locations in South Wales, though, alas, the closure of two nearby rubbish tips in recent years and the cleaning of the beach has meant reduced feeding opportunities for the large number of gulls which frequent the shoreline, and so their numbers have declined.

Species

With some perception, an ornithologist early in 1970 wrote: 'There can be few more likely candidates for future addition to the British and Irish list than the Ring-billed Gull. It is abundant in North America and it winters on the Atlantic coast from New England to the Gulf of Mexico, a migratory pattern that would seem to render it liable to trans-Atlantic vagrancy ... It remains the only gull from the eastern Nearctic still to be recorded in Britain and Ireland' (*British Birds* 66: 115). On 14 March that same year, a birdwatcher painstakingly examining a flock of Common Gulls, its close relative, noted a paler bird that was eventually identified as an adult Ring-billed Gull. This individual, the first record for Great Britain and Ireland, remained until the end of the month. Imagine the surprise when, barely three months later, at the same spot, a first-summer Ring-billed Gull was noted. Individuals were subsequently seen on a regular basis, with up to 11 counted in some years. However, in recent years, as with so many other locations, the occurrence of Ring-billeds has declined and there has been no record here since March 2014.

Even without Ring-billed Gulls present, Blackpill is certainly the place for gull-watching, with all the familiar species present throughout the year. The flock of Common Gulls may number several hundred even in midsummer, despite there being no large colonies closer than the Clyde region of Scotland, while in midwinter 3,000 can be present, and Black-headed Gulls can be double that total. Mediterranean Gulls are now regular here, with the smallest numbers in midwinter, though on occasions up to 40 have been present with colour-ringed birds from colonies in Belgium, Holland and Hungary being identified. Little Gulls may be seen in virtually any month, either here or further west at Mumbles. No other site in Wales boasts such an impressive list of records of rare gulls – all down to the dedication of the 'gull-watchers'. The list includes Ross's, Ivory, Bonaparte's, Laughing, Franklins, Yellow-legged, Iceland, Kumlien's and Glaucous Gulls.

There is a small colony of Kittiwakes at Mumbles Pier, some 141 pairs strong in 2019, and during late summer up to 300 can on occasions can be seen at Blackpill. Skuas regularly enter Swansea Bay on passage, and Pomarine, Arctic and Great Skuas have been seen at Blackpill, or from Mumbles Head. Manx Shearwaters are sometimes sighted close inshore within the bay, and are regular during seawatches at Mumbles Head where Storm Petrels have also been recorded. Black Terns have been seen on a number of occasions, Sandwich, Common, Arctic and Little Terns are regular on both spring and, especially, autumn passage.

The flats at Blackpill are also a good haunt for waders, with 28 species having

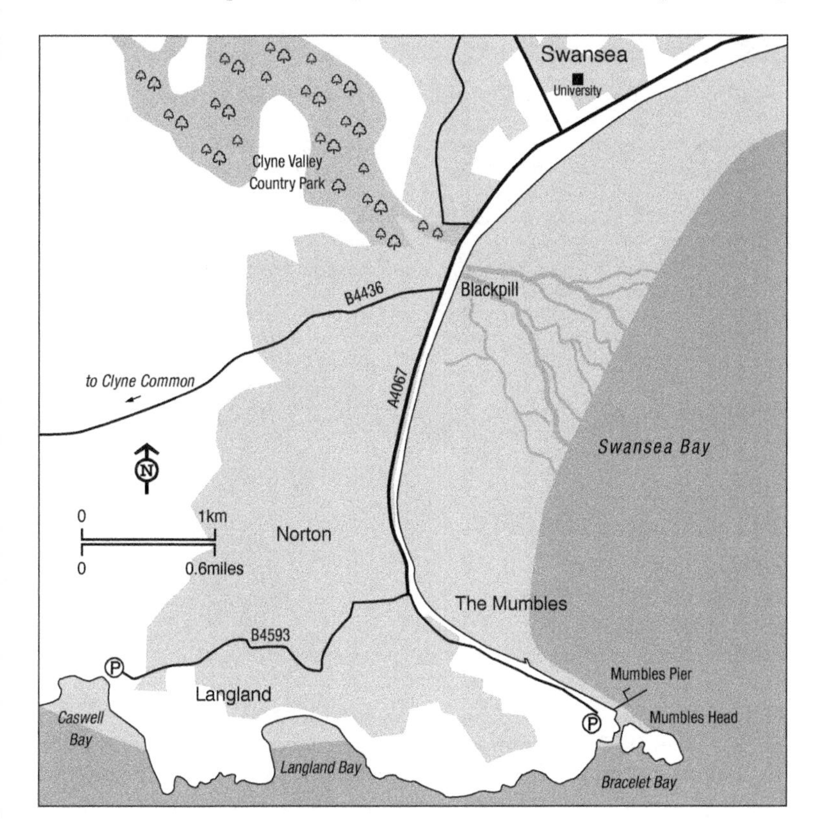

been recorded here. Oystercatchers are present throughout the year and in midwinter up to 3,500 come to feed, many frequenting the nearby Ashleigh Road playing fields when the tide is in. Ringed Plover, Dunlin, Bar-tailed Godwit, Curlew and Redshank occur throughout the year, though in midsummer only a handful may be present. Dunlins are by far the most numerous, with up to 2,300 in January. Other waders regularly seen include Grey Plover, Knot, Sanderling, Curlew Sandpiper, Black-tailed Godwit, Greenshank and Turnstone. Whimbrel come through on spring and autumn passage. Among scarce species have been Avocet, Kentish Plover, Temminck's Stint, White-rumped Sandpiper, Red-necked and Grey Phalaropes.

Timing

It is essential to consult the tide tables and choose a visit for about two hours before or after high-water. At these times the birds are close inshore, whereas at other times they disperse across the flats and viewing is then at long range, for the most part with a telescope.

Access

The A4067 follows the shore west from Swansea to Mumbles. Access is possible along most of its length by crossing the disused line of the old Mumbles railway, now the Swansea bike path.

CALENDAR

Resident: Cormorant, Oystercatcher, Ringed Plover, Dunlin, Bar-tailed Godwit, Curlew, Redshank, Black-headed, Common, Lesser Black-backed, Herring and Great Black-backed Gulls.

December–February: Occasional divers and grebes offshore, Shelduck, Scaup, Goldeneye, Grey Plover, Knot, Sanderling, Greenshank, Turnstone. Wintering gulls may include Kittiwake, Mediterranean and Little with the possibility of Iceland, Glaucous and even Ring-billed.

March–May: Whimbrel, Common Sandpiper, some tern passage. Kittiwakes begin to nest on Mumbles Pier.

June–July: Wader numbers and gull numbers start to increase as July progresses, the first southward-moving terns begin to put in an appearance.

August–November: Autumn passage well underway, including Little Stint, Curlew Sandpiper, Ruff, Spotted Redshank, and as this tails off in late September winter visitors begin to arrive, including occasional Brent Goose, Wigeon, Teal, Pintail. Peregrines regularly hunt here.

141 CWM CLYDACH

OS Landranger Map 159
OS Explorer Map 165
Grid Ref: SN684026
Website: https://www.rspb.org.uk/reserves-and-events/reserves-a-z/cwm-clydach/

Habitat

Some 7 miles (11.2 km) north from the centre of Swansea is the RSPB Cwm Clydach reserve, situated in the narrow valley of the same name through which the Afon Clydach rushes. Cwm Clydach amounts to 215 acres (87 ha) and exists close to a large urban centre, providing many educational possibilities for local schools. The old tramway, a remnant of former industrial glories when up to 1,000 men were employed in the coal mines (the last closed in 1961), forms the main visitor trail through Cwm Clydach Nature Reserve; a smaller trail, the John Nixon Trail, remembers one of the mine owners.

Species

Most woodland species to be expected in Mid- and South Wales will be seen in Cwm Clydach. Birds of prey include Sparrowhawk, Buzzard, Kestrel and Tawny Owl. One normally does not have long to wait before the *kronk* calls of a passing Raven are heard high overhead, while in early spring you may be treated to their splendid acrobatic courtship displays. All three woodpeckers are present, although you will have to search hard for a glimpse of Lesser Spotted Woodpecker. Tree Pipits and Whinchats occupy the fringes of the valley above or at the edge of the woodlands, with Wheatears on the hills. Hole-nesting species such as Redstart and several of the tit family are frequent in the woods,

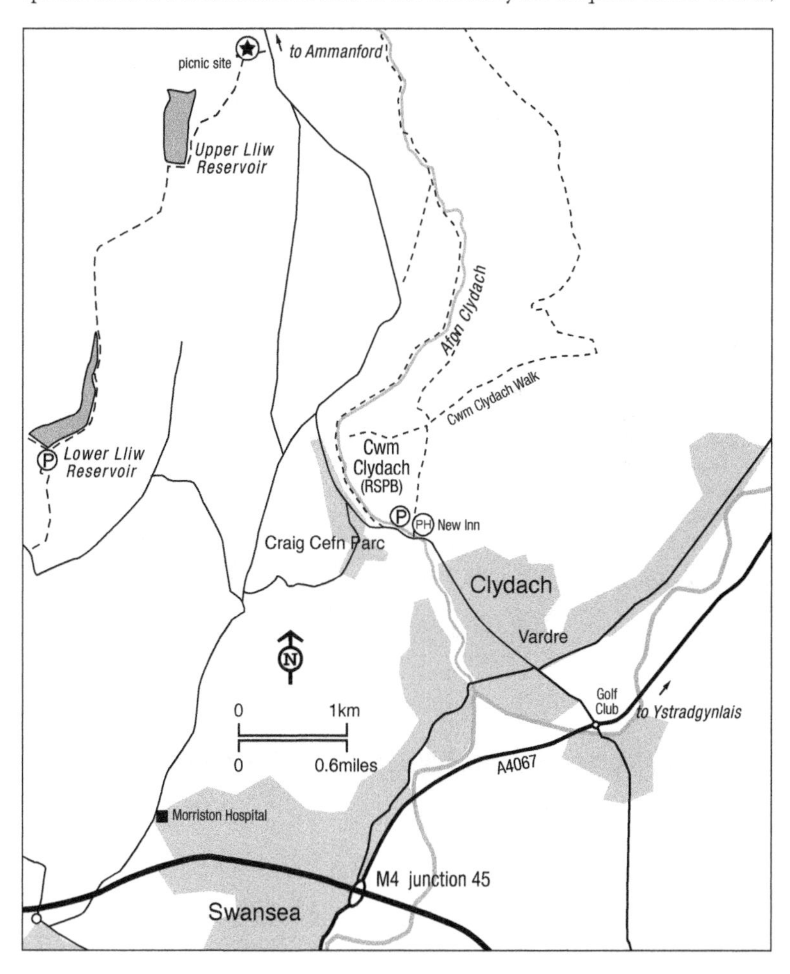

while Pied Flycatcher has declined in recent years. The numerous nest boxes supplement natural nest sites. Summer visitors include Whitethroat, Blackcap, Garden, Wood and Willow Warblers and Chiffchaff. Up to five pairs of Dippers nest along the river together with Grey Wagtails, while Kingfishers occasionally occur. In winter, Snipe and Woodcock are recorded in small numbers and there are usually Fieldfares, Redwings, Siskins and Redpolls about.

Timing

No particular time of day, though early morning is recommended.

Access

Leave M4 at junction 45 and take the A4067 Brecon road, at the crossroads at SN695014, in the centre of Clydach, turn left and follow the unclassified road through the valley until you reach the New Inn pub in Craig Cefn Parc next to which is the car park and entrance to Cwm Clydach Reserve, which is open at all times. The shorter of the trails through the reserve is suitable for wheelchairs. For those using the bus, one from Swansea stops at the reserve entrance. Beyond the reserve the road climbs on to higher ground with a picnic site at SN670068, overlooking the Lliw Reservoirs, which should always be examined and to where small numbers of duck, including Goosander, come in winter. Is Goosander a potential breeding bird on the Clydach, for it has nested on the River Neath since 1987 and at least once on the Tawe?

CALENDAR

Resident: Sparrowhawk, Buzzard, Kestrel, Pheasant, Stock Dove, Woodpigeon, Collared Dove, Tawny Owl, Green, Great Spotted and Lesser Spotted Woodpeckers, Skylark, Meadow Pipit, Grey Wagtail, Dipper, Wren, Dunnock, Robin, Blackbird, Song and Mistle Thrushes, Goldcrest, Long-tailed, Marsh, Blue and Great Tits, Nuthatch, Treecreeper, Jay, Magpie, Raven, Chaffinch, Greenfinch, Goldfinch, Bullfinch.

December–February: Snipe, Woodcock, Fieldfare, Redwing, Brambling, Siskin, Redpoll.

March–May: Chiffchaff usually the first summer visitor to the woodlands and Wheatear to the uplands, both before the end of March, later arrivals include Swallow, Tree Pipit, Redstart, Whinchat, Garden Warbler, Blackcap, Wood and Willow Warblers, Spotted and Pied Flycatchers.

June–July: All breeding species present, though by end of period the woods seem strangely quiet after all the song and bustle of earlier in the season.

August–November: Departure of summer visitors. Some residents, especially tits, form small post-breeding flocks which can be joined by Goldcrest, Treecreeper, even by finches on occasions. Autumn sees the arrival of such visitors as Woodcock, Fieldfare and Redwing.

142 CRYMLYN BURROWS AND CRYMLYN BOG

OS Landranger 159 & 170
OS Explorer 165
Grid References SS717928 & SS694953
Websites:
https://naturalresources.wales/days-out/places-to-visit/south-west-wales/crymlyn-bog-nnr

https://www.swansea.ac.uk/life-on-campus/our-grounds/crymlyn-burrows-sssi/
https://naturalresources.wales/days-out/places-to-visit/south-west-wales/pant-y-sais-nnr

Habitat

Most who speed their way along the A483 to the east of Swansea hardly realise that beyond the screen of roadside trees lies a wildlife gem, often overlooked even by birdwatchers. In pre-industrial times, sand dunes would have stretched westwards the whole 4 miles (6.4 km) between the estuaries Neath and Tawe, both rivers having made their way from the uplands of Fforest Fawr, the hills some 25 miles (40 km) to the north. Now the fragment of dunes – the Burrows – which remains is safeguarded as a Site of Special Scientific Interest, particularly noted for the rich flora including the rare Sea Stock, which seems to have disappeared from a number of its former Welsh locations. Beyond lies Baglan Bay and the estuary of the River Neath, from which a narrow creek with small side channels extends westwards into the saltings. At low-water the shore revealed is extensive and varies from sandy flats to muddy shores with vast quantities of seashells along the strandlines. The Bay Campus of Swansea University is located adjacent to the Burrows, which it is responsible for maintaining.

A short distance inland is another gem, Crymlyn Bog: the most extensive lowland fen in Wales and well described as Swansea's best kept secret. This was an inlet from the sea until about 800 years ago when a violent sandstorm blocked the entrance to the valley and the bog began to slowly form. That it still exists despite encroachment is due to one man, Andrew Lees, who having recognised its importance vigorously campaigned to prevent an extension of the landfill site. His efforts eventually resulted in designation of the bog as a National Nature Reserve and in 1993 as a Ramsar site, while close by Pant y Sais Fen, bordering the Tennant Canal, was compulsorily purchased by the local council and is now a National Nature Reserve. Here you will find a memorial to Lees, who tragically died in 1994, with the words 'At some point I had to stand up and be counted. Who speaks for the butterflies?'

Looking out across Crymlyn Bog is there something missing? Open water! Why, one asks, has Natural Resources Wales not seen fit to devote a small section of the wetland to the creation of a lagoon, a scrape with shallow water and small islets. This is the sort of positive management that has payed mighty dividends elsewhere in Wales for birds and those who watch them.

Species

Do not rush straight onto the Burrows: the woodland edge deserves attention, especially in winter when both Lesser Redpolls and Siskins are likely, while the dunes host Meadow Pipits and Linnets, and where Hen Harrier, Merlin and Short-eared Owl have also been noted. Snipe and Jack Snipe are winter visitors, largely undetected until flushed. Along the strandline Oystercatchers, Curlews and Redshanks are the most numerous, while Grey Plover, Lapwing, Sanderling, Dunlin and Turnstone are also encountered. Wintering Great Crested Grebes have reached 300 in number, in their midst the possibility of Red-breasted Mergansers and smaller grebes. Gull flocks should never be neglected, this after all is part of Swansea Bay, a place renowned for its rare visitors.

The two classic birds of the reedbeds at Crymlyn Bog and Pant y Sais Fen

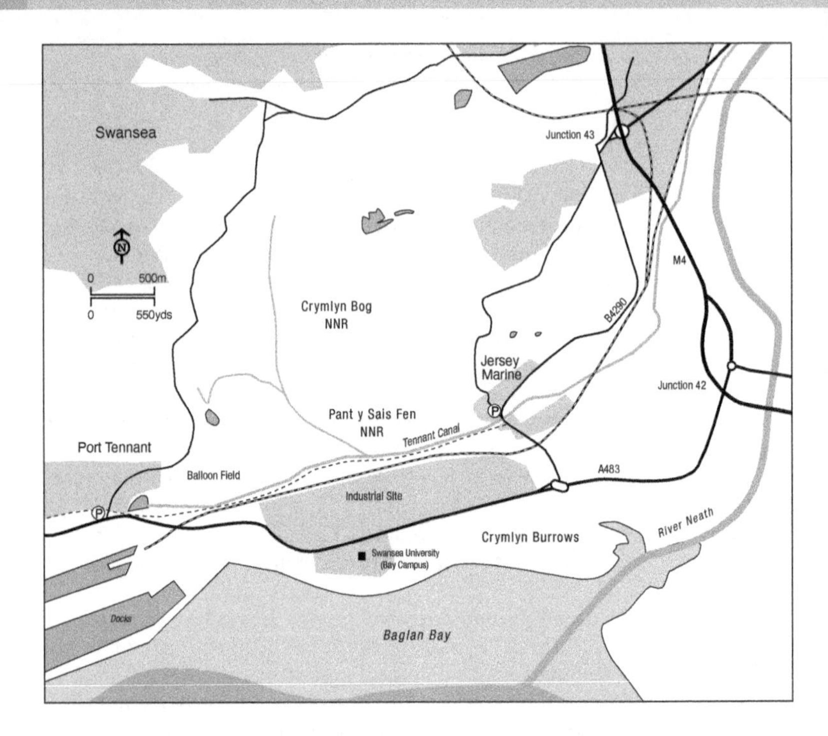

are more often heard than seen – Water Rail and Cetti's Warbler, being joined in summer by Grasshopper, Sedge and Reed Warblers. The latter of these species is described as a 20th-century success story, having strikingly extended its range in Wales; it probably would do so even more if there were more reed-beds. From late July onwards, some of these warblers could well be from breeding locations further north, stopping here in order to fatten up, even to double in weight before a non-stop journey south of some 2,700 miles (4,350 km) to winter quarters in the Sahel. Other summer visitors include Lesser Whitethroat, Blackcap and Willow Warbler, while Reed Buntings are resident. Quite eye-catching when the wind is strong from an easterly quarter are Buzzards, Red Kites and Ravens gliding high over the escarpment above the western boundary of the bog.

Timing

For the Burrows, a glance at the tide tables will prove helpful, ideally on a falling tide. A rising tide, if one is not careful, can mean the creek being impassable, resulting in a lengthy walk westwards before turning east to the car park. Timing is not important at either Crymlyn Bog or Pant y Sais, though if there is a choice early morning is preferable.

Access

For Crymlyn Burrows use the small car park on the south side of the A483 trunk road at the junction with the B4290 for Jersey Marine. Crymlyn Bog is reached from Port Tennant on the eastern fringe of Swansea, from where a minor road heads north skirting on higher ground the bog itself where there are several trails – one crossing the Balloon Field, site of barrage balloons during the Second World War used to deter Luftwaffe attacks on the Llandarcy Oil Refinery, another passes through a section of the wetlands, partly on boardwalks. A section of the Wales Coast Path gives access alongside the Tennant Canal which borders Pant y Sais Fen, where a boardwalk provides a 15-minute walk and is accessible to wheelchairs.

CALENDAR

Resident: Cormorant, Grey Heron, Water Rail, Moorhen, Buzzard, Sparrowhawk, Great Black-backed and Herring Gulls, Kingfisher, Cetti's Warbler – listen for its explosive song from the reedbeds – Reed Bunting.

December–February: Red-throated Diver and Red-breasted Merganser share the open water with Great Crested Grebes. The mixed flocks of gulls along the shore deserve careful scrutiny for scarce visitors. Bearded Tits and Bitterns have occasionally been reported from the bog, the latter breeding here two centuries ago.

March–May: Marsh Harriers have occasionally been reported from the bog. Summer visitors that breed there arrive from early April including Grasshopper, Reed and Sedge Warblers, Whitethroat, Lesser Whitethroat and Willow Warbler, while a visit in early May should result in sighting the first Swifts of the year hawking insects high overhead.

June–July: Shelduck which otherwise are resident depart on their moult migration. Waders begin to return to the shoreline, Oystercatchers, Curlews and Redshanks are often the most numerous. The Swifts will depart for Central Africa in early August.

August–December: Great Crested Grebes winter offshore, over 300 on occasions. Wader passage and the arrival of winter visitors on the estuary really underway by late August including Oystercatcher, Lapwing, Grey Plover, Sanderling, Dunlin, Curlew and Redshank, while by winter both Snipe and Jack Snipe on the saltings, their presence undetected unless accidentally flushed. Water Rails announce their presence from the reedbeds.

APPENDIX A: NATIONAL ORGANISATIONS AND AVIFAUNAS

Welsh Ornithological Society
www.birdsin.wales
www.adar.cymru

The Welsh Ornithological Society (WOS) was formed in 1988 with the aim of coordinating Bird Recording in Wales and providing a vehicle for the publication of material on Welsh ornithology. At first WOS produced the *Welsh Bird Report* and a twice-annual newsletter, but since 1996 the amount of material being written was sufficient to justify the publication of a journal: *Welsh Birds*, now renamed as *Birds in Wales*.

The WOS website contains details of the society, as well as pdf copies of old reports and publications. A subcommittee of WOS is the Welsh Birds Rarities Committee, that adjudicates on records of rare species sighted in Wales, produces an annual report each November (downloadable from the WOS website) and maintains the Welsh bird list.

Membership:
Online or by emailing: wosmembership@btinternet.com

Membership (at the time of printing is £20 a year), payable in January of that year.

Welsh Birds Rarities Committee
A subcommittee of WOS that adjudicates records and descriptions of scarce birds in Wales and maintains the Welsh Bird List. More information on the panel, its list and downloadable record forms and annual reports can be found on the WOS website.

WBRC Secretary: Jon Green, Crud yr Awel, Bowls Road, Blaenporth, Cardigan, SA43 2AR (01239 811561).

email: welshrarebirds@tiscali.co.uk

Royal Society for the Protection of Birds
The RSPB has many nature reserves in Wales. For details contact the reserve direct or one of the two offices coordinating the RSPB's work in Wales.

RSPB Wales Office, Castlebridge 3, 5–19 Castlebridge Road East, Cardiff, CF11 9AB

North Wales Office at RSPB, North Wales Office, Uned 14, Llys Castan, Ffordd Y Parc, Bangor, Gwynedd, LL57 4FH.

National Avifauna

The first complete national avifauna, *Birds in Wales*, was published by Poyser in 1994 by Roger Lovegrove, Graham Williams and Iolo Williams.

An update to this covering the years 1992–2000 by Jonathan Green was published by WOS in 2002.

A completely new complete national avifauna, *The Birds of Wales – Adar Cymru*, was published by Liverpool University Press in July 2021, edited by Rhion Pritchard, Julian Hughes, Ian Spence, Bob Haycock and Anne Brenchley. The avifauna contains information from breeding and ringing studies as well as detailed species accounts, covering the period up to 2019, and is dedicated to all the birdwatchers, past and present, whose records have added to our understanding of birds in Wales, and to the next generation, whose recording and passion should ensure their future.

County Information

Birdwatching details for each county, such as the county recorder, bird groups and further reading is all shown at the beginning of that county's sites.

ANGLESEY (YNYS MÔN)

County Recorder: Steve Culley email: SteCul10@aol.com	**Recent sightings:** https://www.facebook.com/birdwatchingnorthwales/ https://angleseybirdnews.blogspot.com/
Bird Reports: Cambrian Ornithological Society publishes an annual bird report for Anglesey, Caernarfonshire and Meirionnydd. Available from Geoff Gibbs, Fron Wen, Valley Road, Llanfairfechan, LL33 0ET.	**Most up-to-date avifauna:** Walley, P. & Hope-Jones, P. 2004. *Birds of Anglesey*. Brenchley, A., Gibbs, G., Pritchard, R. and Spence, I. M. 2013 *The Breeding Birds of North Wales*. Liverpool University Press.
Bird Groups: None on the island but groups nearby in Bangor and Conwy (see Caernarfonshire)	**Other publications:** Davies, A. & Roberts, O. 2007. *Best Birdwatching Sites in North Wales*. Buckingham Press.
County Trust: North Wales Wildlife Trust https://www.northwaleswildlifetrust.org.uk/	

BRECONSHIRE (BRYCHEINIOG)

County Recorder: Andrew King email: andy@breconbirds.wales	**Recent sightings blog:** https://www.brecknockbirds.co.uk/
Bird Reports: Annual bird reports available from Wildlife Trust South & West Wales, 6 The Bulwark, Brecon, Powys, LD3 7LB.	**Most up-to-date avifauna:** Shrub, M & Peers, M. 1990. *Birds of Breconshire*. Brecknock Wildlife Trust.
Local Bird Groups: Breconshire has not got a formal birdwatching club, but instead there is a local group of the Brecknock Trust that runs events and outings. For more information please contact Pauline Hill on 07957 292235 or email: p.hill@welshwildlife.org	

County Trust: Wildlife Trust of South and West Wales (local group is called the: Brecknock Wildlife Trust) https://www.welshwildlife.org/local-group/brecknock/

CAERNARFONSHIRE

County Recorder: Rhion Pritchard email: rhion678pritchard@gmail.com	**Recent sightings:** https://www.facebook.com/birdwatchingnorthwales/
Bird Reports: Cambrian Ornithological Society publishes an annual bird report for Anglesey, Caernarfonshire and Meirionnydd. Available from Geoff Gibbs, Fron Wen, Valley Road, Llanfairfechan, LL33 0ET. Bardsey Bird Observatory produces an annual bird report, available from Steve Stansfield, Bardsey Bird & Field Observatory, Cristin, Bardsey Island, Gwynedd, LL53 8DE. http://bbfo.blogspot.com/	**Most up-to-date avifauna:** Pritchard R. 2017 *Birds in Caernarfonshire*. Cambrian Ornithological Society Brenchley, A, Gibbs, G., Pritchard, R. and Spence, I.M. 2013 *The Breeding Birds of North Wales*. Liverpool University Press.
Bird Groups: Cambrian Ornithological Society Indoor meetings are normally held at Pensychnant Conservation Centre (01492 592595) at the top of the Sychnant Pass, above Conwy (LL32 8BJ). Membership Secretary: Dr Jeff Williams, Plas Elwy, Llanfair H., Abergele, LL22 8YT. email: jeffreywilliams1@icloud.com. Website: https://www.brnw.cymru/cos Bangor Bird Group The group, which has links with Bangor University, was set up 70 years ago. Organizes a series of talks October–March, with a wide range of speakers and topics. email: bangorbirdgroup@gmail.com	**Other publications:** Davies, A. & Roberts, O. 2007. *Best Birdwatching Sites in North Wales*. Buckingham Press.
County Trust: North Wales Wildlife Trust https://www.northwaleswildlifetrust.org.uk/	

CARMARTHENSHIRE (CAERFYRDDIN)

County Recorder: Gary Harper email: gary.harper3@gmail.com Mobile: 07748 970124	**Recent sightings:** https://carmarthenshirebird.club/ Also on twitter: @CarmsBirdClub

Bird Reports:	Most up-to-date avifauna:
Annual reports available from Carmarthenshire Bird Group, available from Wendell Thomas, 48 Glebe Road, Loughor, Swansea, SA4 6QD.	No recent avifauna since Ingram, G.C.S. and Salmon, H.M. 1954. *A Hand List of the Birds of Carmarthenshire.*
Bird Groups: Carmarthenshire Bird Group Events are organised by Wendell Thomas. Mobile: 07912 577626	
County Trust: The Wildlife Trust South & West Wales: www.welshwildlife.org.uk	

CEREDIGION

County Recorder:	Recent sightings blog:
Russell Jones, Bron y Gan, Talybont, Ceredigion, SY24 5ER. email russell.jones@rspb.org.uk Mobile phone number: 07753 774891	http://ceredigionbirds33.blogspot.com/
Bird Reports: Annual bird reports were produced in paper form until 2019, after which pdf versions are available to download from the WTSWW website (see below).	**Most up-to-date avifauna** Roderick, H. and Davis, P. 2010. *Birds of Ceredigion*. The Wildlife Trust of South and West Wales.
Bird Groups: Ceredigion does not have a formal birdwatching club, but there is a monthly field trip organised by Tim Rayner. email: timgjrayner@live.co.uk Mobile: 07954 012870 Address: Rhydynant, Doldre, Tregaron, SY25 6JT	
County Wildlife Trust: The Wildlife Trust South & West Wales: www.welshwildlife.org.uk	

DENBIGHSHIRE (DINBYCH)

Clwyd Bird Recording Group, part of Bird Recording North Wales (BRNW). Contact: Jacqui Irving email: jacqui970irving@btinternet.com	
County Recorder: Ian Spence. email: ian.spence@zen.co.uk	**Website:** https://www.brnw.cymru/
Bird Reports: North-East Wales Bird reports are produced annually by BRNW and are available to purchase online from the BRNW website.	**Most up-to-date avifauna:** No recent avifauna but the area is included in the *North Wales Bird Atlas*: Brenchley, A., Gibbs, G., Pritchard, R. and Spence, I.M. 2013. *The Breeding Birds of North Wales*. Liverpool University Press.

	Other publications: Davies, A. & Roberts, O. 2007. *Best Birdwatching Sites in North Wales.* Buckingham Press.
County Trust: North Wales Wildlife Trust https://www.northwaleswildlifetrust.org.uk/	

EAST GLAMORGAN (MORGANNWG DWYRAIN)

County Recorder: Phil Bristow email: phlbrstw@gmail.com	**Recent sightings:** Glamorgan Rarities Committee Website: http://grcforum.blogspot.com/ Twitter: @welshbirders
Bird reports: Annual bird reports available from Glamorgan Bird Club.	**Most up-to-date avifauna:** Hurford, C. and Lansdown, P. 1995. *Birds of Glamorgan*. Published by the authors.
Bird Groups: Glamorgan Bird Club https://glamorganbirds.org.uk/ Indoor Meetings: first Tuesday of each of the winter months at The Miners' Welfare Hall, Heol-y-Groes, Bridgend, CF35 5PE. Contact: Paul Denning (01443 202607). Monthly outdoor Meetings: contact John Wilson (johndw1948@gmail.com) or 07999 801645.	**Other publications:** Rosney, A. and Smith, R.G. 2009. *Birding in Glamorgan*. Glamorgan Bird Club.
County Trust: The Wildlife Trust South & West Wales: www.welshwildlife.org.uk	

FLINTSHIRE (FFLINT)

Clwyd Bird Recording Group, part of Bird Recording North Wales (BRNW).	
County Recorder: Ian Spence. email: ian.spence@zen.co.uk	**Website:** https://www.brnw.cymru/
Bird Reports: North-East Wales Bird reports are produced annually by BRNW and are available to purchase online from the BRNW website.	**Most up-to-date avifauna:** No recent avifauna but the area is included in the *North Wales Bird Atlas*: Brenchley, A., Gibbs, G., Pritchard, R. and Spence, I.M. 2013 *The Breeding Birds of North Wales*. Liverpool University Press.
Local Bird Groups: Wrexham Birdwatchers Group: indoor meetings on first Friday of each month at Gresford Memorial Hall, LL12 8PS. Outdoor meetings: throughout the year. Deeside Naturalists: http://www.deenats.org.uk/ Programme of events and access to Connah's Quay Reserve.	**Other publications:** Morris, G. 2015. *The Birds of Connah's Quay Nature Reserve and Oakenholt Marsh.* Davies, A. & Roberts, O. 2007. *Best Birdwatching Sites in North Wales.* Buckingham Press.

County Wildlife Trust: North Wales Wildlife Trust: https://www.northwaleswildlifetrust.org.uk/
Other useful websites: Dee Estuary Birding: http://www.deeestuary.co.uk/

GWENT

County Recorder: Darryl Spittle: countyrecorder@gwentbirds.org.uk	**Recent sightings:** http://www.gwentbirds.org.uk/
Bird Reports: Produced annually by Gwent Ornithological Society and can be purchased from their website.	**Most up-to-date avifauna:** Venables, W.A. et al. 2008. *The Birds of Gwent*. Christopher Helm.
Bird Groups: Gwent Ornithological Society: www.gwentbirds.org.uk email: secretary@gwentbirds.org.uk Fortnightly indoor Meetings are held at Goytre Village Hall, Newtown Road, Penperlleni, NP4 0AR. Contact Hannah Daniels indoor@gwentbirds.org.uk Regular outdoor meetings – contact Dave Brassey: outdoor@gwentbirds.org.uk	**Other publications:** Venables, W. et al. 2013. *Bird Watching Walks in Gwent*. Gwent Ornithological Society. Tyler, Stephanie et al. 1987. *The Gwent Atlas of Breeding Birds*. Gwent Ornithological Society.
County Trust: Gwent Wildlife Trust www.gwentwildlife.org	

MEIRIONNYDD

County Recorder: Jim Dustow email: meinirowen@live.co.uk	**Recent sightings:** https://www.facebook.com/birdwatchingnorthwales/
Bird Reports: Cambrian Ornithological Society publishes an annual bird report for Anglesey, Caernarfonshire and Meirionnydd. Available from Geoff Gibbs, Fron Wen, Valley Road, Llanfairfechan, LL33 0ET.	**Most up-to-date avifauna:** Pritchard, R. 2012. *Birds in Meirionnydd*. Cambrian Ornithological Society. Brenchley, A., Gibbs, G., Pritchard, R. and Spence, I.M. 2013. *The Breeding Birds of North Wales*. Liverpool University Press.
Bird Groups: None in the county but groups nearby in Bangor and Conwy (see Caernarfonshire).	**Other publications:** Davies, A. & Roberts, O. 2007. *Best Birdwatching Sites in North Wales*. Buckingham Press.
County Trust: North Wales Wildlife Trust https://www.northwaleswildlifetrust.org.uk/	

MONTGOMERYSHIRE (TREFALYDWYN)

County Recorder: Simon Boyes: montbird@gmail.com	**Recent sightings blog:** http://montgomerybirdblog.blogspot.com/
Bird Reports: http://montgomerybirdblog.blogspot.com/p/county-reports.html	**Most up-to-date avifauna:** Holt, B. and Williams, G. 2008. *Birds of Montgomeryshire.*
Local Bird Groups: Montgomery Wildlife Trust Bird Group meets on the third Wednesday of each month September to April at Welshpool Methodist Hall. Contact: Montgomeryshire Wildlife Trust (01938 555654).	**Other publications:** Davies, A. & Roberts, O. 2007. *Best Birdwatching Sites in North Wales.* Buckingham Press.
County Trust Montgomeryshire Wildlife Trust (MWT) https://www.montwt.co.uk/	

PEMBROKESHIRE (SIR PENFRO)

County Recorder: Jon Green and Steve Berry, c/o Crud yr Awel, Bowls Road, Blaenporth, Cardigan, SA43 2AR. email: jonrg@tiscali.co.uk	**Recent sightings blog:** https://pembsbirds.blogspot.com/
Bird Reports: Bird Reports are produced annually and are freely available to download from the website above. Skokholm Bird Observatory also produces an annual report, which is also freely downloadable from the island's website: http://skokholm.blogspot.com/	**Most up-to-date avifauna:** Online version via Pembrokeshire Bird Group website or https://pembsavifauna.co.uk/ Donovan, J. and Rees, G. 1994. *Birds of Pembrokeshire: Status and Atlas of Pembrokeshire's Birds.* Dyfed Wildlife Trust.
Bird Group: Pembrokeshire Bird Group, meets during the autumn and winter in Haverfordwest, details are on their website: www.pembrokeshirebirdgroup.blogspot.com	**Other publications:** Rees, G., Haycock, A., Haycock, B., Hodges, J., Sutcliffe, S., Jenks, P. and Robbins, R. 2009. *Atlas of Breeding Birds in Pembrokeshire 2003–2007.* Pembrokeshire Bird Group. Green, J. and Roberts, O. 2004. *Birding in Pembrokeshire.* Available from Jon Green, email: welshrarebirds@tiscali.co.uk
County Trust: The Wildlife Trust South & West Wales: www.welshwildlife.org.uk	

RADNORSHIRE (MAESYFED)

County Recorder: Pete Jennings: email: radnorshirebirds@hotmail.com	**Recent sightings blog:** https://radnorshirebirds.wordpress.com/ Facebook: https://www.facebook.com/radnorbirdblog/
Bird Reports: No recent annual bird reports.	**Most up-to-date avifauna:** Jennings, J. 2014. *The Birds of Radnorshire.* Ficedula Books.
Local Bird Groups: No formal bird group, although there are local Wildlife Trust groups that run meetings.	
County Trust: Radnorshire Wildlife Trust https://www.rwtwales.org/	

WEST GLAMORGAN & GOWER

County Recorder: Rob Jones e-mail: goweros10@gmail.com	**Recent sightings:** https://www.gowerbirds.org.uk/ Glamorgan Rarities Committee Website: http://grcforum.blogspot.com/ Twitter: @welshbirders
Bird reports: Annual reports are available from Gower Ornithological Society, see https://www.gowerbirds.org.uk.	**Most up-to-date avifauna:** Hurford, C. and Lansdown, P. 1995 *Birds of Glamorgan.* Published by the authors.
Bird Groups: Gower Ornithological Society – The Bird Club for Gower, Swansea & Neath Port Talbot Indoor meetings are held monthly at Environment Centre, Pier Street, Swansea, AS1 1RY – see website for details. Outdoor Meetings are also held every month – see https://www.gowerbirds.org.uk.	**Other publications:** Rosney, A. and Smith, R.G. 2009. *Birding in Glamorgan.* Glamorgan Bird Club Thomas, D.K. 1992. *An Atlas of Breeding Birds in West Glamorgan.* Gower Ornithological Society.
County Trust: The Wildlife Trust South & West Wales: www.welshwildlife.org.uk	

APPENDIX B: ALPHABETICAL LIST OF SPECIES WITH WELSH AND SCIENTIFIC NAMES

This list includes all birds mentioned in the text. The Welsh names are based on those in *Birds of Wales* 2021.

Accentor	Alpine	Prunella collaris	*Prunella collaris*
Albatross	Black-browed	Albatros Aelddu	*Diomedea melanophris*
Auk,	Little	Carfil Bach	*Alle alle*
Avocet		Cambig	*Recurvirostra avosetta*
Bee-eater		Gwybedog y Gwenyn	*Merops apiaster*
Bittern		Aderyn y Bwm	*Botaurus stellaris*
	American	Aderyn-bwm America	*Botaurus lentiginosus*
	Little	Aderyn-bwm Lleiaf	*Ixobrychus minutus*
Blackbird		Mwyalchen	*Turdus merula*
Blackcap		Telor Penddu	*Sylvia atricapilla*
Bluetail	Red-flanked	Cynffonlas Ystlysgoch	*Tarsiger cyanurus*
Bluethroat		Bronlas Smotyn Coch	*Luscinia svecica*
Bobolink		Bobolinc	*Dolichonyx oryzivorus*
Brambling		Pinc y Mynydd	*Fringilla montifringilla*
Bullfinch		Coch y Berllan	*Pyrrhula pyrrhula*
Bunting	Black-headed	Bras Penddu	*Emberiza melanocephala*
	Cirl	Bras Ffrainc	*Emberiza cirlus*
	Corn	Bras yr Ŷd	*Miliaria calandra*
	Cretzschmar's	Bras Cretzschmar	*Emberiza caesia*
	Indigo	Bras Dulas	*Passerina cyanea*
	Lapland	Bras y Gogledd	*Catcarius lapponicus*
	Little	Bras Lleiaf	*Emberiza pusilla*
	Ortolan	Bras y Gerdd	*Emberiza hortulana*
	Reed	Bras y Cyrs	*Emberiza schoeniclus*
	Rock	Bras y Craig	*Emberiza cia*
	Rustic	Bras Gwledig	*Emberiza rustica*
	Snow	Bras yr Eira	*Plectrophenax nivalis*
	Yellow-breasted	Bras Bronfelen	*Emberiza aureola*
Buzzard		Bwncath	*Buteo buteo*
	Rough-legged	Bod Bacsiog	*Buteo lagopus*
Catbird	Grey	Cathaderyn Llwyd	*Dumetella carolinensis*
Chaffinch		Ji-binc	*Fringilla coelebs*
Chiffchaff	Common	Siff-siaff	*Phylloscopus collybita*
	Iberian	Siff-siaff Iberia	*Phylloscopus ibericus*
Chough		Bran Goesgoch	*Pyrrhocorax pyrrhocorax*
Coot		Cwtiar	*Fulica atra*

Cormorant		Mulfran	*Phalacrocorax carbo*
Corncrake		Rhegen yr Ŷd	*Crex crex*
Crake	Baillon's	Rhegen Baillon	*Porzana pusilla*
	Spotted	Rhegen Fraith	*Porzana porzana*
Crane		Garan	*Grus grus*
Cream-coloured Courser		Rhedwr y Twyni	*Cursorius cursor*
Crossbill	Common	Gylfin Groes	*Loxia curvirostra*
	Two-barred	Gylfingroes Adeinwyn	*Loxia leucoptera*
Carrion	Carrion	Bran Dyddyn	*Corvus corone*
	Hooded	Bran Lwyd	*Corvus cornix*
Cuckoo		Cog	*Cuculus canorus*
Curlew		Gylfinir	*Numenius arquata*
Dipper		Bronwen y Dŵr	*Cinclus cinclus*
Diver	Black-throated	Trochydd Gyddfddu	*Gavia arctica*
	Great Northern	Trochydd Mawr	*Gavia immer*
	Red-throated	Trochydd Gyddfgoch	*Gavia stellata*
	White-billed	Trochydd Pigwyn	*Gavia adamsii*
Dotterel		Hutan y Mynydd	*Charadrius morinellus*
Dove	Collared	Turtur Dorchog	*Streptopelia decaocto*
	Rock (Feral Pigeon)	Colomen y Graig	*Columba livia*
	Stock	Colomen Wyllt	*Columba oenas*
	Turtle	Turtur	*Streptopelia turtur*
Dowitcher	Long-billed	Giach Gylfin-hir	*Limnodromus scolopaceus*
Duck	Black	Hwyaden Ddu	*Anas rubripes*
	Falcated	Hwyaden Grymanblu	*Mareca falcata*
	Ferruginous	Hwyaden Lygadwen	*Aythya nyroca*
	Long-tailed	Hwyaden Gynffon-hir	*Clangula hyemalis*
	Mandarin	Hwyaden Gribog	*Aix galericulata*
	Ring-necked	Hwyaden Dorchog	*Aythya collaris*
	Ruddy	Hwyaden Goch	*Oxyura jamaicensis*
	Tufted	Hwyaden Gopog	*Aythya fuligula*
Dunlin		Pibydd y Mawn	*Calidris alpina*
Dunnock		Llwyd y Gwrych	*Prunella modularis*
Eagle	Golden	Eryr Euraid	*Aquila chrysaetos*
	White-tailed	Eryr y Môr	*Haliaeetus albicilla*
Egret	Great White	Creyr Mawr Gwyn	*Ardea alba*
	Little	Creyr Bach	*Egretta garzetta*
Eider		Hwyaden Fwythblu	*Somateria mollissima*
	King	Hwyaden Fwythblu'r Gogledd	*Somateria spectabilis*
Falcon	Gyr	Hebog y Gogledd	*Falco rusticolus*
	Red-footed	Cudyll Troedgoch	*Falco vespertinus*
Fieldfare		Socan Eira	*Turdus pilaris*
Firecrest		Dryw Penfflamgoch	*Regulus ignicapillus*
Flycatcher	Collared	Gwybedog Torchog	*Ficedula albicollis*
	Pied	Gwybedog Brith	*Ficedula hypoleuca*

	Red-breasted	Gwybedog Brongoch	*Ficedula parva*
	Spotted	Gwybedog Mannog	*Muscicapa striata*
Frigatebird		Ffrigad	*Fregata sp.*
Fulmar		Aderyn-Drycin y Graig	*Fulmarus glacialis*
Gadwall		Hwyaden Lwyd	*Anas strepera*
Gannet		Hugan	*Sula bassana*
Garganey		Hwyaden Addfain	*Anas querquedula*
Godwit	Bar-tailed	Rhostog Gynffonfrith	*Limosa lapponica*
	Black-tailed	Rhostog Gynffonddu	*Limosa limosa*
Goldcrest		Dryw Eurben	*Regulus regulus*
Goldeneye		Hwyaden Lygad-aur	*Bucephala clangula*
Goldfinch		Nico	*Carduelis carduelis*
Goosander		Hwyaden Ddanheddog	*Mergus merganser*
Goose	Barnacle	Gwŷdd Wyran	*Branta leucopsis*
	(Taiga) Bean	Gwŷdd y Llafur	*Anser fabalis*
	Brent	Gwŷdd Ddu	*Branta bernicla*
	Canada	Gwŷdd Canada	*Branta canadensis*
	Greylag	Gwŷdd Wyllt	*Anser anser*
	Lesser White fronted	Gwŷdd Dalcen-wen Leiaf	*Anser erythropus*
	Pink-footed	Gwŷdd Droedbinc	*Anser brachyrhynchus*
	White-fronted	Gwŷdd Dalcen-wen	*Anser albifrons*
Goshawk	Gwalch Marth	Accipter gentilis	
Grebe	Black-necked	Gwyach Yddfddu	*Podiceps nigricollis*
	Great Crested	Gwyach Fawr Gopog Podiceps cristatus	
	Little	Gwyach Fach	*Tachybaptus ruficollis*
	Pied-billed	Gwyach Gylfinfraith	*Podilymbus podiceps*
	Red-necked	Gwyach Yddfgoch	*Podiceps grisegena*
	Slavonian	Gwyach Gorniog	*Podiceps auritus*
Greenfinch	Llinos Werdd	Carduelis chloris	
Greenshank	Pibydd Coeswerdd	Tringa nebularia	
Grosbeak	Rose-breasted	Gylfindew Brongoch	*Pheucticus ludovicianus*
Grouse	Black	Grugiar Ddu	*Tetrao tetrix*
	Red	Grugiar	*Lagopus lagopus*
Guillemot		Gwylog	*Uria aalge*
	Black	Gwylog Ddu	*Cepphus grylle*
Gull	Black-headed	Gwylan Benddu	*Larus ridibundus*
	Bonaparte's	Gwylan Bonaparte	*Larus philadelphia*
	Caspian	Gwylan Bontaidd	*Larus cachinnans*
	Common	Gwylan y Gweunydd	*Larus canus*
	Franklin's	Gwylan Franklin	*Larus pipixcan*
	Glaucous	Gwylan y Gogledd	*Larus hyperboreus*
	Glaucous-winged	Gwylan Adeinlas	*Larus glaucescens*
	Great Black-backed	Gwylan Gefnddu Fwyaf	*Larus marinus*
	Herring	Gwylan y Penwaig	*Larus argentatus*
	Kumlien's	Gwylan Kumlien	*Larus glaucoides kumliena*

	Iceland	Gwylan yr Arctig	*Larus glaucoides*
	Ivory	Gwylan Ifori	*Pagophila eburnea*
	Laughing	Gwylan Chwerthinog	*Larus atricilla*
	Lesser Black-backed	Gwylan Gefnddu Leiaf	*Larus fuscus*
	Little	Gwylan Fechan	*Larus minutus*
	Mediterranean	Gwylan Môr y Canoldir	*Larus melanocephalus*
	Ring-billed	Gwylan Fodrwybig	*Larus delawarensis*
	Ross's	Gwylan Ross	*Rhodostethia rosea*
	Sabine's	Gwylan Sabine	*Larus sabini*
	Yellow-legged	Gwylan Goesfelen	*Larus michahellis*
Harrier	Hen	Bod Tinwen	*Circus cyaneus*
	Marsh	Bod y Gwerni	*Circus aeruginosus*
	Montagu's	Bod Montagu	*Circus pygargus*
	Pallid	Boda Llwydwyn	*Circus macrourus*
Hawfinch		Gylfinbraff	*Coccothraustes coccothraustes*
Heron	Green	Creyr Gwyrdd	*Butorides virescens*
	Grey	Creyr Glas	*Ardea cinerea*
	Purple	Creyr Porffor	*Ardea purpurea*
	Squacco	Creyr Melyn	*Ardeola ralloides*
Hobby		Hebog yr Ehedydd	*Falco subbuteo*
Hoopoe		Copog	*Upupa epops*
Honey Buzzard		Bod y Mêl	*Pernis apivorus*
Ibis	Glossy	Ibis Du	*Plegadis falcinellus*
Jackdaw		Jac-y-do	*Corvus monedula*
Jay		Ysgrech y Coed	*Garrulus glandarius*
Junco	Dark-eyed	Jynco Llwyd	*Junco hyemalis*
Kestrel		Cudyll Coch	*Falco tinnunculus*
Killdeer		Cwtiad Torchog Mawr	*Charadrius vociferus*
Kingfisher		Glas y Dorlan	*Alcedo atthis*
Kite	Black	Barcud Du	*Milvus migrans*
	Red	Barcud	*Milvus milvus*
Kittiwake		Gwylan Goesddu	*Rissa tridactyla*
Knot		Pibydd yr Aber	*Calidris canutus*
Lapwing		Cornchwiglen	*Vanellus vanellus*
Lark	Black	Ehedydd Du	*Melanocorypha yeltoniensis*
	Crested	Ehedydd Copog	*Galerida cristata*
	Shore	Ehedydd y Traeth	*Eremophila alpestris*
	Short-toed	Ehedydd Llwyd	*Calandrella brachydactyla*
	Sky	Ehedydd	*Alauda arvensis*
	Wood	Ehedydd y Coed	*Lullula arborea*
Linnet		Llinos	*Carduelis cannabina*
Magpie		Pioden	*Pica pica*
Mallard		Hwyaden Wyllt	*Anas platyrhynchos*
Martin	Crag	Gwennol y Clogwyn	*Ptyonoprogne rupestris*
	House	Gwennol y Bondo	*Delichon urbica*
	Sand	Gwennol y Glennydd	*Riparia riparia*
Merganser	Red-breasted	Hwyaden Frongoch	*Mergus serrator*

Merlin		Cudyll Bach	*Falco columbarius*
Moorhen		Iar Ddŵr	*Gallinula chloropus*
Night-Heron		Creyr y Nos	*Nycticorax nycticorax*
Nighthawk	Common	Cudylldroellwr	*Chordeiles minor*
Nightingale		Eos	*Luscinia megarhynchos*
	Thrush	Eos Fronfraith	*Luscinia luscinia*
Nightjar	Troellwr Mawr	Caprimulgus europaeus	
Nuthatch	Telor y Cnau	Sitta europaea	
Oriole	Golden	Euryn	*Oriolus oriolus*
	Northern	Euryn y Gogledd	*Icterus galbula*
Osprey		Gwalch y Pysgod	*Pandion haliaetus*
Ouzel	Ring	Mwyalchen y Mynydd	*Turdus torquatus*
Owl	Barn	Tylluan Wen	*Tyto alba*
	Little	Tylluan Fach	*Athene noctua*
	Long-eared	Tylluan Gorniog	*Asio otus*
	Scops	Tylluan Scops	*Otus scops*
	Short-eared	Tylluan Glustiog	*Asio flammeus*
	Snowy	Tylluan yr Eira	*Bubo scandiacus*
Tawny		Tylluan Frech	*Strix aluco*
Oystercatcher		Pioden y Môr	*Haematopus ostralegus*
Parakeet	Ring-necked	Paracit Torchog	*Psittacula krameri*
Partridge	Red-legged	Petrisen Goesgoch	*Alectoris rufa*
Pelican	White	Pelican Gwyn	*Pelecanus onocrotalus*
Peregrine		Hebog Tramor	*Falco peregrinus*
Petrel		Pedryn Torchog	*Pterodroma leucoptera*
	Leach's	Pedryn Gynffon-fforchog	*Oceanodroma leucorrhoa*
	Storm	Pedryn Drycin	*Hydrobates pelagicus*
	Wilson's	Pedryn Wilson	*Oceanites oceanicus*
	Zino's/Fea's/ Desertas	Pedryn Madeira/Cabo Verde/Desertas	*Pterodroma madeira/feae/ deserta*
Phalarope	Grey	Llydandroed Llwyd	*Phalaropus fulicarius*
	Red-necked	Llydandroed Gyddfgoch	*Phalaropus lobatus*
	Wilson's	Llydandroed Wilson	*Phalaropus tricolor*
Pheasant		Ffesant	*Phasianus colchicus*
Pintail		Hwyaden Lostfain	*Anas acuta*
Pipit	Buff-bellied	Corhedydd Melynllwyd	*Anthus rubescens*
	Meadow	Corhedydd y Waun	*Anthus pratensis*
	Olive-backed	Corhedydd Cefnwyrdd	*Anthus hodgsoni*
	Red-throated	Corhedydd Gyddfgoch	*Anthus cervinus*
	Richard's	Corhedydd Richard	*Anthus novaeseelandiae*
	Rock	Corhedydd y Graig	*Anthus petrosus*
	Tawny	Corhedydd Melyn	*Anthus campestris*
	Tree	Corhedydd y Coed	*Anthus trivialis*
	Water	Corhedydd y Dŵr	*Anthus spinoletta*
Plover	American Golden	Corgwtiad Aur	*Pluvialis dominca*
	Golden	Cwtiad Aur	*Pluvialis apricaria*
	Greater Sand	Cwtiad y Tywod Mwyaf	*Charadrius leschenaultii*

	Grey	Cwtiad Llwyd	*Pluvialis squatarola*
	Kentish	Cwtiad Caint	*Charadrius alexandrinus*
	American Golden	Corgwtiad Aur	*Pluvialis dominica*
	Little Ringed	Cwtiad Torchog Bach	*Charadrius dubius*
	Pacific Golden	Corgwtiad y Môr Tawel	*Pluvialis fulva*
	Ringed	Cwtiad Torchog	*Charadrius hiaticula*
	Sociable	Cwtiad Heidiol	*Chettusia gregaria*
Pochard		Hwyaden Bengoch	*Aythya ferina*
	Red-crested	Hwyaden Gribgoch	*Netta rufina*
Pratincole	Black-winged	Cwtiadwennol Aden-ddu	*Glareola nordmanni*
	Collared	Cwtiadwennol Dorchog	*Glareola pratincola*
Puffin		Pal	*Fratercula arctica*
Quail		Sofliar	*Coturnix coturnix*
Rail	Water	Rhegen y Dŵr	*Rallus aquaticus*
	Sora	Rhegen Sora	*Porzana carolina*
Raven		Cigfran	*Corvus corax*
Razorbill		Llurs	*Alca torda*
Redhead			*Aythya americana*
Redpoll	Arctic	Llinos Bengoch yr Arctig	*Carduelis hornemanni*
Lesser	Llinos Bengoch	Carduelis flammea	
Redshank	Pibydd Coesgoch	Tringa totanus	
	Spotted	Pibydd Coesgoch Mannog	*Tringa erythropus*
Redstart	Tingoch	Phoenicurus phoenicurus	
	Black	Tingoch Du	*Phoenicurus ochruros*
	Moussier's	Tingoch Moussier	*Phoenicurus moussieri*
Redwing	Coch Dan-aden	Turdus iliacus	
Robin	Robin Goch	Erithacus rubecula	
	American	Robin America	*Turdus migratorius*
	White-throated	Gyddfwyn Robin Goch	*Irania gutturalis*
Roller	Rholydd	Coracias garrulus	
Rook	Ydfran	Corvus frugilegus	
Rosefinch	Common	Llinos Goch	*Carpodacus erythrinus*
Ruff		Pibydd Torchog	*Philomachus pugnax*
Sanderling		Pibydd y Tywod	*Calidris alba*
Sandpiper	Baird's	Pibydd Baird	*Calidris bairdii*
	Broad-billed	Pibydd Llydanbig	*Limicola falcinellus*
	Buff-breasted	Pibydd Bronllwyd	*Tryngites subruficollis*
	Common	Pibydd y Dorlan	*Actitis hypoleucos*
	Curlew	Pibydd Cambig	*Calidris ferruginea*
	Green	Pibydd Gwyrdd	*Tringa ochropus*
	Least	Pibydd Bychan	*Calidris minutilla*
	Marsh	Pibydd y Gors	*Tringa stagnalis*
	Pectoral	Pibydd Cam	*Calidris melanotos*
	Purple	Pibydd Du	*Calidris maritima*
	Semi-palmated	Pibydd Llwyd	*Calidris pusilla*
	Sharp-tailed	Pibydd Gynffonfain	*Calidris acuminata*
	Spotted	Pibydd Brych	*Actitis macularia*

	Terek	Pibydd Terek	*Xenus cinereus*
	Upland	Pibydd Cynffonir	*Bartramia longicauda*
	White-rumped	Pibydd Tinwen	*Calidris fuscicollis*
	Wood	Pibydd y Graean	*Tringa glareola*
Scaup		Hwyaden Benddu	*Aythya marila*
	Lesser	Hwyaden Benddu Fechan	*Aythya affinis*
Scoter	Black	Môr-hwyaden America	*Melanitta americana*
	Common	Môr-hwyaden Ddu	*Melanitta nigra*
	Surf	Môr-hwyaden yr Ewyn	*Melanitta perspicillata*
	Velvet	Môr-hwyaden y Gogledd	*Melanitta fusca*
Serin		Llinos Frech	*Serinus serinus*
Shag		Mulfran Werdd	*Phalacrocorax aristotelis*
Shearwater	Cory's	Aderyn-Drycin Cory	*Calonectris diomedea*
	Great	Aderyn-Drycin Mawr	*Puffinus gravis*
	Little	Aderyn-Drycin Bach	*Puffinus assimilis*
	Manx	Aderyn-Drycin Manaw	*Puffinus puffinus*
	Balearic	Aderyn-Drycin Mediterranean	*Puffinus yelkouan*
	Sooty	Aderyn-Drycin Du	*Puffinus griseus*
Shelduck		Hwyaden yr Eithin	*Tadorna tadorna*
	Ruddy	Hwyaden Goch yr Eithin	*Tadorna ferruginea*
Shoveler	Hwyaden Lydanbig	Anas clypeata	
Shrike	Daurian/ Turkestan	Cigydd Dawria/Tyrcestan	*Lanius isabellinus/ phoenicuroides*
	Great Grey	Cigydd Mawr	*Lanius excubitor*
	Lesser Grey	Cigydd Glas	*Lanius minor*
	Red-backed	Cigydd Cefngoch	*Lanius collurio*
	Woodchat	Cigydd Pengoch	*Lanius senator*
Siskin	Pila Gwyrdd	Carduelis spinus	
Skua	Arctic	Sgiwen y Gogledd	*Stercorarius parasiticus*
	Great	Sgiwen Fawr	*Stercorarius skua*
	Long-tailed	Sgiwen Lostfain	*Stercorarius longicaudus*
	Pomarine	Sgiwen Frech	*Stercorarius pomarinus*
Smew		Lleian Wen	*Mergus albellus*
Snipe		Gïach Gyffredin	*Gallinago gallinago*
	Great	Gïach Mawr	*Gallinago media*
	Jack	Gïach Fach	*Lymnocryptes minimus*
Sparrow	House	Aderyn y To	*Passer domesticus*
	Song	Llwyd Persain	*Melospiza melodia*
	Spanish	Golfan Sbaen	*Passer hispaniolensis*
	Tree	Golfan y Mynydd	*Passer montanus*
	White-throated	Llwyd Gyddfwyn	*Zonotrichia albicollis*
Sparrowhawk		Gwalch Glas	*Accipiter nisus*
Spoonbill		Llwybig	*Platelea leucorodia*
Starling		Drudwen	*Sturnus vulgaris*
	Rose-coloured	Drudwen Wridog	*Sturnus roseus*
Stilt	Black-winged	Hirgoes	*Himantopus himantopus*

Stint	Little	Pibydd Bach	*Calidris minuta*
	Temminck's	Pibydd Temminck	*Calidris temminckii*
Stonechat		Clochdar y Cerrig	*Saxicola torquata*
	Siberian	Clochdar Siberia	*Saxicola torquata maura and stejnegeri*
Stone-curlew		Rhedwr y Moelydd	*Burhinus oedicnemus*
Stork	Black	Ciconia Du	*Ciconia nigra*
	White	Ciconia Gwyn	*Ciconia ciconia*
Swallow		Gwennol	*Hirundo rustica*
	Red-rumped	Gwennol Dingoch	*Hirundo daurica*
Swan	Bewick's	Alarch Bewick	*Cygnus columbianus*
	Mute	Alarch Dof	*Cygnus olor*
Whooper		Alarch y Gogledd	*Cygnus cygnus*
Swift		Gwennol Ddu	*Apus apus*
	Alpine	Gwennol Ddu'r Alpau	*Apus melba*
	Chimney	Coblyn y Simddu	*Chaetura pelagica*
	Little	Gwennol Ddu Fach	*Apus affinis*
	Pallid	Gwennol Ddu Weiw	*Apus pallidus*
Tanager	Summer	Euryn yr Haf	*Piranga rubra*
Tattler	Grey-tailed	Pibydd Gynffonlwd	*Heteroscelus brevipes*
Teal		Corhwyaden	*Anas crecca*
	Blue-winged	Corhwyaden Asgell-las	*Anas discors*
	Green-winged	Corhwyaden Asgell-Werdd	*Anas crecca carolinensis*
Tern	Arctic	Morwennol y Gogledd	*Sterna arctica*
	Black	Corswennol Ddu	*Chlidonias niger*
	Bridled	Morwennol Ffrwynog	*Sterna anaethetus*
	Caspian	Morwennol Fwyaf	*Sterna caspica*
	Common	Morwennol Gyffredin	*Sterna hirundo*
	Elegant	Morwennol Gain	*Thalasseus elegans*
	Forster's	Morwennol Forster	*Sterna forsteri*
	Gull-billed	Morwennol Ylfinbraff	*Gelochelidon nilotica*
	Little	Morwennol Fechan	*Sterna albifrons*
	Roseate	Morwennol Wridog	*Sterna dougallii*
	American Royal	Morwennol Fawr	*Sterna maxima*
	Sandwich	Morwennol Bigddu	*Sterna sandvicensis*
	Sooty	Morwennol Fraith	*Onychoprion fuscata*
	Whiskered	Corswennol Farfog	*Chlidonias hybridus*
	White-winged Black	Corswennol Adeinwen	*Chlidonias leucopterus*
Thrush	Black-throated	Brych Gyddfdywll	*Turdus ruficollis atrogularis*
	Dusky	Brych Tywyll	*Turdus naumanni*
	Grey-cheeked	Bronfraith Fochlwyd	*Catharus minimus*
	Mistle	Brych y Coed	*Turdus viscivorus*
	Rock	Brych y Graig	*Monticola saxatilis*
	Song	Bronfraith	*Turdus philomelos*
	Swainson's	Carfonfraith	*Catharus ustulatus*
Tit	Bearded	Titw Barfog	*Panurus biarmicus*
	Blue	Titw Tomas Las	*Parus caerulus*

	Coal	Titw Penddu	*Parus ater*
	Great	Titw Mawr	*Parus major*
	Long-tailed	Titw Gynffon-hir	*Aegithalos caudatus*
	Marsh	Titw'r Wern	*Parus palustris*
	Penduline	Titw Pendil	*Remiz pendulinus*
	Willow	Titw'r Helyg	*Parus montanus*
Treecreeper		Y Dringwr Bach	*Certhia familiaris*
Turnstone		Cwtiad y Traeth	*Arenaria interpres*
Twite		Llinos y Mynydd	*Carduelis flavirostris*
Vireo	Red-eyed	Telor Melyn	*Vireo olivaceous*
Wagtail	Citrine	Siglen Sitraidd	*Motacilla citreola*
	Eastern Yellow	Siglen Felen y Dwyrain	*Motacilla tschutschensis*
	Grey	Siglen Llwyd	*Motacilla cinerea*
	Pied	Siglen Fraith	*Motacilla alba*
	White	Siglen Wen	*Motacilla alba alba*
	Yellow	Siglen Felen	*Motacilla flava*
Warbler	Aquatic	Telor y Dŵr	*Acrocephalus paludicola*
	Arctic	Telor yr Arctig	*Phylloscopus borealis*
	Barred	Telor Rhesog	*Sylvia nisoria*
	Blackburnian	Telor Blackburnian	*Dendroica fusca*
	Blackpoll	Telor Tinwen	*Dendroica striata*
	Black-and-White	Telor Brith	*Mniotilta varia*
	Booted	Telor Bacsiog	*Hippolais caligata*
	Cetti's	Telor Cetti	*Cettia cetti*
	Dartford	Telor Dartford	*Sylvia undata*
	Dusky	Telor Tywyll	*Phylloscopus fuscatus*
	Eastern Subalpine	Telor Brongoch y Dwyrain	*Curruca cantillans*
	Garden	Telor yr Ardd	*Sylvia borin*
	Grasshopper	Troellwr Bach	*Locustella naevia*
	Great Reed	Telor Mawr y Cyrs	*Acrocephalus arundinaceus*
	Greenish	Telor Gwyrdd	*Phylloscopus trochiloides*
	Hume's Leaf	Telor Hume	*Phylloscopus humei*
	Icterine	Telor Aur	*Hippolais icterina*
	Lanceolated	Telor Gwaywog	*Locustella lanceolata*
	Marmora's	Telor Marmora	*Curruca sarda*
	Marsh	Telor y Twerni	*Acrocephalus palustris*
	Melodious	Telor Per	*Hippolais polyglotta*
	Myrtle	Telor Tinfelyn	*Setophaga coronata*
	Pallas's	Telor Pallas	*Phylloscopus proregulus*
	Paddyfield	Telor Padi	*Acrocephalus agricola*
	Radde's	Telor Radde	*Phylloscopus schwarzi*
	Reed	Telor y Cyrs	*Acrocephalus scirpaceus*
	River	Telor yr Afon	*Locustella fluviatilis*
	Rüppell's	Telor Rueppell	*Sylvia rueppelli*
	Sardinian	Telor Sardinia	*Sylvia melanocephala*
	Savi's	Telor Savi	*Locustella luscinioides*
	Sedge	Telor yr Hesg	*Acrocephalus schoenobaenus*

	Western Orphean	Telor Llygad Arian y Gorllewin	*Curruca hortensis*
	Western Bonelli's	Telor Bonelli y Gorllewin	*Phylloscopus bonelli*
	Western Subalpine	Telor Brongoch y Gorllewin	*Curruca iberiae*
Willow		Telor yr Helyg	*Phylloscopus trochilus*
	Wood	Telor y Coed	*Phylloscopus sibilatrix*
	Yellow	Telor Melyn	*Dendroica petechia*
	Yellow-browed	Telor Aelfelen	*Phylloscopus inornatus*
Wheatear		Tinwen y Garn	*Oenanthe oenanthe*
	Desert	Tinwen y Diffaethwch	*Oenanthe deserti*
	Isabelline	Tinwen Isabella	*Oenanthe isabellina*
	Pied	Tinwen Fraith	*Oenanthe pleschanka*
Western Black-eared		Tinwen Clustiog Du	*Oenanthe hispanica*
Whimbrel		Coegylfinir	*Numenius phaeopus*
	Hudsonian	Coegylfinir yr Hudson	*Numenius hudsonicus*
	Little	Coegylfinir Bach	*Numenius minutus*
Whinchat	Crec yr Eithin	Saxicola rubetra	
Whitethroat	Llwydfron	Sylvia communis	
	Lesser	Llwydfron Fach	*Sylvia curruca*
Wigeon	Chwiwell	Anas penelope	
	American	Chwiwell America	*Anas americana*
Woodcock		Cyffylog	*Scolopax rusticola*
Woodpecker	Great Spotted	Cnocell Fraith Fwyaf	*Dendrocopos major*
	Green	Cnocell Werdd	*Picus viridis*
	Lesser Spotted	Cnocell Fraith Leiaf	*Dendrocopos minor*
Woodpigeon		Ysguthan	*Columba palumbus*
Wren		Dryw	*Troglodytes troglodytes*
Wryneck		Pengam	*Jynx torquilla*
Yellowhammer		Bras Melyn	*Emberiza citrinella*
Yellowlegs	Lesser	Melyngoes Bach	*Tringa flavipes*
	Greater	Melyngoes Mawr	*Tringa melanoleuca*
Yellowthroat	Common	Gyddf-felyn	*Geothlypis trichas*

INDEX TO SPECIES

Figures denote site number. Those sites marked in bold are of more unusual species, where observers are most likely to see that species, while those in italics are sites where that species is most often seen, at the right time of the year.